ENERGY-EFFICIENT MOTOR SYSTEMS:
A Handbook on Technology, Program, and Policy Opportunities

Other Titles on Energy Conservation and Energy Policy

Energy Efficiency in Buildings: Progress and Promise

Financing Energy Conservation

Energy Efficiency: Perspectives on Individual Behavior

Electric Utility Planning and Regulation

Residential Indoor Air Quality and Energy Efficiency

State of the Art of Energy Efficiency: Future Directions

Efficient Electricity Use: A Development Strategy for Brazil

Energy Efficiency and the Environment: Forging the Link

ENERGY-EFFICIENT MOTOR SYSTEMS:
A Handbook on Technology, Program, and Policy Opportunities

STEVEN NADEL

R. NEAL ELLIOTT

MICHAEL SHEPARD

STEVE GREENBERG

GAIL KATZ

ANIBAL T. DE ALMEIDA

American Council for an Energy-Efficient Economy
Washington, D.C.
Second Edition
2002

Energy-Efficient Motor Systems:
A Handbook on Technology, Program, and Policy Opportunities

Published by the American Council for an Energy-Efficient Economy
1001 Connecticut Avenue, N.W., Suite 801, Washington, D.C. 20036.

Cover art: M.C. Escher's "Waterfall" © 2001 Cordon Art B.V.—Baarn—Holland. All rights reserved.

Cover and book design by Chuck Myers, American Labor Education Center

Library of Congress Cataloging-in-Publications Data

Energy-efficient motor systems : a handbook on technology, program, and policy opportunities / Steven Nadel ... [et al.].—2nd ed.
p. cm.
Includes bibliographical references and index.

ISBN 0-918249-40-6
1. Electric motors—Energy consumption—Handbooks, manuals, etc.
2. Electric power—Conservation—Handbooks, manuals, etc. I. Nadel, Steven, 1957- II. American Council for an Energy-Efficient Economy.

TK2781 .E54 2002

333.79'68—dc21   2001060300

NOTICE

Printed on recycled paper.

Acknowledgments

A book of this scope requires the help of many individuals and organizations. Thanks to the dozens of people, too numerous to list here, who provided us with data and information, much of it not previously published. It is due to their assistance that this book is able to break new ground.

Thanks also to our project sponsors—organizations who recognized the usefulness of this project and agreed to provide funds. For this second edition, project sponsors are the Energy Center of Wisconsin, National Grid Corp., New York State Energy Research and Development Authority, and Pacific Gas & Electric Company.

Many people reviewed drafts and provided insightful comments. For this second edition, substantial assistance was provided by Johnny Douglas, Washington State University; Fred Gordon, Pacific Energy Associates; Roger Lawrence, Electric Power Research Institute; Aimee McKane, Lawrence Berkeley National Laboratory; Bruce Meberg, USA Data; Jim Raba, U.S. Department of Energy; and Mitch Rosenberg, XENERGY, Inc.

Other important contributors include Carl Blumstein, the series editor, who helped originally conceive this book, and the production team at ACEEE, including Glee Murray, Renee Nida, Susan Ziff, Cathy Romanowski, and Liz Brown.

Finally, we would like to thank our families, who put up with many evenings and weekends of meeting deadlines and reviewing manuscripts.

Dedication

This book is dedicated to our valued colleague and friend, Gail Katz, whose practical perspective enriched this book and helped keep us on track. Her tragic death near the end of writing the first edition of this book was a blow to all who have worked with her.

Gail was nationally recognized for her pioneering work in energy conservation engineering in industrial and commercial facilities. A memorial scholarship fund has been established in Gail's name to benefit women engineering students working in the fields of energy conservation, renewable energy, and appropriate technology. Contributions and inquiries should be directed to the Gail Katz Memorial Fund, Society of Women Engineers, 345 East 47th St., New York, N.Y. 10017

Contents

List of Figures

List of Tables

Overview and Summary

A sizable percentage (15–25%) of U.S. electricity (see calculations in Chapter 7) can be saved by optimizing the performance of electric motors and their associated wiring, power-conditioning equipment, controls, and transmission components. These networks of devices are also known as motor systems.

Electric motors are remarkable machines: rugged, reliable, and far more efficient than the animals and steam-powered equipment that motors have replaced over the past century. A well-designed and well-maintained electric motor can convert over 90% of its input energy into useful shaft power, 24 hours a day, for decades. The popularity of motors attests to their effectiveness: they provide more than four-fifths of the nonvehicular shaft power in the United States and use upward of 60% of the nation's electricity as input. It is this popularity that makes electric motor systems such an important potential source of energy savings: because more than half of all electricity flows through them, even modest improvements in their design and operation can yield tremendous dividends.

Touring a Motor System

The key to making motor systems more efficient and economical is to take advantage of high-performance technologies and the synergism among the various system components. To illustrate, let's take a brief tour of a system. Starting from the point at which electricity enters the facility, we will move downstream through the wiring, power-conditioning equipment, and controls to the motor. Finally, we will continue through the transmission hardware to the driven devices.

Along the way, we will identify some of the major opportunities for savings.

In theory, electricity arrives at a customer's facility as perfectly balanced and synchronized single- or three-phase power of constant voltage, free of harmonics and other kinds of distortion. In reality, this ideal condition is almost never reached. Phases are often slightly out of balance, voltages may dip and rise, and various kinds of distortion commonly occur. This less-than-perfect power provision is subject to further unbalance and distortions from equipment inside customers' facilities (e.g., welders, lighting ballasts, arc furnaces, and variable-frequency motor controls). Sometimes problems can arise from a poor arrangement of equipment, such as the uneven distribution of single-phase and three-phase devices on a circuit. Such deviations from the pure, ideal electric waveform can reduce the efficiency, performance, and life of motors and other electric equipment.

Avoiding and correcting such problems requires careful monitoring of power quality, repair of faulty devices, and, in some cases, installation of specialized power-conditioning equipment. Some analysts believe that such tune-ups may be among the largest reservoirs of untapped drivepower savings, although the scanty data available allow only rough estimates of the overall potential. Field studies suggest that the effort and expense of electrical tune-ups can be worthwhile due to reduced energy costs, better equipment performance, improved process control, and reduced downtime from damaged equipment. Further details of some major opportunities in this area are discussed in Chapter 3.

Just as it pays to streamline the power flowing through the wires, so too it is important to optimize the efficiency of the wires themselves. In most facilities, distribution wiring is sized according to the National Electrical Code, which principally addresses safety, not energy efficiency. Wires that are larger than the minimum size requirement of the code have lower resistance to the flow of electricity, and hence fewer energy losses. Therefore, in new installations or major renovations, it often pays to exceed code standards. Unfortunately, the benefits of doing so are not widely appreciated by architects, designers, electricians, and facility managers, so considerable amounts of energy and money are being wasted through in-plant distribution losses, before the electricity even does any work. Details on wire sizing are covered in Chapter 3.

Motor-driven processes frequently require some form of control over the motor's start-up, speed, or torque (rotational force). For example, fan-, compressor-, and pump-driven systems moving gaseous or liquid loads may require frequent changes in the rate of flow. This

2

is the case for fans and chillers for ventilation and cooling of commercial buildings, pumps for hydronic heating and/or cooling systems, fans and feed water pumps for industrial and power plant boilers, and municipal water and wastewater pumps. Modern adjustable-speed drives (ASD), discussed in Chapter 4, allow the motor's speed to be precisely controlled, which can significantly reduce energy consumption. This device precisely controls the speed of alternating-current (AC) motors, eliminating the need for wasteful throttling devices in fluid flow applications and rendering many traditional controls and uses for direct-current (DC) motors obsolete. ASDs yield sizable energy savings (15–40% in many cases) and extend equipment life by allowing for gentle start-up and shutdown.

Most systems with variable flow, however, have not been updated and continue to use mechanical devices such as inlet vanes, outlet dampers, or throttling valves to control fluid flow while the motor continues to run at full speed. These techniques are analogous to driving a car with the accelerator pushed to the floor while controlling the vehicle's speed with the brake. Such methods yield imprecise control and waste a lot of energy.

The electronic ASD is not the only new control technology, although it may be the most important one. Other technologies include microprocessor-based controllers that monitor system variables and adjust motor load accordingly, and power-factor controllers that can trim the energy use of small motors driving grinders, drills, and other devices that idle at nearly zero loading most of the time. There are also application-specific controls such as those that sequence the operation of multiple compressors in a compressed-air system.

Other developments enlarge the range of control applications. For instance, advanced sensors are allowing ASDs to be used in applications (lumber-drying kilns, for example) where they previously would not work due to limitations in sensing or in matching the response time required by a control loop. Electronic advances also are allowing lumber mills to control cuts better and to mill more product from raw stock without increasing energy use. These developments and others in the controls area represent the largest slice of the drivepower savings pie and are discussed in Chapter 4.

In other kinds of loads requiring varying speed or torque—winders, mills, conveyors, elevators, cranes, and servodrivers—motor users have employed various kinds of mechanical, electromechanical, or hydraulic speed controls in conjunction with AC motors or have used DC motors where the speed can be easily controlled. However, most of these speed-control options have pitfalls, including high cost, low efficiency, or poor reliability. New motor technologies, discussed in Chapter 2, are emerging that may address these applications' needs while improving energy efficiency at the same time.

Motors are available in a range of efficiencies, as discussed in Chapter 2. Higher-efficiency motors are available for most applications. These motors are typically 2 to 10 percentage points more efficient than standard-efficiency motors, with smaller motors at the high end of this range and larger motors at the low end. Due principally to their better materials, high-efficiency units cost 10–30% more but tend to last longer than standard models. While a few percentage points of efficiency do not sound like much, such an improvement can add up to sizable savings over the life of a motor. A heavily used motor can easily have electricity bills ten times its purchase price each year. If cars were comparable, a $10,000 car would use $100,000 worth of gasoline annually. With so much of the life-cycle cost in operating expense, each increment of efficiency is extremely valuable. Therefore, the payback on the added cost of high-efficiency motors is often very attractive. However, these more efficient motors have been a small part of the market. As presented in Chapter 6, efficient motors accounted for 16% of 1- to 200-horsepower (hp) motor sales on a unit basis and 32% on a value basis in 1997.

The most important recent development has been the implementation of the minimum efficiency standards for industrial motors that were in the Energy Policy Act of 1992 (EPAct), which went into effect in 1997. As discussed in Chapter 2 and Appendix B, this law eliminated the least efficient industrial motors from the new motors market. However, efficient motors made up only 9.1% of the integral motor stock in U.S. manufacturing plants in 1997. Consequently, significant economically attractive opportunities exist for replacing less efficient motors now in service with new, more efficient motors.

While EPAct eliminated the least efficient products from the market, a range of efficiencies above the minimum levels continue to exist. In many cases, choosing these *premium-efficiency* motors (PEMs) is attractive when a motor is bought for a new application or to replace a failed motor. In some cases, the retrofit of an operating motor can be justified. Unfortunately, these motors are not well labeled, as is discussed in Chapter 2. This lack of labeling has resulted in market confusion and made it more difficult for motor purchasers to identify the most efficient products on the market.

As we replace older, less efficient motors with more efficient models, we can capture savings bonuses by correcting for two problems endemic to the existing motor stock: oversizing and rewind damage. Many motors are oversized for their applications, and because motor efficiency drops off sharply below about 40% of rated load, oversized motors often run far below their nameplate efficiency. In addition, many motors are repaired at least once, and often several times, before they are discarded. While quality repair practices can maintain the efficiency of a motor, less attention to detail can reduce the motor's efficiency and life significantly. The proper sizing of new motors and either the use of quality rewind practices or the adoption of replace-instead-of-rewind policies can thus add significant savings. These matters are covered in Chapters 2 and 3.

Energy enters a motor as electricity and emerges as mechanical power in the form of a rotating shaft. To put that energy to use often requires a transmission, provided typically by belts, gears, or chains. Such devices are often overlooked in efficiency analyses. They also typically receive unsophisticated installation and maintenance. This neglect is unfortunate because, as discussed in Chapter 3, the proper selection, installation, and maintenance of transmission hardware can profoundly affect the performance and efficiency of a motor system. For example, too loose a belt will slip, wasting energy. Too tight a belt can place extreme loads on a bearing, causing it to fail prematurely and lead to costly downtime. Such problems can be avoided in some applications by using synchronous belts, which run on toothed sprockets and are generally more efficient than V-belts, which run on smooth pulleys.

Optimized drivetrains are also important because they are far downstream in the drivepower system. Even modest improvements can ripple back through the system to yield significant savings. For instance, a unit of energy saved in the drivetrain means the motor doesn't have to work as hard, so it draws less energy, which reduces losses in the distribution wiring, and so on, back to the power plant. An additional, potentially large bonus comes in the form of indirect savings from reduced building cooling load due to lower current flow and less heat dissipation from the more efficient equipment.

The shaft of the motor drives some types of equipment, such as fans, pumps, compressors, and conveyors. No matter how efficient the system is up to that point, if the system does unnecessary work, significant amounts of energy can be wasted. In Chapter 5, we discuss what is needed to optimize the motor-driven system. Savings approaching 50% can often be realized at little cost just by matching the operation of the system to the end-use requirements.

The need for careful, ongoing monitoring and maintenance applies to the entire motor system. A high-efficiency system will only stay that way if given proper care, from simple cleaning and lubrication to sophisticated troubleshooting of power quality problems. While the energy savings from top-notch maintenance are substantial, the greatest dividend comes in the form of more reliable, trouble-free operation and extended equipment life. When equipment downtime can mean thousands of dollars per hour in lost production, quality maintenance is worthwhile.

We have completed our tour of the motor system and touched on some of the major technical areas that later chapters will deal with in greater depth. If nothing else, this brief survey is designed to emphasize the notion of a motor *system* and to underscore the critical importance of the interactions and synergism among the various system components.

A Note on Lost Opportunities

Most of the efficiency options discussed here are more economical in new installations than in retrofits. These options are termed "lost opportunity" resources because if they are not implemented during new construction or renovation, they are much more costly to install later. In some cases, however, it makes economic sense to replace and upgrade operating equipment rather than to wait for it to fail. Where load factors are very high, for instance, it often pays to scrap standard-efficiency motors and replace them with efficient models. As described in Chapter 2, Stanford University did this with 73 motors, with average paybacks of less than 3 years. Energy conservation program planners and facility managers should remember this distinction between new and retrofit efficiency opportunities as they implement programs.

Barriers to Drivepower Savings

If the potential savings are so large, why are so few motor users aggressively pursuing them? The answer lies in a maze of barriers to investment in energy efficiency in general and to drivepower improvements in particular. Some of the most important of these barriers are highlighted below and discussed in detail in Chapter 8.

Aversion to Downtime

In many businesses, particularly in industry, shutting down equipment for upgrading or replacement can mean losing thousands of dollars per hour in forgone production. Such penalties may induce an understandable aversion to downtime. Because of this, many facility managers shy away from new, unfamiliar technology that they fear might be less reliable than the equipment they are used to. Furthermore, if a high-efficiency substitute for a failed motor is not stocked by the distributor, in order to save time the user is likely to buy a standard replacement or simply repair the old motor.

Purchase Practices

Existing equipment is usually replaced or repaired without engineering analysis and is often replaced with the same size, brand, and model number. Only in the case of large motors (over approximately 250 hp) with high operating costs does an engineering or economic analysis usually precede decisions concerning replacement equipment.

Customers commonly believe that motors under approximately 200 hp and other drivepower components are commodity items, meaning that models produced by different manufacturers are interchangeable. While this is true from the functional perspective, it could not be further from the truth from an energy efficiency perspective. For many customers, purchase decisions are based primarily on reliability, price, and availability, not on efficiency. Consequently, energy cost saving is a factor in decisions, but not a primary concern. Some large companies (and a few smaller ones) have formal motor-purchase policies that address motor efficiency; however, most do not.

Repair Shops Compete on Speed and Price

When motors fail, most end-users replace small motors and repair large ones because repairing is generally more expensive than replacing a small motor and less expensive than replacing a large one. Repair-or-replace decisions are generally made at the plant level, although a few large corporations have established guidelines for their plants. End-users select repair shops primarily on the basis of price and speed of service. Most motor repair shops do not provide the customer with any evaluation of the motor to be repaired or recommendations on replacement options unless the motor is severely damaged. To encourage competition and responsiveness, most end-users use more than one repair shop. Unless consistent reliability problems are encountered, the quality of the shops' repairs is not considered.

Maintenance Practices

Motor maintenance practices are generally limited to what is needed to keep equipment running rather than attempting to optimize performance and save energy. Most industrial plants and large commercial firms have full-time maintenance staff who regularly lubricate (and often overlubricate) motors, listen for bearing noise (a sign of wear or misalignment), and check and tighten belts as needed. Few firms do any more sophisticated monitoring or maintenance work on motor systems. According to some industrial observers, the time available for maintenance is becoming even more limited in some firms due to industrial company downsizing over the past decade, so the situation is likely to deteriorate.

Other Factors Influencing Decision-Making

Several other factors, in addition to those related specifically to motor systems, influence most efficiency-related investment. Some of the more important ones are discussed below.

- *Limited Information.* As noted above, most maintenance managers and other decision-makers are very busy, leaving little time to research new opportunities, including opportunities to save energy. This lack of time generally causes knowledge of energy-saving options to be limited. Only among large companies were the majority of decision-makers aware of the availability of premium-efficiency motors or decision-assisting tools. Adding to this confusion is publicity surrounding the EPAct motor standards, leading many users to mistakenly conclude that all motors are efficient and that they no longer need to pay attention to efficiency.

  To our knowledge, similar survey data are not available for other energy-saving measures, such as optimization of fan, pump, and compressed-air systems. Given the fact that these other opportunities are usually more complicated than purchasing improved-efficiency motors, the lack of information is likely to be even more of a problem for these other opportunities.

- *Limited Access to Capital.* The average end-user is more restrictive with capital than with operating funds. Generally, capital expenses are closely scrutinized and require approval at multiple levels in a company. To minimize capital outlay, companies tend to choose the least expensive equipment that will do the job satisfactorily.

  Operating funds, on the other hand, are relatively easy to obtain, since they are required for production. Operating budgets are typically based on expenses in previous years and are only seriously examined when out of line with expectations. Moreover, unlike capital costs, operating costs are paid with pretax dollars.

- *Payback Gap.* It is a curious fact that most firms look for a simple payback period of 2–3 years or less on energy projects and other operations and maintenance investments, even though longer paybacks are often considered when investing in new product lines. This difference, known as the payback gap, makes it difficult to implement all but rapid-payback energy-saving measures, although measures with longer paybacks will sometimes be considered as part of a major facility upgrade designed to improve the long-term competitiveness of the firm. The payback gap is most pronounced when viewed from the societal perspective—individual firms pass up energy-saving investments with paybacks of 3–4 years, while utilities invest in distribution lines with economic returns equivalent to 10- to 20-year paybacks.

- *Low Priority Assigned to Energy Matters.* For the average industrial firm, energy costs represent only a small percentage of total costs; labor and material costs are usually far greater. For example, in 1998 the U.S. Census's Annual Survey of Manufacturers estimated that, on average, electricity accounts for a little over 1% of manufacturing costs. Since motors make up about 70% of manufacturing electricity use (see Chapter 6), they make up about 1% of total costs for the average industrial firm. Since energy costs represent a small proportion of an average end-user's total operating costs, motor and other energy-related operating costs are rarely examined in reviews of operating expenses.

- *Transaction Costs.* Contributing to the low priority that energy matters take is the fact that many energy-saving measures, including motor measures, have substantial transaction costs. Comparing equipment or optimizing a system takes time, which is a commodity in short supply in many firms. For larger projects, outside engineers can be brought in to help with project design and implementation, but for small projects, if existing staff are short on time, decisions are commonly made based on expediency rather than economic merit.

- *Misplaced Program Emphasis.* Since they generally have full-time maintenance staff or energy managers, large firms are more likely to be interested in energy efficiency. Even in firms with energy managers, however, motor systems historically have not received much attention because of (often incorrect) perceptions that motor system improvements have high capital expense, low rates of return, and low percentage savings. Energy managers tend to focus on low capital cost measures with high savings. While this approach is reasonable during the start-up stages of an energy management effort, many firms have not moved beyond high-savings, low-cost measures. Moreover, many drivepower-saving measures are relatively inexpensive.

- *Lack of Internal Incentives.* For many companies, energy bills are paid by the company as a whole and not allocated to individual departments. This practice gives maintenance and engineering staff little incentive to pursue energy-saving investments because the savings in energy bills show up in a corporate-level account where the savings provide little or no benefit to maintenance and engineering decision-makers. As is discussed in Chapter 10, mechanisms to improve internal incentives have been put into place in some facilities.

This listing of the barriers to motor system improvement is by no means exhaustive. It does cover, however, enough of the major impediments to clarify the nature of the challenge. Fortunately, there are many ways to remove or lower these hurdles to sound investment. Some of the more important options are outlined briefly below and are covered in greater depth in Chapter 9.

Overcoming the Hurdles

In the intervening decade since the first edition of this book was published, significant progress has been made in improving motor system efficiency. We have made many steps toward improving the quality and availability of information on motors and motor system efficiency. Utilities, energy agencies, manufacturers, universities, and private organizations have developed publications, videos, seminars, and design and calculation aids. These products have been used across the country in programs discussed in Chapter 9. These products and programs have begun to have a significant impact on the motor market.

While significant steps have been made, more is needed. We discuss the perspectives and needs of these various players in the motor market in detail in Chapter 8.

With EPAct, we have minimum efficiency and motor labeling standards in place in the United States Now educational efforts are needed to make the market aware of these standards and to assist motor owners in making sound motor decisions. While EPAct eliminated the least efficient industrial motors from the market, motors significantly more efficient that EPAct levels are available. These more efficient products are cost-effective in most replacement applications and many retrofit applications, as discussed in Chapter 2. What is needed now is a brand to easily identify these products in the marketplace. National Electrical Manufacturers Association (NEMA), motor manufacturers, and voluntary programs, such as ENERGY STAR®, need to step up and implement a national premium-efficient branding program.

Financial incentives have proven useful in certain instances to overcome the perverse effects of the payback gap and motor users' limited access to capital. The impacts of these programs have been modest but have yielded important visibility for motor efficiency. We have also learned important lessons that are presently leading to improved programs. Recently, programs have shifted their focus from rebates for individual motor purchases to strategically moving the motor marketplace toward products and practices that are more efficient. Chapter 9 covers the experience to date with motor system programs.

In addition, the programs for increasing drivepower efficiency need to be broader in scope. Most drivepower efficiency programs have focused only on efficient motors instead of on the entire motor-decision process. A good program would address repair-versus-replace decisions, the implementation of life-cycle analysis of new motor purchase decisions, and the importance of demanding quality motor repairs.

Improved motor repair practices have long been identified as significant opportunities for energy efficiency. Unfortunately, we have only begun to see the first, tentative steps toward implementing programs to realize these savings. Research discussed in Chapter 2 has provided us with a foundation upon which programs can be built. We need to now focus on implementing programs that raise the standard of practice to the level of the best shops, which can restore a motor to near its original efficiency. Such programs need to work with repair shops to assist them in improving the quality of their services and also work with repair shop customers to help these customers understand why and how they can obtain quality repairs.

A number of programs were motivated by the opportunity created by ASDs, and have attempted to focus on motor-driven systems, particularly fan, pump, and compressed-air systems. As discussed in Chapter 7, the largest opportunities for cost-effective saving are in improved optimization of these systems. The success of these programs has been mixed to date, largely because of the site-specific effort required to identify and implement projects. However, some recent efforts that build on the successes and failures in this area show promise and provide a foundation for new motor system program designs that can help capture huge savings potential in this area. This process is addressed in Chapter 9.

Finally, most programs have ignored other efficiency-related topics, such as motor sizing, rewinding, and controls other than ASDs. Few programs that we know of have addressed the savings available from electrical tune-ups, better selection and maintenance of drivetrains and bearings, better system monitoring, and the upsizing of distribution wires in new installations. While the savings from these measures may appear incremental, they are frequently among the most cost-effective, and they also offer significant nonenergy benefits in the form of improved reliability and productivity.

Motor Technologies

M otors produce useful work by causing a shaft to rotate. The twisting force (torque) applied to the shaft is produced by the interaction of two magnetic fields, one produced by the fixed part of the motor (stator) and the other produced by the rotating component of the motor (rotor). The forces developed in a motor resemble the force between two magnets held close together: similar poles repel each other; dissimilar poles attract. If one of the magnets is mounted on a shaft, the attracting and repelling forces create torque (see Figure 2-1).

Figure 2-1

Torque Generation in a Motor

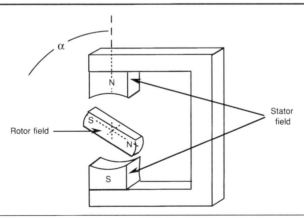

Note: The generated torque is proportional to the strength of each magnetic field and depends on the angle (α) between the two fields. Mathematically, torque equals $B_{rotor} \times B_{stator} \times \sin \alpha$, where B refers to a magnetic field.

Figure 2-2

Estimated Distribution of Input Energy by General Type of Motor, Based on Motor Sales in the United States

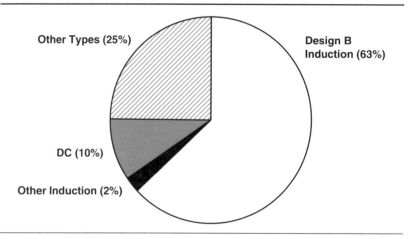

Other Types (25%)

Design B Induction (63%)

DC (10%)

Other Induction (2%)

Note: See Chapter 6 for further discussion.

A magnetic field can be generated either by a permanent magnet, in which case the field is constant, or by a winding in which an electric current flows. In the latter case, the magnetic field is generally proportional to the number of turns of wire in the winding and to the amount of current. The iron in the motor provides an easy path for the magnetic field in the same way that copper provides a low-resistance path for the electric current. A wire with a low resistance to current flow has high conductivity; a material, like iron, with a low resistance to a magnetic field has high permeability. Using a highly permeable material in the magnetic circuits of the rotor and stator reduces the amount of current required to produce a given magnetic field.

There are three basic types of electric motors: AC induction/asynchronous; AC synchronous; and DC. A detailed breakdown of motor types by horsepower and end-use appears in Chapter 6. Figure 2-2 shows the relative shares of electrical input used by different motor types. Because more than 90% of energy input goes to AC induction motors, this type is discussed in more detail than the others.

Principles of Induction Motors

Induction motors can be categorized by whether they run on single- or three-phase power. Houses are usually supplied with single-phase electricity. As a result, household appliances such as refrigerators, washers, dryers, heat pumps, and furnaces use single-phase motors. Utility companies provide most commercial and industrial facilities with three-phase service, which is used to run most motors larger than 1 hp. The overwhelming majority of motors are single-phase. Because of their relatively small size, however, single-phase motors account for less than 20% of the total drivepower energy input in the United States.

Many single-phase motors are integrated with the equipment they drive so when the motor fails, the equipment must be replaced. Three-phase motors are typically separate from equipment and can be easily replaced. Three-phase motors are emphasized in this section because they use more energy and are more readily replaced with high-efficiency models.

Rotating Field and Synchronous Speed

Three-phase induction motors, also called polyphase asynchronous motors, have three stator windings symmetrically arranged 120° apart in a cylinder surrounding the rotor. When supplied with three-phase power, also offset by 120° (see Figure 2-3), the windings act as electromagnets, creating a rotating magnetic field, which starts and drives the motor.

Figure 2-3

AC Sinusoidal Voltage for Single-Phase and Three-Phase Systems

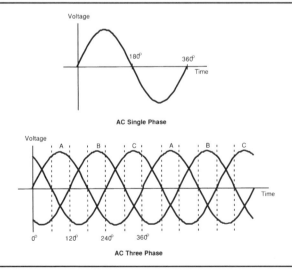

Figure 2-4

**Schematic of a Single-Phase Induction Motor
(Capacitor-Start Design)**

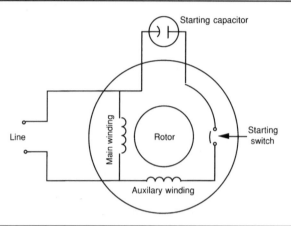

Note: In addition to the main winding there is an auxiliary winding offset by 90°, normally connected in a series with a capacitor. The sum of the magnetic fields generated by the main and the auxiliary windings is a rotating field of north and south magnetic poles that revolve around or move around the stator. The changing magnetic field from the stator induces a current in the rotor conductors, in turn creating the rotor magnetic field. Magnetic forces in the rotor tend to follow the stator magnetic fields, producing rotary motor action.

Source: Andreas 1982

Because single-phase motors do not have a three-phase field, they require a special starting system that employs an auxiliary winding, offset 90° from the main winding, which is normally connected in series with a capacitor. In some designs, the auxiliary winding and capacitor are disconnected after the motor starts, by a centrifugal or thermal switch; such machines are commonly known as capacitor-start motors. Motors that do not disconnect the capacitor are known as permanent split-capacitor (PSC) motors. There are also motors that combine the two designs, using one capacitor for starting and another for normal operation; these are known as capacitor-start, capacitor-run motors. The basic circuitry of a single-phase motor is shown in Figure 2-4.

The speed of the rotating magnetic field in an induction motor, known as the synchronous speed, depends on the frequency of the supplied voltage and the number of pole pairs in the motor. This is expressed as the following equation:

$$\text{synchronous speed (rpm)} = \frac{\text{Frequency of applied voltage (Hz)} \times 60}{\text{number of pole pairs}}$$

Thus, when a motor with two poles (one pole pair) is supplied by a 60-cycle-per-second (hertz [Hz]) supply, the synchronous speed is 3,600 rpm. A four-pole motor supplied with 60 Hz power has a synchronous speed of 1,800 rpm, and a six-pole motor has a synchronous speed of 1,200 rpm.

Induction Motor Slip

Induction motors are referred to as asynchronous motors because they operate slightly below synchronous speed. For example, a motor with four poles and a synchronous speed of 1,800 rpm will actually spin between 1,725 and 1,790 rpm.

The difference between the synchronous and actual speeds of an induction motor is called the motor slip. Slip is expressed either as a percentage of synchronous speed or as revolutions per minute. For example, a four-pole induction motor with a synchronous speed of 1,800 rpm operating at 1,750 rpm has a slip of 2.8%, or 50 rpm. The full-load motor slip ranges from 4% in small motors to 1% in large motors (see Figure 2-5).

Figure 2-5

Full-Load Revolutions per Minute vs. Horsepower for Four-Pole Induction Motors

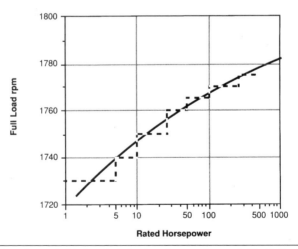

Note: The stepped curve is data from typical motors; the smooth curve is fitted to the data. For four-pole induction motors, 1,800 rpm is the synchronous speed (approximately the speed under no load). The full-load speed is less than the synchronous speed; this difference (or "slip") is smaller for larger motors.

Source: Nailen 1987

Figure 2-6

**Relationship between Active, Reactive, and Total Current for
(A) High Power Factor (90%) and (B) Low Power Factor (45%)**

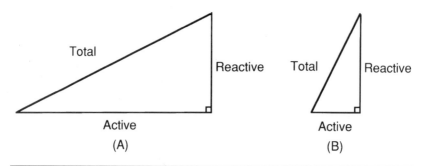

Note: The active current is 90% of the total in (A) and 45% of the total in (B). Active, reactive, and total power follow the same relationship as the current.

Power Factor

The current in an induction motor has two components: active and reactive. The active component is responsible for the torque and work performed by the motor; the active part of the current is small with no load and rises as the load increases. The reactive component creates the rotating magnetic field and is almost always constant from no load to full load, as is the magnetic field.

Although the reactive component does not perform useful work, it is required to excite the motor and must be supplied by the power network. The ratio of active to total current is called the power factor (see Figure 2-6).

When the motor is operating at no load, the energy it absorbs is limited to the power losses (motor inefficiencies). Therefore, the active component is small, and the power factor can be as low as 10%. At full load, however, the active component is at its maximum with a power factor that is typically 70–95% for a three-phase motor. A high power factor is desirable since it implies a low reactive-power component. (Power factor is commonly expressed as a percentage or a decimal fraction.) A poor power factor has the following effects:

- Higher losses in the cables and transformers, and thus higher energy bills for a given amount of useful work output

- A reduced available capacity of transformers, circuit breakers, and cables because their output depends on the total current; the capacity falls linearly as the power factor decreases: a 1,000-kilovolt-ampere (kVA) transformer supplying loads with a 70% power factor is only able to supply 700 kilowatts (kW)

- Higher voltage drops, yielding problems associated with undervoltage, as discussed in Chapter 3

These effects have caused most utility companies to penalize consumers whose power factor is below a threshold level, typically in the range of 85–95%. Thus, when consumers improve the power factor, they reduce both the energy bill and the reactive-power bill. As discussed in Chapter 3, the savings from avoided utility penalties are typically larger than the energy savings from power-factor correction. Measures to improve power factor are discussed in Chapters 3 and 4.

Types of Induction Motors

According to the rotor configuration, induction motors are classified as either squirrel-cage or wound-rotor. Squirrel-cage induction motors are the most common and are either three-phase or single-phase.

Squirrel-Cage Induction Motors

Most induction motors contain a rotor in which the conductors, made of either aluminum or copper, are arranged in a cylindrical format resembling a "squirrel cage" (see Figure 2-7). Squirrel-cage induction motors are used in the vast majority of commercial and industrial applications because they are relatively simple, inexpensive, reliable, and efficient.

Squirrel-cage induction motors have no external electrical connections to the rotor, which is made of solid, uninsulated aluminum or copper bars short-circuited at both ends of the rotor with solid rings of the same metal. The rotor and stator are connected by the magnetic field that crosses the air gap. This simple construction results in relatively low maintenance requirements.

The relationship between torque and speed in squirrel-cage motors is largely dependent on rotor resistance. As the rotor resistance decreases, the performance speed improves and the starting torque decreases. The smaller the slip for a given load, the higher the efficiency, because the induced rotor currents and their associated rotor losses are also smaller.

Three-phase squirrel-cage induction motors dominate applications above 1 hp. Single-phase squirrel-cage induction motors are more common in sizes below 1 hp and in large home appliances. Single-phase motors are larger and more expensive, with a lower efficiency than

Figure 2-7

Operation of a Four-Pole Squirrel-Cage Induction Motor

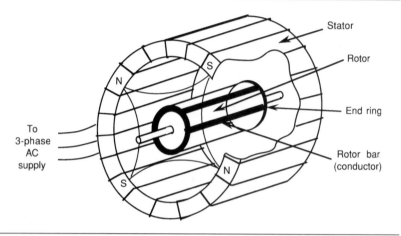

Note: The rotating magnetic field is created in the stator by AC currents carried in stator windings. A three-phase voltage source results in the creation of north and south magnetic poles that revolve around or move around the stator. The changing magnetic field from the stator induces current in the rotor conductors, in turn creating the rotor magnetic field. Magnetic forces in the rotor tend to follow the stator magnetic fields, producing rotary motor action.

Source: Lawne 1987

three-phase motors that have the same power and speed ratings. For example, the full-load efficiency of a 2 hp, 1,800 rpm, three-phase, standard-efficiency motor is 72% with a power factor of 62%, whereas the efficiency of a 2 hp single-phase motor from the same manufacturer is 66.2% with a power factor of 62.1%. Additionally, three-phase motors are more reliable since they do not need special starting equipment. Thus, they are typically used whenever a three-phase supply is available. In commercial and industrial installations involving a large number of small motors, single-phase models have the further disadvantage of causing voltage unbalance if they are unevenly distributed on the three phases (see the discussion of voltage unbalance in Chapter 3).

Shaded-Pole Motors

Another type of induction design, the shaded-pole motor, is most commonly used in packaged equipment applications below 0.17 (1/6) hp, such as computers, small fans found in portable heaters, and small condensing units for air conditioning and refrigeration. Although shaded-pole motors are cheaper than single-phase squirrel-cage motors, their efficiency is poor (below 20%) and their use should be restricted to low-power applications with a limited number of operating hours. For low-power applications with longer operating hours, higher-efficiency single-phase motors should be used, such as the permanent split-capac-itor (a type of squirrel-cage motor discussed previously) or permanent-magnet units (discussed later in this chapter).

Wound-Rotor Induction Motors

Wound-rotor induction motors are sometimes used in industrial ap-plications, typically 20 hp or larger, where the starting current, torque, and speed must be precisely controlled. As the name suggests, these motors feature insulated copper windings in the rotor similar to those in the stator. The rotor windings are fed with power using slip rings and brushes. This rotor construction is substantially more expensive, with higher maintenance requirements, than the squirrel-cage type.

Factors to Consider in Selecting Induction Motors

Some of the factors described in this section apply to all kinds of mo-tors, including noninduction designs. Other factors, like the National Elec-trical Manufacturers Association design classes, apply exclusively to squir-rel-cage induction motors, particularly three-phase versions. The technical information required to apply a motor can be found on the nameplate (see Figure 2-8), and most is also available in manufacturers' catalogs.

Figure 2-8

Sample Motor Nameplate Showing Nominal and Minimum Efficiency

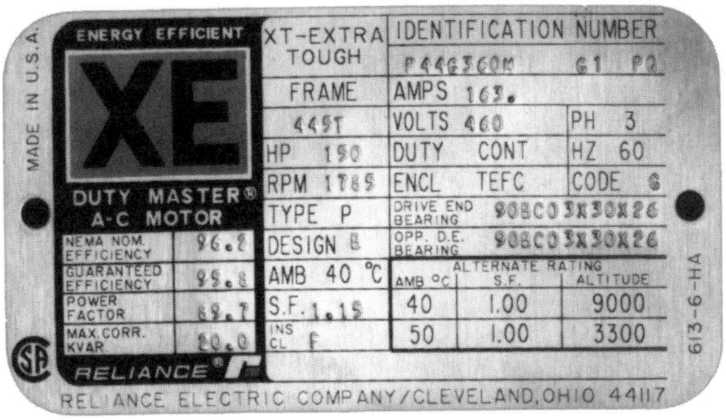

Source: Reprinted with permission from Reliance Electric

NEMA Designs

The type of load determines the type of motor chosen to drive it. NEMA has defined standards for different types of squirrel-cage designs to meet the needs of different operating conditions. NEMA standard designs fall into five categories: A, B, C, D, and E.

- *Design B motors* are the dominant type on the market and are used for most applications, including fans, pumps, some compressors, and many other types of machinery. "Normal" torque is defined by that which is produced by a Design B motor. The torque peaks at approximately 80% of the synchronous speed. Design B units have a "normal" starting current of approximately five times the full-load current. In manufacturers' literature, general purpose motors are Design B motors.

- *Design A motors* are similar to Design B motors except that the maximum torque is 15–25% higher. The starting current is six to seven times the full-load current because of the design tradeoffs necessary to increase peak torque.

- *Design C motors* are characterized by high torque. Substantial starting torques make them useful for machines that can start with a full load (such as conveyors or some compressors).

- *Design D motors* have high starting torque and high slip. High slip allows the motor speed to vary somewhat from the rated speed. As a result, Design D motors are normally used where there is the potential for a shock load on the motor, as in punch presses and shears, since the motors' ability to adjust their speed will act as a shock absorber and protect the driven equipment. Design D motors have the lowest efficiency for a given size and speed because of their high slip.

- *Design E motors* have lower starting torque, high starting current, and low slip. As a result of the high inrush current, Design E motors require special starting wiring, motor control, and other related equipment (NEMA 1999). They are the most efficient induction motor class, but as a result of design compromises necessary to achieve high efficiency, they may not be capable of starting under a significant load. Design E motors are used predominantly in fan applications, where the lower starting torques are not a problem. Design E motors have not achieved a significant market and may be discontinued in the future (Bonnett 1999).

Figure 2-9 shows the available torque as a function of speed for NEMA Design A, B, C, D, and E motors.

Figure 2-9

General Shape of Torque vs. Speed Curves for Induction Motors with NEMA Designs A, B, C, D, and E

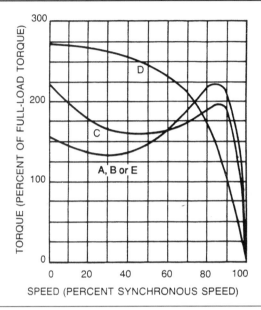

Note: The torque at zero speed is the starting torque. The full-load torque occurs at speeds somewhat below the synchronous speed (i.e., at the full-load speed) (NEMA 1994). See Appendix C for further discussion of torque.

Available Speeds

As mentioned earlier, the number of pole pairs is determined by the synchronous speed of the motor. Induction motors are available with synchronous speeds of 3,600, 1,800, 1,200, 900, 720, 600, 450, and 300 rpm when operated at 60 Hz. (When operated at 50 Hz, the corresponding synchronous speeds are 3,000, 1,500, 1,000, 750, 600, 500, 375, and 250 rpm.) Actual speeds are slightly lower because of motor slip. For a given horsepower, as the speed decreases, the number of poles increase, costs increase, and the efficiency and power factor are reduced (see Figure 2-10).

The 1,800 rpm motor probably accounts for more than 50% of the motor population. Both 1,200 and 3,600 rpm motors are popular enough to be stocked by distributors and manufactured in large quantities. Those slower than 1,200 rpm are often treated as special orders. For fixed speeds lower than 300 rpm, it is generally more economical to use a motor with a

Figure 2-10

Typical Full-Load Efficiencies and Power Factors

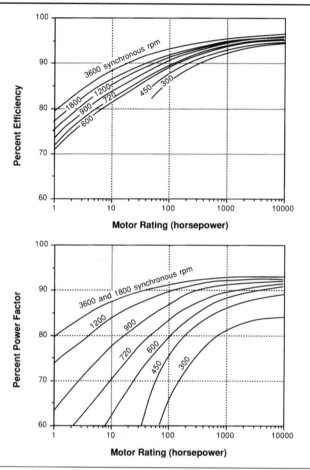

Note: Typical full-load efficiencies (top) and power factors (bottom) for standard-efficiency NEMA Design B motors (normal-torque, low-starting-current, three-phase induction motors) with synchronous speeds from 300 to 3,600 rpm. These are general trends; values for particular motors may vary.

Source: Smeaton 1987

speed of 300 rpm or more, combined with a mechanical transmission system like gears or belts, to achieve the desired speed reduction.

Enclosures

Motor enclosures are designed to match the motor to its operating environment. An open enclosure allows heat to dissipate readily, leading to better motor cooling. It offers less protection, however, against the entry of potentially damaging foreign objects such as dirt, metal pieces, and water. NEMA Standard MG 1-1978 (NEMA 1978)

Motor Enclosures

The *open-enclosure* types normally used are as follows:

- *Open drip-proof (ODP),* in which the ventilation openings are positioned to keep out liquid or solid particles falling at any angle between 0° and 15° from the vertical. This enclosure is not adequate for harsh environments but is common in applications driving fans and pumps in indoor heating, ventilating, and air conditioning (HVAC) applications.

- *Splash-proof,* which is the same as drip-proof except that the downward angle of the vents is increased to 100° so that liquid or solid particles arriving at a slightly upward angle will not enter the motor.

- *Guarded,* in which all openings giving direct access to rotating or electrically live metal parts are limited in size by screens, baffles, grilles, or other barriers to the entry of objects larger than 0.75 inch in diameter. Thus, insects and dirt are not prevented from entering. The purpose of guarded fittings is more to protect personnel than the motor.

- *Weather protected,* for outdoor use, include guarded enclosures and ventilation passages designed to minimize the entrance of rain, snow, and airborne particles.

Totally enclosed machines are designed to prevent the free exchange of air between the inside and outside. The most common types are as follows:

- *Totally enclosed fan-cooled (TEFC),* in which the motor is equipped with a fan for external cooling. Normally the external fan is mounted on the shaft opposite the load and equipped with a guard to improve safety and aerodynamics.

- *Explosion-proof (EXP),* in which the enclosure is designed to withstand the explosion of a specified gas or vapor within and prevent ignition of a specified external gas or vapor by sparks, flashes, or explosions that may occur inside the motor casing. These enclosures may be fan-cooled (EXPFC) or nonventilated (EXPNV).

- *Dust-ignition-proof,* which are designed to exclude dust, ignitable or not, that might affect performance or rating, while preventing external dust from being ignited by arcs, sparks, or heat generated from within.

Figure 2-11

Motor Enclosure Types

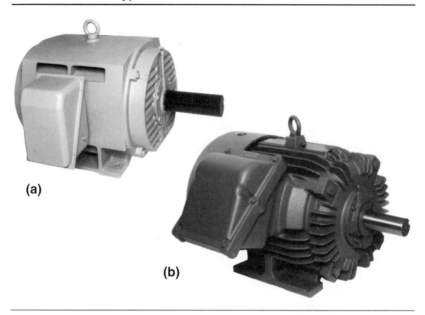

(a)

(b)

Note: The most common motor enclosures are (a) ODP and (b) TEFC. In ODP enclosures such as the one shown, internal fans bring cooling air into the motor through openings in one or both ends and then discharge it through openings in the side. Another ODP design brings in air through one end and discharges it through the other. In either case, the cooling air flows directly through the motor. In TEFC designs, there is no air exchange between the inside and outside of the motor. A fan, driven by an extension of the motor shaft, and shown here in the smooth housing on the left end of the motor, pulls air through slots in its housing and then blows it over the exterior of the motor, which is usually made with fins (as shown) for cooling. Figure 2-20 is a cutaway view of a TEFC motor; the internal arrangement of both motor types is basically the same.

Source: Reprinted with permission from Toshiba

defines twenty types of enclosures clustered into two basic groups: open and totally enclosed (see box). The most common enclosures used in commercial and industrial facilities are open drip-proof (ODP); totally enclosed fan-cooled (TEFC); and explosion-proof (EXP), a type of TEFC motor (see Figure 2-11). Each of these three basic types of enclosures has subsets listed in catalogs for special environments (agricultural, corrosive, or wet conditions, for example).

Temperature Ratings and Classes of Insulation

Motor losses are transformed into heat, which increases the temperature of the motor. Table 2-1 shows the insulation class required to

Table 2-1

Allowable Temperature Increases (°C) for Various Insulation Classes (NEMA Standards)

	Insulation Class			
	A[a]	B	F	H
Open or TEFC motors with 1.0 service factor, rise at rated load (°C)	60	80	105	125
All motors having service factor, rise at 115% rated load (°C)	70	90	115	135[b]

[a] Of historical interest only. Class A insulation is no longer used in integral-horsepower motors. Classes C, D, E, and G were never used.
[b] Not NEMA standard but common industry practice.

Source: Nailen 1987

withstand different temperature rises according to NEMA Standard MG 1-1998 (NEMA 1999). Class A is no longer made. Class B is the most common. Classes C, D, E, and G were never used. Classes F and H are used in applications with high ambient temperatures in order to allow a larger reserve margin for overload conditions or to enable the design of smaller, less expensive motors for intermittent-duty operation. Most efficient motors use Class F insulation as part of the general package to upgrade performance.

Service Factor

The service factor specifies the capacity of the motor to withstand prolonged overload conditions. When the service factor is 1.0, prolonged operation above full load can damage the insulation and cause the motor to fail. If the service factor is 1.15, the motor can work at 1.15 times its rated horsepower without failing, although insulation life may be reduced (typically by 50% when compared with the same motor working at full load). Standard service factors for 3,600 and 1,800 rpm motors range as high as 1.35 for 0.5 hp and smaller, 1.25 for 0.5 and 0.75 hp, and 1.15 for 1 hp and above. Motors running at 1,200 and 900 rpm generally have lower service factors. However, service factors of 1.5 or more are available on special order. In general, the class of insulation on the motor windings determines the service factor. Motors above 1 hp with Class B insulation have a service factor of 1.0, whereas motors with Class F insulation have a service factor of 1.15.

Frame Size

The frame size defines the shape and size of the motor and depends on horsepower, speed, voltage, and duty requirements. Motors built prior to 1952 did not use industry-wide standard frame sizes. Then the U-frame was standardized, and all motors with the same code, such as 254U, had the same frame size. With the advent of new high-temperature insulation, NEMA authorized smaller, lighter T-frames in 1964, which remain the prevalent type for new three-phase motors. Most small (under 1 hp) and very large (over 300 hp) motors use frames other than T- or U-designs. Standard-efficiency U-frame designs are still made for replacing worn-out or damaged U-frames because replacing a U-frame motor with a standard- or high-efficiency T-frame unit typically requires modification of the mounting hardware and is therefore not practical for all applications. Thus, not all existing U-frame motors can be replaced with high-efficiency T-frame models.

Frame size is an important determinant of motor efficiency and performance. To reduce production cost, manufacturers often try to fit a motor into the smallest possible frame but, in so doing, they must limit the service factor to ensure that the motor will not overheat under the reduced cooling of the smaller frame. Thus, some manufacturers build 5 hp motors in frames typically used for 3 hp motors: to meet cooling requirements, they will have a service factor no higher than 1.0 (Gilmore 1990).

Supply Voltage

Most three-phase motors in the United States are designed to operate at 460 volt (V), 60 Hz, which allows some voltage drop from the nominal 480 V supply commonly used in newer large commercial and industrial facilities. Smaller commercial facilities often use either 230 V or 208 V three-phase power. Older facilities (both commercial and industrial) sometimes use 575 V three-phase power. To decrease the distribution losses in cables and transformers, large motors (200 hp and above) can be specified with a supply voltage over 600 V. Fractional-horsepower single-phase motors most commonly run on 120 V power.

Before we continue, we should clarify some terms. The Energy Policy Act of 1992 (EPAct) is an important piece of legislation for efficiency because it established minimum-efficiency levels for electric motors manufactured or imported after October 1997. EPAct, which was based on NEMA standards, defined a number of terms, including what constitutes an *energy-efficient* motor. This concept will be described in greater detail later in this chapter (see Table 2-9) and in Appendix B. In the wake of the ruling, industry began manufacturing motors that exceeded EPAct standards and became alert to labeling and marketing these more efficient motors. To accommodate these motors, the Consortium for Energy Efficiency (CEE)—a nonprofit group in which utilities, public interest groups, and government agencies such as the Department of Energy have representatives—established *premium-efficiency* motor levels. These specifications are also outlined in Table 2-9.

Specialty motors are available to run on two voltages (230/460 V, for example) or on a range of voltages along with some other, typically

---

How to Read a Motor Catalog

The specific information presented in a motor catalog and the format for that information vary among manufacturers. Figures 2-12 and 2-13 show sample pages from one catalog.

Most catalogs cluster information by specific motor types. For example, all ODP EPAct-compliant motors are listed in a single table that contains generic information on the type of mounting system, the housing, the materials of construction, the insulation class, the service factor, and the design rating. EPAct requires a listing of the nominal full-load efficiency for the motor, and catalogs may include a special symbol ("ee") for those motors that meet the EPAct minimum requirements (see Appendix B). In most catalogs, the general listing will also specify whether the motor line is EPAct or premium-efficiency. Some also list whether the motor meets the CEE premium-efficiency specification.

In addition to generic information, other tables will list size, speed, frame number, full-load amps, and list price for individual motors. Most motors actually sell for 30–70% of the list price.

Most catalogs outline in a separate table motor efficiencies and power factors at full, 0.75, and 0.5 load. This table also typically contains data on motor torque.

Finally, most motor catalogs include a table of dimensional data organized by the motor frame number. In the NEMA system, all motors with the same frame number are the same size. Dimensional data are generally used to determine the changes required in the mounting system or drive shaft when downsizing a motor or converting from a U-frame motor to a T-frame motor.

Figure 2-12

Typical Motor Application Data from a Major Manufacturer's Catalog

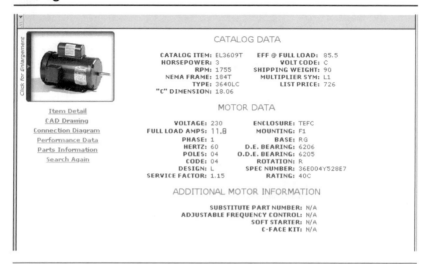

CATALOG DATA

CATALOG ITEM: EL3609T	EFF @ FULL LOAD: 85.5
HORSEPOWER: 3	VOLT CODE: C
RPM: 1755	SHIPPING WEIGHT: 90
NEMA FRAME: 184T	MULTIPLIER SYM: L1
TYPE: 3640LC	LIST PRICE: 726
"C" DIMENSION: 18.06	

MOTOR DATA

VOLTAGE: 230	ENCLOSURE: TEFC
FULL LOAD AMPS: 11.8	MOUNTING: F1
PHASE: 1	BASE: RG
HERTZ: 60	D.E. BEARING: 6206
POLES: 04	O.D.E. BEARING: 6205
CODE: 04	ROTATION: R
DESIGN: L	SPEC NUMBER: 36E004Y528E7
SERVICE FACTOR: 1.15	RATING: 40C

ADDITIONAL MOTOR INFORMATION

SUBSTITUTE PART NUMBER: N/A
ADJUSTABLE FREQUENCY CONTROL: N/A
SOFT STARTER: N/A
C-FACE KIT: N/A

Item Detail
CAD Drawing
Connection Diagram
Performance Data
Parts Information
Search Again

Source: Baldor 2001

Figure 2-13

Typical Motor Performance Data from a Major Manufacturer's Catalog

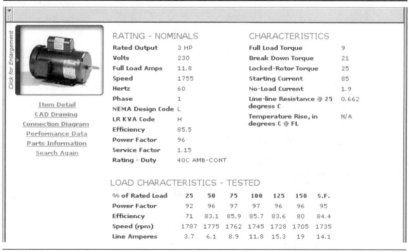

RATING - NOMINALS

Rated Output	3 HP
Volts	230
Full Load Amps	11.8
Speed	1755
Hertz	60
Phase	1
NEMA Design Code	L
LR KVA Code	H
Efficiency	85.5
Power Factor	96
Service Factor	1.15
Rating - Duty	40C AMB-CONT

CHARACTERISTICS

Full Load Torque	9
Break Down Torque	21
Locked-Rotor Torque	25
Starting Current	85
No-Load Current	1.9
Line-line Resistance @ 25 degress C	0.662
Temperature Rise, in degrees C @ FL	N/A

Item Detail
CAD Drawing
Connection Diagram
Performance Data
Parts Information
Search Again

LOAD CHARACTERISTICS - TESTED

% of Rated Load	25	50	75	100	125	150	S.F.
Power Factor	92	96	97	97	96	96	95
Efficiency	71	83.1	85.9	85.7	83.6	80	84.4
Speed (rpm)	1787	1775	1762	1745	1728	1705	1735
Line Amperes	3.7	6.1	8.9	11.8	15.3	19	14.1

Source: Baldor 2001

doubled, value (such as 208–230/460 V). In recent years it has become common to use motors rated at 230 V for 208 V applications. Although these motors will operate at any voltage in this range, at the reduced operating voltage their efficiency, life, and torque will decrease, while slip increases. For NEMA Design B motors, slip will typically increase 30%, torque will decrease 20–30%, and efficiency will decline by 3%. Design E motors are less affected by reduced-voltage operation (Bonnett 1999).

Motors designed to operate at 60 Hz can be satisfactorily operated at 50 Hz if the voltage and horsepower rating are reduced by five-sixths. All other performance characteristics will be essentially the same as at 60 Hz (NEMA 1999).

It is common to see the actual voltage and frequency supplied to the motor significantly differ from the design values because of problems in the electric supply system, as discussed in Chapter 3. Frequent operation at off-voltage can have a significant impact on motor performance. Steps should be taken to operate the motor at as close to the design voltage as possible (Bonnett 1999).

General, Special, and Definite Purpose Induction Motors

Most NEMA Design A and B motors are *general purpose*. NEMA defines a general purpose motor as an open or closed motor, 500 hp or less, rated for continuous duty, without special mechanical construction, that can be used in *usual* service conditions without restrictions to a particular application or type of application. If a standard-rating or construction motor is designed to operate under conditions other than *usual* or in a particular application, it is classed as *definite purpose*. Examples are motors designed to occasionally be submerged in water when not running, or in a salt spray environment. Some applications require special mechanical construction, or operating specifications, or both, designed for a particular application. These motors are referred to as *special purpose*. Examples are motors designed to operate in a vertical shaft position that requires the use of thrust bearings, and motors with windings that are encapsulated to operate in a corrosive environment (NEMA 1999). In general, motors with these special features are more costly than general purpose motors, and some of the features may reduce the energy efficiency of the motor compared with that of a similar design that does not incorporate the features. Appendix B discusses features that characterize a motor as general or special or definite purpose and that are likely to reduce efficiency.

Other Types of Motors

Although induction models use more than 90% of all motor input energy, they are not appropriate for all applications. The following varieties of motors are also important.

Synchronous Motors

Synchronous motors have a stator similar to that of induction motors, with three windings that produce a rotating field. The rotor contains a winding to produce the rotor field and a starting winding similar to the rotor of a squirrel-cage induction motor; the connection from the power supply to the rotor field winding is made through slip rings and brushes. Because of their complex rotors, synchronous motors are more expensive to build and maintain than induction motors.

The starting winding makes the motor act like an induction motor at up to about 95% of the synchronous speed. At that point, the rotor field winding is switched on, and the rotor quickly catches up to the rotating field, reaching the synchronous speed. For further information on the operation of synchronous motors, refer to Nailen 1987, Fitzgerald 1983, or Smeaton 1987.

Figure 2-14

General Areas of Application of Synchronous Motors and Induction Motors

---

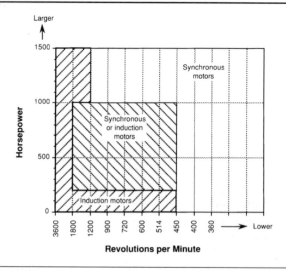

Source: Nailen 1987

Synchronous motors can run at lower speeds and are slightly more efficient than induction motors, especially at low speeds. They also have the virtue of being able to generate or absorb reactive power, whereas induction motors only absorb reactive power. A large synchronous motor can thus correct the overall power factor of an entire plant by generating the reactive power absorbed by the induction motors in the plant. Figure 2-14 shows the typical speed and horsepower ranges of induction and synchronous motors. Synchronous motors are used in applications in which fixed constant speed is required, such as in the textile fiber industry, or in large-power, low-speed applications, where the motor's additional cost is offset by its higher efficiency and capability for power-factor compensation. Synchronous motors tend to be large and in operation most of the time. There are few in use, and their number has been decreasing in recent years. Their percentage contribution to energy and power demands is therefore small.

Direct-Current Motors

Direct-current motors normally have windings in both the stator and the rotor. As the name implies, DC motors are fed by a DC voltage, which may change in magnitude but not in polarity. Thus, the magnetic field produced by the stator has a constant orientation, though its size may change as a function of the voltage applied to the terminals. DC motors are often used for applications in which speed control is required since varying the voltage changes the motor speed.

In a DC motor, electricity reaches the windings in the rotor via a ring of electrically isolated copper bars, a device known as a commutator (see Figure 2-15). Corresponding contacts known as brushes are connected to the power supply and ride against this commutator. DC motor rotors are complex, expensive to manufacture, and unreliable because of wear on the brushes and commutator caused by sparking and friction as the rotor turns. The wear creates the need for frequent inspection and replacement of the brushes. In addition, the commutator must be repaired or replaced at longer intervals. These motors have additional drawbacks in larger sizes: they are bulky; cannot sustain high speeds; and are less efficient than AC motors of similar size.

There is one type of DC motor, in which the rotor and stator windings are connected in series, that can be used with AC voltages (because the torque in series motors maintains the same direction even if the polarity of the voltage is reversed). They are known as universal motors and are generally found in small portable appliances and power tools. For fractional-horsepower sizes, the universal motor has a superior power-to-weight ratio. These motors generally operate for very limited periods so their energy use is not significant.

Figure 2-15

Typical DC Motor Design Showing Major Parts

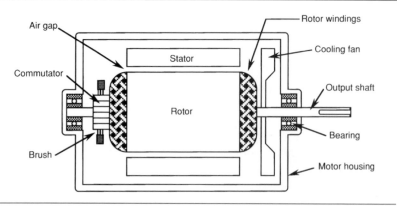

Note: The stator windings and connections to the brushes and stator are not shown.

Source: Bodine 1978

Besides being used in low-energy-use applications such as small appliances and power tools, DC motors are still being sold for industrial applications requiring very high starting torque or inexpensive speed regulation. However, with the advent of high-performance AC drives, their market share is dwindling to below 5%.

Permanent-Magnet Motors

In some small DC models, a permanent magnet replaces the stator winding, although the rotor is still fed by a conventional brush-and-commutator system. A more important type of permanent-magnet (PM) motor has a stator with three windings producing a rotating field, as in induction and synchronous motors. The rotor consists of one or more permanent magnets that interact with the rotating field so as to align the poles in the rotor with the poles of the rotating field. Thus, the speed of the motor is the speed of the rotating field. Because there is no rotor current and the rotor magnetic field is constant, there are no losses in the rotor, helping to make PM motors more efficient (by 5 to 10 percentage points in small sizes) than induction motors.

The most common form of PM motor is the brushless DC motor, also known as an electronically commutated motor (see Figure 2-16). Electronically commutated permanent-magnet motors (ECPMs) consist of a rotor with multiple permanent magnets bonded to it and a stator made of electrical windings that create a varying magnetic field to drive

Figure 2-16

Schematic of Electronically Commutated Permanent-Magnet Motor

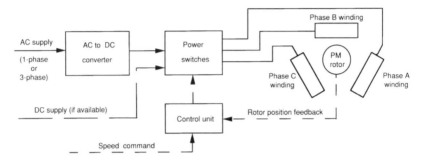

Note: The motor (on the right) is composed of three sets of stator windings arranged around the PM rotor. AC power is first converted to DC and then switched to the windings according to the signals provided by the control unit, which responds to both the desired speed ("speed command") and rotor position feedback. If a DC supply is available, it can be used in place of the AC supply and converter. The function of the commutator and brushes in the conventional DC motor is replaced by the control unit and power switches. The PM rotor follows the rotating magnetic field created by the motor windings. The speed of the motor is easily changed by varying the speed of switching.

Figure 2-17

Compared Efficiencies of 10 hp AC Induction Motors with ASD and Brushless DC Motors as a Function of Speed

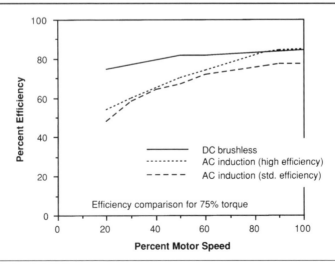

Source: Lovins et al. 1989

the rotor. The stator field is driven electronically using solid-state power devices and feedback from angular-position sensors. This arrangement eliminates rotor resistive losses, brush friction, and maintenance associated with conventionally commutated motors. Other advantages are precise speed control, lower operating temperature, and higher power factor than for induction motors. ECPM efficiency cannot match induction motor efficiency for fixed-speed, full-load operation but has a significant advantage at reduced speeds. Under these part-load conditions, induction motor efficiency drops significantly, while ECPM efficiency remains flat (Nadel et al. 1998).

Typical induction motor/adjustable-speed drive combinations have a range of full-load efficiencies between 85% and 90%, but this falls off 15–20 points at light load. ECPMs, in comparison, can maintain their efficiencies at part load within 5 points, with full-load efficiency as high as 95% in 100 hp sizes (E Source 1999). Figure 2-17 shows the comparative performance of AC induction motors with ASD and brushless DC motors, rated at 10 hp. Cost premiums for ECPMs currently are on the order of $50/hp (Nadel et al. 1998).

ECPMs are available from many manufacturers in sizes ranging from fractional to 60 hp. Powertec Industrial Corporation produces integral-horsepower motors in NEMA frame sizes to compete directly with induction motors. GE, Emerson, and A.O. Smith produce small fractional-horsepower integral ECPMs for use in HVAC equipment. In addition to high part-load efficiency and variable speeds, these motors have several unique features that make them particularly attractive to HVAC equipment manufacturers, including the ability to maintain a constant air flow and to ramp up to speed slowly (Nadel et al. 1998). Other kinds of PM motors include (1) small DC motors that have brushes, a commutator, and a wound rotor plus a PM stator; and (2) a type of AC synchronous motor available on special order from Siemens U.S. and Reliance Electric (Lovins et al. 1989).

PM motors have in the past been limited to fractional-horsepower sizes because they are bulkier and much more expensive than induction motors. The magnets in most fractional-horsepower PM motors, such as those used in residential appliances, are made from ferrite, primarily because of its low cost. In recent years, however, the performance of PM materials has improved dramatically (see Figure 2-18). In particular, neodymium-iron-boron alloys feature high energy density at moderate cost. Such PM materials allow the design of compact and high-efficiency motors in larger sizes up to 600 hp (E Source 1999). Improved materials can also yield very high efficiencies. A 50 hp PM motor with an efficiency of 97% has been developed (EPRI 1989).

Figure 2-18

The Evolution of Permanent-Magnet Materials, Showing the Increasing Magnetic Energy Density ("Energy Product")

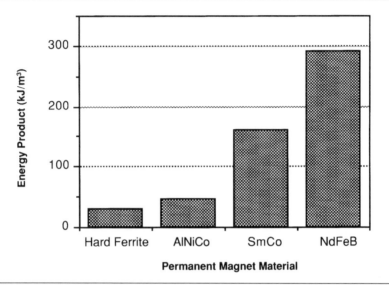

Permanent Magnet Material

Note: Ferrites were developed in the 1940s; AlNiCos (aluminum, nickel, and cobalt) in the 1930s. The "rare earth" magnets were developed beginning in the 1960s (samarium-cobalt) and in the 1980s (neodymium-iron-boron). The higher the energy density, the more compact the motor design can be for a given power rating.

Sources: Baldwin 1989; Krupp-Widia 1987

Since PM motors have neither rotor windings nor slip rings, they can be as robust and reliable as induction motors. They must be totally enclosed, however, to avoid attracting iron particles. The magnets can be demagnetized by high temperature, and therefore the motor cannot be overheated. For example, the neodymium-iron-boron magnets can be demagnetized if the temperature exceeds 302°F (150°C). However, the low losses of PM motors mean that their operating temperature is well below 300°F at the rated power.

PM motors coupled with electronic speed controls are already being used in cordless power tools, as well as residential air conditioners, furnaces, and heat pumps. Refrigerators and freezers are likely candidates for PM motor applications. Because of their high efficiency and reliability, and the recent availability of high-performance magnetic materials at reasonable costs, PM motors offer great promise as a general purpose motor.

Reluctance Motors

Reluctance designs are another promising family of motors that are synchronous but do not require electrical excitation of the rotor. Losses are lower and efficiencies are generally higher than in induction motors because no current is induced in the rotor. The shaft power of reluctance motors is smaller than in similar-sized PM motors. Reluctance motors are well established in very low power applications such as clocks, timers, and turntables, in which an inexpensive, low-power, constant-speed motor is required.

A variation known as the switched-reluctance (SR) motor shows great promise as a future competitor to the induction motor, especially in adjustable-speed applications. The switched-reluctance drive is a compact and efficient brushless, electronically commutated AC motor with high efficiency and torque, variable-speed regulation, and simple construction. Available in virtually any size, the SR motor offers the advantage of variable-speed capability (very low to very high) and precision control. Switched reluctance is an old idea that has advanced recently with progress in solid-state electronics and software that allows precision control. The motor comes as a package integrated with a controller (Nadel et al. 1998).

SR motors with control systems are competing to supplant induction motors with variable-speed drives in a number of applications. Both are attractive in new and OEM installations because they come as a motor-controller package. The SR motor is now being used in the Maytag Neptune line of horizontal-axis residential and commercial washing machines. Other likely applications include residential heating and cooling and commercial HVAC fans and pumps. Most SR research and application in the United States is in fractional-horsepower printer, copier, and precision-motion tasks and appliances. Other potential applications include fans, machine (servo) control, and electric vehicles (Wallace 1998). SR motors could potentially replace 20–50% of the existing general purpose motors in service today (Albers 1998; OIT 1998).

The rugged rotor of an SR motor is much simpler than that of other motors since it has no field coils or embedded magnetic materials. This design enables some models to operate at speeds as low as 50 rpm and as high as 100,000 rpm (E Source 1999). The coils and magnets attached to the rotor are subjected to very high stresses, necessitating more complex designs (Albers 1998). Because of its simplicity, the SR motor in mass production should theoretically cost no more than, and perhaps less than, mass-produced induction motor/ASD packages of comparable size, as discussed in Chapter 4.

However, at this time, automating the manufacturing of integral-horsepower and larger fractional-horsepower SR motors is proving difficult, and it is uncertain whether the hoped-for price reductions will materialize (Albers 1998; Wallace 1998).

Currently, SR motors and their associated controls, starter, and enclosure cost about 50% more than comparably sized and equipped induction motors with variable-speed controls (Albers 1998; Wallace 1998)—or about a $2,000 premium for a 20 hp installation in 1998.

Written-Pole Motors

The written-pole (WP) motor is a single-phase AC motor that acts like an induction motor during start-up, then like a synchronous motor on reaching full operating speed. Much like a computer hard drive, which records data onto a disk, the WP "writes" the number of poles and their locations electronically onto the rotor. This allows the WP motor to obtain higher energy efficiency and a lower start-up inrush current. The lower inrush inherent in the WP design may extend the expected life of the motor by reducing the inrush stresses (Nadel et al. 1998).

Single-phase motors have historically not been available in sizes over 16 hp because of the high inrush currents (six to seven times the nominal operating current) they create (EPRI 1994). Single-phase WP motors are now available in 15, 20, 40, and 60 hp sizes. The WP motor could potentially replace 4% of the integral-horsepower general purpose motors in service (Bannerjee 1998; OIT 1998). The WP's main advantage is not so much energy efficiency but rather that it allows a higher-horsepower single-phase motor to be used in applications for which only three-phase motors were available in the past. The motor also offers some power-outage ride-through capability that is of use in some industrial applications.

WP motors were originally intended to replace three-phase motors that use phase converters so the motors can operate on single-phase power systems, particularly in rural applications such as drying fans, conveyors, and irrigation pumps. In these cases, efficiency was not considered to be a significant issue.

WP motors are appropriate for new and OEM installations because they come as motor-controller packages. WP motors are now being used for irrigation pumps, conveyor motors, water pumps, food-processing air dryers, and process stirring. At this time, only one manufacturer, Precise Power Company of Bradenton, Florida, produces WP motors (Bannerjee 1998). WP motor research and application in the United States are limited to the 15–100 hp size range. WP motors have been used in less than 100 commercial installations to date (Morash 1998), but the potential U.S. residential, commercial, and industrial general purpose motor market that WP motors could replace annually is estimated to be about $140 million (EPRI 1994).

The WP motor is not complicated to manufacture, but costs are still high because of lack of production volume (Precise Power 1998). The installed cost of a 20 hp WP motor and controller package is about 60% higher than for a conventional induction motor with controller (Morash 1998), although the WP motor is more of a niche product and not completely comparable. Once the WP motor reaches full production levels, the cost premium is expected to drop by 50% (Nadel et al. 1998). In 1998, the WP cost $6,500 for a 20 hp unit with controller, starter, and enclosure package. The comparable induction motor and controller package cost $4,000.

The WP motor product line has relatively flat efficiency curves with maximum efficiencies of 92% for 40 hp and below, and 93–94% for units of 60 hp or more, at load levels as low as 70% (Precise Power 1998).

The primary barriers facing WP motor technology are its limited market niche, high initial cost, and lack of product understanding by the motor-buying public. Utilities, Electric Power Research Institute (EPRI), Precise Power Company, and OEMs are working to identify more opportunities to place the WP motor into finished goods. Demonstrations and educational programs are needed in the near term to raise awareness of the ways in which WP technology can deliver superior performance in certain kinds of applications (Nadel et al. 1998).

Characteristics of Commercially Available Motors

We have now discussed the principal types of commercially available motors. Their major characteristics are summarized in Table 2-2. Other types of motors exist in various stages of commercialization (for an overview, see E Source 1999). In the next section we will focus on the elements of, and trends in, motor efficiency.

Motor Efficiency

Motor efficiency has a slightly different definition than most other efficiency measurements because motor ratings are based on

Table 2-2

Classification of Common Motor Types, with Their General Applications and Special Characteristics

AC			
	Induction	Squirrel-Cage	Three-phase (general purpose, >0.5 hp, low cost, high reliability)
			Single-phase (low [typically <5] hp range, high reliability)
		Wound-Rotor	(special purpose for torque and starting current regulation, typically >20 hp, greater maintenance required than for squirrel-cage)
	Synchronous	Wound-Rotor	(high efficiency and reliability, very large sizes, greater maintenance requirement than for squirrel-cage)
		Reluctance	Standard (small motors, reliability, synchronous speed)
			Switched (rugged, high efficiency, good speed control, high cost)
		Brushless Permanent-Magnet	(high efficiency, high-performance applications, high reliability)
DC	Wound-Rotor	(limited reliability, relatively high maintenance requirements)	Series (traction and high-torque applications)
			Shunt (good speed control)
			Compound (high torque with good speed control)
			Separated (high-performance drives [e.g., servos])

Note: Brushless permanent-magnet motors overlap between AC and DC types.

output rather than input. Efficiency, η, is expressed as:

$$\eta = \frac{\text{output}}{\text{input}}$$

or

$$\eta = \frac{dW}{dW + l}$$

where dW is the energy output of the motor and l is the sum of the losses. This variation on the normal efficiency calculation is often not recognized and has led to confusion and calculation errors. The focus on motor losses stems from the approach used to design efficiency in a motor. We first discuss the sources of these losses and how motor efficiency is measured. We then turn to how design tradeoffs affect efficiency, concluding with a discussion of efficiency labeling and standards.

Motor Losses

There are four basic kinds of loss mechanisms in a motor: electrical; magnetic (core); mechanical (windage and friction); and stray. The relative contributions of these losses vary with motor load and are depicted in Figure 2-19.

Whenever current flows through a conductor, power is dissipated. These electrical losses are a function of the square of the current times the resistance, and thus are termed I^2R losses, where I is the symbol for current and R is the symbol for resistance. In a motor, I^2R losses occur in the stator and rotor. Because they rise with the square of the current, such losses increase rapidly with the motor's load. By using more, and sometimes better, lower-resistance materials (switching from aluminum to copper, for instance), manufacturers have reduced the I^2R losses in efficient motors.

Magnetic losses occur in the steel laminations of the stator and rotor and are due to eddy currents and hysteresis (see Appendix C).

Figure 2-19

Variation of Losses with Load for a 10 hp Motor

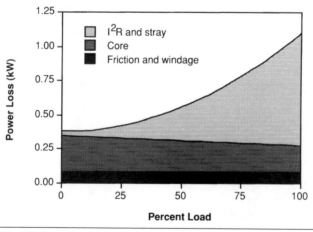

Source: Nailen 1987

Figure 2-20

An Efficient Induction Motor Cutaway View Showing Important Features and Construction

External fan (windage)   Stator windings (I^2R)   Stator laminations
(hysteresis and eddy current)

Armature conductors
(beneath surface) (I^2R)

Armature laminations
(hysteresis and eddy
current)

Armature fan
(windage)

Ball bearings
(friction)

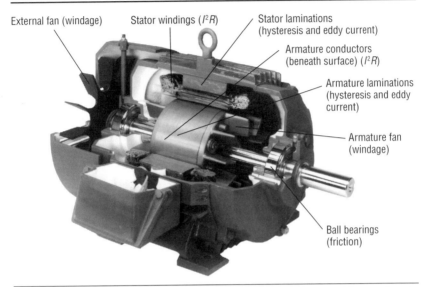

Note: Labeling indicates the major components that contribute to motor losses and (in parentheses) the type of loss that takes place. *Armature* is another name for rotor. Windage losses can be reduced through improved fan design. Hysteresis and eddy current (magnetic losses) can be reduced through the use of larger cross-sections, thinner laminations, and special steel alloys. I^2R losses can be reduced through the use of conductors with larger cross-sections (e.g., bigger wire).

Source: Reprinted with permission from Reliance Electric

Use of larger cross-sections of iron in the stator and rotor, thinner laminations, and improved magnetic materials can decrease the magnetic losses. Figure 2-19 illustrates that magnetic losses in a given motor decrease slightly as the load increases.

Mechanical losses occur in the form of bearing friction and "windage" created by the fans that cool the motor. Windage losses can be decreased through improved fan design. Mechanical losses are relatively small in open, low-speed motors but may be substantial in large high-speed motors or TEFC motors.

Stray losses are miscellaneous losses resulting from leakage flux, nonuniform current distribution, mechanical imperfections in the air gap, and irregularities in the air gap flux density. They typically represent 10–15% of the total losses and increase with the load. Stray losses can also be decreased by optimal design and careful manufacturing.

Figure 2-20 shows a cutaway of an efficient motor with the areas of efficiency improvement and types of loss minimized by each measure. In the integral-horsepower sizes, premium motors are 1–3 percentage points more efficient than required by EPAct standards, with the low end of this range applied to large models. In fractional-horsepower, single-phase models, the spread between standard- and high-efficiency units can be greater than 10 points.

Measuring Motor Efficiency

Different standards for testing motors have been developed by various organizations in several countries. Because of these differences, the test results for a given motor may vary depending on the procedure that is carried out. Users need to be aware of these differences so that they will compare motors as much as possible on the basis of uniform test methods. The principal testing methods are briefly described below, with comments on the relative efficiency rating produced by each procedure.

As mentioned at the outset of this chapter, motor efficiency is the ratio of the mechanical output and the electrical input. Although the definition is simple, there are difficulties associated with its accurate measurement. In the United States, the basic motor test measure is Institute of Electrical and Electronics Engineers (IEEE) Standard 112-1996, entitled "Standard Test Procedure for Polyphase Induction Motors and Generators," which comprise of five testing methods. IEEE Standard 112, Method B, is the most accurate, but also the most time-consuming and expensive (IEEE 1996). Using the basic definition of motor efficiency, it directly measures the mechanical output and electrical input to determine the efficiency. This standard is now substantially harmonized with Canadian Standards Association (CSA) Standard C-390-93 (CSA 1993). Both

Table 2-3

Comparison of the Efficiencies of Typical Motors, Tested According to Different Standards

Standard	Full-Load Efficiency (%)	
	7.5 hp	20 hp
CSA C-390-93	80.3	86.9
IEEE 112, Method B	80.3	86.9
IEC-34.2	82.3	89.4
JEC-37	85.0	90.4

Note: CSA = Canadian Standards Association
IEC = International Electrotechnical Commission (in Europe);
IEEE = Institute of Electrical and Electronic Engineers; and
JEC = Japanese Electrotechnical Commission.

Source: B.C. Hydro 1988

the IEEE and CSA standards account for stray losses by measuring them indirectly.

Other standards used in the international market provide a less accurate estimate of motor efficiency. International Electrotechnical Commission (IEC) Standard 34.2, used in Europe and some other parts of the world, allows for a tolerance in the efficiency and does not calculate the stray losses, assuming they are fixed at 0.5% of the full-load power (Control Engineering 1998). Japanese Electrotechnical Commission (JEC) Standard 37 ignores stray losses altogether, giving even less credible results (Control Engineering 1998).

As can be seen in Table 2-3, the efficiency of motors tested according to the different standards varies significantly. The assumptions of IEC-34.2 and JEC-37 are especially optimistic in small- and medium-horsepower motors. Considering that stray losses represent typically 10–15% of the motor losses at full load, in a model

Table 2-4

Proposed Default Values for Stray Losses Being Considered by IEC

Motor Size		Assumed Stray Losses (percentage of full-load input power)	
(hp)	(kW)	Current IEC	Proposed IEC
1	0.7	0.50	3.00
1.3	1	0.50	3.00
1.5	1.1	0.50	2.99
2	1.5	0.50	2.99
3	2.2	0.50	2.98
5	3.7	0.50	2.97
7.5	5.6	0.50	2.96
10	7.5	0.50	2.94
15	11.2	0.50	2.92
20	15	0.50	2.89
25	19	0.50	2.86
30	22	0.50	2.84
40	30	0.50	2.78
50	37	0.50	2.72
60	45	0.50	2.66
75	56	0.50	2.58
100	75	0.50	2.44
125	93	0.50	2.30
150	112	0.50	2.16
200	149	0.50	1.88
250	187	0.50	1.60
268	200	0.50	1.50

Source: de Almeida 1999

whose efficiency is 85%, stray losses represent 1.5–2.25% of the full-load power, not the 0.5% as assumed by the IEC, or 0% as assumed by the JEC.

Groups in the IEC have recognized this problem and are currently considering the adoption of a new test procedure. One proposal under consideration would allow a manufacturer to measure the losses directly, which is essentially identical to IEEE/CSA methodology, or to use a default value for the stray load losses. These proposed values (see Table 2-4) represent a near-worst-case scenario. Most manufacturers would therefore benefit from direct measurement of these losses. It is unclear when revisions are likely to be implemented (de Almeida 1999).

Table 2-5

Approximate Estimation of Comparable Efficiency Levels Using JEC, IEC, and IEEE Test Methods

Motor Size		Motor Efficiency (%)		
hp	kW	IEEE 112-B	IEC[a]	JEC[b]
1	0.7	76.8	78.8	79.6
2	1.5	81.1	83.1	83.8
3	2.2	81.4	83.4	84.1
5	3.7	83.9	85.9	86.5
7.5	5.6	84.8	86.8	87.3
10	7.5	85.6	87.6	88.1
15	11.2	87.4	89.4	89.9
20	15	88.3	90.3	90.7
25	19	88.9	90.4	90.8
30	22	89.8	91.3	91.7
40	30	90.4	91.9	92.3
50	37	91.0	92.0	92.4
60	45	91.5	92.5	92.8
75	56	92.0	93.0	93.3
100	75	92.0	93.0	93.3
125	93	92.2	92.7	93.0
150	112	92.8	93.3	93.6
200	149	93.8	94.3	94.6

Note: Estimates of IEC and JEC values are calculated for the specified IEEE 112 levels. These calculated values are subject to a substantial band of uncertainty as the relation between JEC, IEC, and IEEE 112-B efficiency varies with motor design and the calculations shown here are based on very limited comparative data.

[a] Adjusted for differences between IEC and IEEE 112-B test procedures based on limited comparative test data.

[b] Based on the following formula: $JEC = (1.05 \times IEC\ effic.)/(1 + 0.05 \times IEC\ effic.)$.

Source: ERM 1999

It would be convenient to convert the efficiency measure determined using one test procedure into values for the other test procedures without actually retesting the motor. Unfortunately, as noted above, the percentage of stray losses for any given motor varies with design and material selection so an exact equivalence in not achievable. If some simplifying assumptions are made, a rough estimate of comparisons for 1 to 200 hp motors can be developed (see Table 2-5).

Designing Motor Efficiency

The efficiency of a motor is determined by a series of design decisions. Efficiency is not the sole design parameter, and thus the design process involves a series of tradeoffs among the various parameters. The designer uses three broad strategies to achieve an efficient motor. First, extra time can be invested in the design of the motor. Although this approach is challenging, it usually involves only a modest unit cost. Second, the motor can be built to tighter tolerances. This approach requires additional capital investment in manufacturing, offers benefits in reducing variation between individual motors, and usually produces only modest incremental increases in unit costs. Finally, the design can make use of higher-quality materials, which can increase efficiency but can also significantly increase the unit cost of the motor. In actuality, the designer uses a combination of all these approaches to reach the price and performance goals for the particular target market.

Motors that have a range of efficiencies have always been available on the market. The average efficiency has fluctuated with market conditions, although it did increase from the end of World War II to the mid-1950s as new materials and technology became available. From the mid-1950s until the mid-1970s (a period of inexpensive energy), efficiency

Figure 2-21

Historical Efficiency of Standard and Energy-Efficient Motors

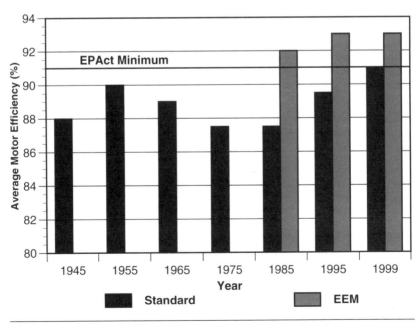

Source: Derived from Van Son 1994

declined as manufacturers built relatively inefficient motors that minimized the use of copper, aluminum, and steel (see Figure 2-21). These motors were designed with a focus on lower initial costs, unlike their predecessors, and used more energy because of their inefficiency (Van Son 1994).

The less efficient and more compact motors were made possible by the development of insulation materials that could withstand higher temperatures. Inefficiency is manifested in a motor as heat. Paper, cotton, enamels, and varnishes used in the 1940s deteriorated rapidly above 210°F (99°C), whereas the synthetic insulating materials developed since can tolerate operating temperatures up to 390°F (199°C). Thus it was possible to design motors that could accommodate higher losses without damaging the insulation and reducing motor lifetime.

Using less material in the magnetic and electrical circuits led to designs that were more compact. The development of improved steel with a higher permeability allowed for a further reduction of the magnetic circuits. The superior electrical characteristics of the new insulation materials allowed windings to be packed tighter in the slots, thereby reducing volume requirements even further. A combination of these factors resulted in smaller motor frames, as shown in Figure 2-22. This reduction in size and efficiency occurred mainly in motors

---

Figure 2-22

Relative Diameters of a 7.5 hp, 1,725 rpm Three-Phase Motor Reflecting NEMA Standards in Recent Years

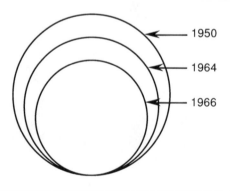

1950

1964

1966

Note: The 1964 size is the U-frame design. The 1966 size (the presently used T-frame) utilizes higher-temperature insulation and is allowed a higher temperature rise than previous motors.

Source: Lloyd 1969

smaller than 100 hp. In larger motors, the amount of heat that could be dissipated determined the motor size while limiting the minimum efficiency that could be tolerated.

By the mid-1970s, electricity prices began escalating rapidly. Consequentially, the majority of the large manufacturers introduced a line of energy-efficient motors in addition to their standard-efficiency models. Energy-efficient motors feature optimized design, more generous electrical and magnetic circuits, and a higher quality of materials. As can be seen in Figure 2-21, the average efficiency of energy-efficient motors has risen since their introduction.

Variations in Motor Efficiency

The efficiency of different units of the same motor model will vary. These variations can be attributed to differences in raw materials and multiple random factors in the manufacturing processes, as well as to dissimilarities in the results of efficiency testing (see NEMA 1993, Section 12.58.2). A 10% difference in the iron core losses, which is within the tolerance of magnetic steel manufacture, can by itself produce a 0.3% change in the efficiency of a 10 hp motor. Mechanical variations can also affect efficiency by altering the size of the air gap (a 10% difference in air gap size is not uncommon), which results in the increase in the stray losses. As noted above, precision machining of motor parts is more costly, and motor manufacturers settle for a trade-off between precision and cost when purchasing the equipment used in the production line.

The determination of efficiency is further complicated by variations due to uncertainty in test results. Efficiency determination is a complex and demanding exercise, and a significant difference can be introduced by variations in the technician's practices as well as measurement errors. In a study of different motors of the same model, losses varied often by 10% and sometimes by as much as 19%, corresponding to efficiency reductions of one to two percentage points (NEMA 1999).

Labeling of Motor Efficiency

In the late 1970s and early 1980s, NEMA established a labeling program for the most common types and sizes of motors ranging from 1 to 125 hp. Under this program, the nominal and minimum efficiency ratings for a motor are listed on its nameplate (where nominal efficiency is analogous to the average efficiency of a sample of motors of the same design and minimum efficiency roughly represents the fifth percentile of the sample). Since variations in materials, manufacturing processes, and testing result in motor-to-motor efficiency variations, NEMA specified a standard procedure for labeling efficiencies. The standard assumes that the distribution of efficiencies for a population

Table 2-6

Allowable Efficiency Levels for Labeling of NEMA Design A, B, and E Motors

Nominal Efficiency (%)	Minimum Efficiency Based on 20% Loss Difference (%)	Nominal Efficiency (%)	Minimum Efficiency Based on 20% Loss Difference (%)
99.00	98.80	91.00	89.50
98.90	98.70	90.20	88.50
98.80	98.60	89.50	87.50
98.70	98.50	88.50	86.50
98.60	98.40	87.50	85.50
98.50	98.20	86.50	84.00
98.40	98.00	85.50	82.50
98.20	97.80	84.00	81.50
98.00	97.60	82.50	80.00
97.80	97.40	81.50	78.50
97.60	97.10	80.00	77.00
97.40	96.80	78.50	75.50
97.10	96.50	77.00	74.00
96.80	96.20	75.50	72.00
96.50	95.80	74.00	70.00
96.20	95.40	72.00	68.00
95.80	95.00	70.00	66.00
95.40	94.50	68.00	64.00
95.00	94.10	66.00	62.00
94.50	93.60	64.00	59.50
94.10	93.00	62.00	57.50
93.60	92.40	59.50	55.00
93.00	91.70	57.50	52.50
92.40	91.00	55.00	50.50
91.70	90.20	52.50	48.00
		50.50	46.00

Note: The nominal efficiency listed is the lowest value for each range; the minimum efficiency corresponds to 20% higher losses than the nominal values. The actual average motor full-load efficiency is used to determine the NEMA efficiency number. For example, a motor with a full-load average efficiency of 93.5% would have nameplate efficiency of 93.0%.

Source: NEMA 1999, Table 12.9

of a given motor is normal. The motor should be labeled with a value from a table of allowable values that is less than or equal to the nominal value of the sample population (NEMA 1999). When NEMA first adopted these numerical ratings, there were 22 values for rounding. NEMA subsequently redefined the list to 43 preset ranges, later expanding it slightly to 51 (see Table 2-6).

The strength of the labeling program is that it embodies the natural variation in individual motors and provides a standard measure of motor performance that makes comparison between different manufacturers' products easy. The weakness of the program has been that there has been no certification of the manufacturers' reported efficiency values. The EPAct motor law addresses this problem by establishing an efficiency certification process and requiring that all motors be labeled with the certified value (see Appendix B for further details).

As can be seen in Figure 2-23, there is significant overlap in the distribution between adjacent nameplate efficiency values, so two nameplate steps are required for there to be a statistically significant difference

Figure 2-23

Bell-Shaped Probability Curve Showing the General Population Distribution of Three Nameplate Motor Efficiency Values (90.2%, 91%, and 91.7%)

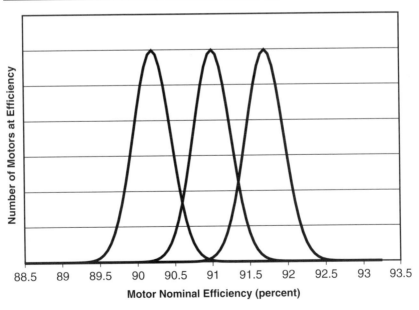

in efficiency. The average is considered the nominal efficiency of the motor and is used to predict the power requirements for a given installation. The minimum efficiency represents a near-worst-case combination of raw materials and manufacturing tolerances. However, 5% of the motors in a population may, depending on the manufacturer, have efficiencies lower than the minimum (NEMA 1999).

The NEMA labeling method currently defines the minimum full-load efficiency of a motor as that level corresponding to 20% higher losses than the listed nominal value. Some experts have voiced concerns that this definition allows for a sizable discrepancy between the efficiency that motor purchasers think they are getting and what they may actually receive. These experts have called on NEMA to tighten the range so that the minimum efficiency of a motor will correspond to losses 10% (instead of 20%) higher than the nominal efficiency. However, the NEMA committee has shown no interest in changing this aspect of MG 1.

In earlier versions of NEMA MG 1-1989 (NEMA 1989), a letter on the nameplate designated the efficiency level of the motor. Although this convention is no longer used, many motors in the operating inventory still bear these designations. Table 2-7 lists the nominal and minimum efficiencies for the index letters.

For meaningful comparisons to be made, it is essential that manufacturers measure efficiency in accordance with IEEE Standard 112,

Table 2-7

Correspondence of NEMA Motor Nameplate Index Letters and NEMA Nominal and Minimum-Efficiency Levels

Index Letter	Nominal	Minimum	Index Letter	Nominal	Minimum
A	—	>95.0	M	78.5	75.5
B	95.0	94.1	N	75.5	72.0
C	94.1	93.0	P	72.0	68.0
D	93.0	91.7	R	68.0	64.0
E	91.7	90.2	S	64.0	59.5
F	90.2	88.5	T	59.5	55.0
G	88.5	86.5	U	55.0	50.5
H	86.5	84.0	V	50.5	46.0
K	84.0	81.5	W	—	46.0
L	81.5	78.5			

Source: NEMA 1989, Table 12.53

Method B, preferably with the additional restrictions imposed by NEMA Standard MG 1-1998, Section 12.53 (a) and (b) (NEMA 1999) or CSA Standard C-390-93 (CSA 1993). Because of the range in efficiencies of the NEMA labeling system, it is better to obtain the actual average or minimum efficiencies for the motor model under consideration, and to determine whether that minimum is guaranteed. The source of this information can be either a motor catalog or the manufacturer.

Depending on one's perspective, it may make more sense to use either nominal or minimum efficiencies for analysis. For example, if numerous motors are being considered for the same application, the nominal values will give a good indication of what the overall energy usage and savings will be. On the other hand, if the application of a single motor is being analyzed, and it is important for the usage to be no greater than a particular value, then it would make more sense to use the minimum efficiency.

Energy-Efficient Motors

The terms used to describe the efficiency of motors have caused significant confusion in the motor marketplace. *High-efficiency, energy-efficient,* and *premium-efficiency* are commonly used by manufacturers to label motors. Only energy-efficient motors have an established definition, which was instituted by NEMA and is used in EPAct.

In 1989, NEMA first developed and adopted energy-efficient performance values in its Standard MG 1-1989 (NEMA 1989). These values were relatively weak. In 1991, NEMA adopted a more stringent complementary set of values labeled "suggested standard for future design" to help guide development of new high-efficiency products. In 1992, Congress used these latter values as the basis for EPAct, which established minimum-efficiency levels for all new general purpose electric motors manufactured or imported after October 1997 and required the labeling of motors with a certified efficiency value (U.S. Congress 1992). The law covers three-phase; general purpose; two-, four-, and six-pole Design A and B motors from 1 to 200 hp. These values were subsequently incorporated by NEMA into Table 12-10 of Standard MG 1-1993 (see Table 2-9) (NEMA 1993).

The term *general purpose* is somewhat vague from a regulatory perspective. The U.S. Department of Energy, as part of the final EPAct implementation rule (Federal Register 1999), has interpreted general purpose to mean any motor that can be used to replace another motor, in a broad range of common applications. The rule identified mechanical and electrical modifications that determine whether a motor is considered "covered product" or "covered equipment" (see Appendix B for a more detailed discussion of the EPAct rule and what products are covered by the law). With the implementation of the EPAct minimum-efficiency levels in October 1997, manufacturers' product offerings changed radically. A significant volume of new, qualifying products became available and will continue to be introduced over the next few years as new designs and toolings defuse to products not covered by EPAct.

Most motor manufacturers only have product meeting EPAct available in all motor sizes and enclosures. Table 2-8 shows current data on the most efficient three-phase motors available by manufacturer, size, and enclosure. This table reflects general trends, but motor-specific data are preferable when making selection decisions. As mentioned earlier, the best source of motor-specific data is the *MotorMaster+*® database (WSU 1999).

The range of efficiencies is greatest in the smaller motors, particularly 1,800 rpm TEFC motors. These 1,800 rpm motors have the highest sales volume. It is anticipated that the market share for premium-efficiency motors will continue to grow and that the average efficiencies of all motors will increase as newer designs are phased in across product lines and competition encourages further design optimization.

In addition, most manufacturers currently offer motors that significantly exceed the EPAct minimum-efficiency levels. A new set of efficiency levels for this class of motors (see Table 2-9) has been developed under the leadership of the Consortium for Energy Efficiency, a nonprofit coalition of utilities, public interest groups, and government (CEE 1996). Several utilities are now adopting these premium-efficiency levels for their incentive programs.

In late 2000, in response to the CEE premium-efficiency specification and the initiation of the development of an ENERGY STAR label by the U.S. Environmental Protection Agency (EPA), NEMA's Motor Generator Committee developed a *NEMA Premium*™ specification for motors (NEMA 2000). The NEMA program scope is significantly broader than the EPAct and CEE specifications, which apply only to motors that are general purpose; single-speed; polyphase; from 1 to 200 hp; two-, four-, and six-pole; open and enclosed squirrel-cage induction motors; NEMA Design A or B; continuous rated; and rated for operation at 230 and/or 460 V. The NEMA definition extends to all low-voltage motors rated for operation at or below 600 V and covers special and definite purpose motors in addition to general purpose motors. The size range covered is also extended up to 500 hp. The NEMA definition also includes medium-voltage products, rated at or below 5,000 V.

The nominal energy-efficiency levels that motors must meet or exceed to be labeled as *NEMA Premium*™ are presented in Table 2-10 for low-voltage motors and Table 2-11 for medium-voltage motors. For 1 to 200 hp motors that are covered under both EPAct and the CEE premium specification, the four-pole motor levels are identical to the CEE levels. For the six-pole product, the NEMA levels are two efficiency

Table 2-8

Nominal Full-Load Efficiencies for the Most Efficient Three-Phase, 1,800 rpm, NEMA Design B, General Purpose Induction Motors Available: 1999

	Open Drip-Proof										
	Horsepower										
Manufacturer	1	2	3	5	7.5	10	25	50	75	100	200
Baldor	85.5	86.5	89.5	89.5	91.0	91.0	94.1	94.5	95.0	95.4	95.8
Dayton	85.5	86.5	90.2	89.5	91.7	91.7	94.1	94.5	NA	95.4	NA
GE	NA	NA	86.5	88.5	91.0	91.0	93.6	94.5	94.5	95.4	95.8
Lincoln	85.5	86.5	89.5	89.5	91.0	91.0	94.1	94.5	95.0	95.4	96.2
MagnaTek	NA	NA	89.5	89.5	91.7	91.7	93.6	94.5	94.1	95.4	95.4
Marathon	84.0	85.5	89.5	89.5	91.0	91.0	91.7	93.0	94.5	95.0	95.8
Reliance	87.4	88.8	89.3	89.3	89.5	91.3	93.0	93.6	95.0	95.4	95.8
Siemens	85.5	85.5	87.5	88.5	90.2	90.2	93.0	93.6	94.5	94.1	95.0
Teco/Westinghouse	84.0	86.5	87.5	88.5	91.0	91.0	93.6	94.5	95.0	94.5	95.0
Toshiba	84.0	86.5	89.5	89.5	91.7	91.7	94.1	94.5	95.0	95.4	96.2
U.S. Motors	85.5	86.5	90.2	89.5	91.7	91.7	93.6	95.0	95.4	95.4	96.2
WEG	82.5	84.0	86.5	87.5	88.5	89.5	91.7	93.0	94.1	94.1	95.0
Average Premium	86.2	87.3	89.7	89.5	88.9	91.7	94.0	94.5	95.0	95.4	96.0

	Totally Enclosed Fan-Cooled										
	Horsepower										
Manufacturer	1	2	3	5	7.5	10	25	50	75	100	200
Baldor	85.5	86.5	89.5	90.2	89.5	89.5	93.6	94.1	95.4	95.4	96.2
Dayton	86.5	84.0	87.5	90.2	90.1	91.0	93.6	94.5	94.1	94.5	95.0
GE	NA	NA	89.5	89.5	91.0	91.7	93.6	94.1	94.5	95.0	95.8
Lincoln	85.5	85.5	89.5	89.5	90.1	91.0	94.1	95.0	95.0	95.8	96.5
Marathon	82.5	85.5	87.5	90.2	91.7	91.7	93.6	94.1	94.5	95.8	95.4
MagnaTek	84.5	86.5	88.5	89.5	91.0	91.0	93.0	94.1	95.4	95.4	95.8
Reliance	NA	88.8	90.3	88.5	90.2	90.2	93.6	94.5	95.0	95.4	96.2
Siemens	84.0	86.5	89.5	89.5	89.5	89.5	93.0	93.6	94.1	94.5	95.8
Teco/Westinghouse	86.5	86.5	89.5	89.5	91.7	91.7	93.6	94.5	95.4	95.4	96.2
Toshiba	85.5	86.5	89.5	89.5	91.0	91.0	93.6	94.1	95.4	95.4	96.2
U.S. Motors	86.5	86.5	89.5	90.2	91.7	91.7	93.6	94.1	95.4	95.4	96.2
WEG	85.5	86.5	89.5	89.5	91.7	92.4	93.0	94.5	94.5	95.0	96.2
Average Premium	85.9	86.9	89.5	89.7	91.7	91.7	93.7	94.5	95.4	95.5	96.2

Note: These motors represent the most efficient 1,800 rpm NEMA Design B general purpose motors that can be operated at 460 V, as listed in the October 1999 release of the *MotorMaster+*® database. These manufacturers' offerings are subject to change. These data are provided to reflect general trends in product offerings. Motors meeting or exceeding the CEE premium-efficiency definition are set in **boldface** type.

Table 2-9

Minimum Full-Load Efficiencies for NEMA Energy-Efficient and CEE Premium-Efficiency Open and Enclosed Motors

Efficiency Levels for ODP Motors

	1,200 rpm		1,800 rpm		3,600 rpm	
Horsepower	EPAct Energy-Efficient (%)	CEE Premium-Efficiency (%)	EPAct Energy-Efficient (%)	CEE Premium-Efficiency (%)	EPAct Energy-Efficient (%)	CEE Premium-Efficiency (%)
1	80	82.5	82.5	85.5	N/A	80
1.5	84	86.5	84	86.5	82.5	85.5
2	85.5	87.5	84	86.5	84	86.5
3	86.5	89.5	86.5	89.5	84	86.5
5	87.5	89.5	87.5	89.5	85.5	89.5
7.5	88.5	91.7	88.5	91	87.5	89.5
10	90.2	91.7	89.5	91.7	88.5	90.2
15	90.2	92.4	91	93	89.5	91
20	91	92.4	91	93	90.2	92.4
25	91.7	93	91.7	93.6	91	93
30	92.4	93.6	92.4	94.1	91	93
40	93	94.1	93	94.1	91.7	93.6
50	93	94.1	93	94.5	92.4	93.6
60	93.6	95	93.6	95	93	94.1
75	93.6	95	94.1	95	93	94.5
100	94.1	95	94.1	95.4	93	94.5
125	94.1	95.4	94.5	95.4	93.6	95
150	94.5	95.8	95	95.8	93.6	95.4
200	94.5	95.4	95	95.8	94.5	95.4

Efficiency Levels for TEFC Motors

	1,200 rpm		1,800 rpm		3,600 rpm	
Horsepower	EPAct Energy-Efficient (%)	CEE Premium-Efficiency (%)	EPAct Energy-Efficient (%)	CEE Premium-Efficiency (%)	EPAct Energy-Efficient (%)	CEE Premium-Efficiency (%)
1	80	82.5	82.5	85.5	75.5	78.5
1.5	85.5	87.5	84	86.5	82.5	85.5
2	86.5	88.5	84	86.5	84	86.5
3	87.5	89.5	87.5	89.5	85.5	88.5
5	87.5	89.5	87.5	89.5	87.5	89.5
7.5	89.5	91.7	89.5	91.7	88.5	91
10	89.5	91.7	89.5	91.7	89.5	91.7
15	90.2	92.4	91	92.4	90.2	91.7
20	90.2	92.4	91	93	90.2	92.4
25	91.7	93	92.4	93.6	91	93
30	91.7	93.6	92.4	93.6	91	93
40	93	94.1	93	94.1	91.7	93.6
50	93	94.1	93	94.5	92.4	94.1
60	93.6	94.5	93.6	95	93	94.1
75	93.6	95	94.1	95.4	93	94.5
100	94.1	95.4	94.5	95.4	93.6	95
125	94.1	95.4	94.5	95.4	94.5	95.4
150	95	95.8	95	95.8	94.5	95.4
200	95	95.8	95	96.2	95	95.8

Note: Reported levels are nominal efficiencies, representing the median efficiency of a population of motors of a given design as determined by IEEE Method 112-B.

Sources: CEE 1996; NEMA 1999, MG 1, Section 12-10

bands above EPAct, which is slightly lower for some sizes and enclo-sures than the CEE levels. For the two-pole product, efficiency is one efficiency band above EPAct. This level is significantly lower than the CEE levels, which are at least two bands above EPAct. NEMA chose this level, rather than the higher CEE levels, so that IEEE 841 motors

Table 2-10

Nominal Efficiencies for *NEMA Premium*™ Induction Motors (Low Voltage, Rated 600 V or Less, Random Wound)

Horsepower	Open Drip-Proof			Totally Enclosed Fan-Cooled		
	6-pole	4-pole	2-pole	6-pole	4-pole	2-pole
1	82.5	85.5	77[a]	82.5	85.5	77
1.5	86.5	86.5	84	87.5	86.5	84
2	87.5	86.5	85.5	88.5	86.5	85.5
3	88.5	89.5	85.5	89.5	89.5	86.5
5	89.5	89.5	86.5	89.5	89.5	88.5
7.5	90.2	91	88.5	91	91.7	89.5
10	91.7	91.7	89.5	91	91.7	90.2
15	91.7	93	90.2	91.7	92.4	91
20	92.4	93	91	91.7	93	91
25	93	93.6	91.7	93	93.6	91.7
30	93.6	94.1	91.7	93	93.6	91.7
40	94.1	94.1	92.4	94.1	94.1	92.4
50	94.1	94.5	93	94.1	94.5	93
60	94.5	95	93.6	94.5	95	93.6
75	94.5	95	93.6	94.5	95.4	93.6
100	95	95.4	93.6	95	95.4	94.1
125	95	95.4	94.1	95	95.4	95
150	95.4	95.8	94.1	95.8	95.8	95
200	95.4	95.8	95	95.8	96.2	95.4
250	95.4	95.8	95	95.8	96.2	95.8
300	95.4	95.8	95.4	95.8	96.2	95.8
350	95.4	95.8	95.4	95.8	96.2	95.8
400	95.8	95.8	95.8	95.8	96.2	95.8
450	96.2	96.2	95.8	95.8	96.2	95.8
500	96.2	96.2	95.8	95.8	96.2	95.8

[a] The value of 77 for the 2-pole ODP 1 hp motor is based on the Natural Resources Canada (NRCAN) requirement of 75.5% for an energy-efficient motor since NEMA MG 1 and EPAct do not contain any value for this rating.

Source: NEMA 2000

Table 2-11

Nominal Efficiencies for *NEMA Premium*™ Induction Motors (Medium Voltage, Rated 5,000 V or Less, Form Wound)

	Open Drip-Proof			Totally Enclosed Fan-Cooled		
hp	6-pole	4-pole	2-pole	6-pole	4-pole	2-pole
250	95	95	94.5	95	95	95
300	95	95	94.5	95	95	95
350	95	95	94.5	95	95	95
400	95	95	94.5	95	95	95
450	95	95	94.5	95	95	95
500	95	95	94.5	95	95	95

Source: NEMA 2000

could be labeled as premium. IEEE 841 motors have a more restrictive allowable inrush current than is allowed for NEMA Design B motors. This restriction prevents manufacturers from making two-pole IEEE 841 motors that meet the CEE levels (Kline 2001).

In December 2000, NEMA and CEE motor committee members met and tentatively agreed to adopt these *NEMA Premium*™ tables as the common definition for premium motors. Both organizations recommended to EPA that it also use the *NEMA Premium*™ as the basis for its ENERGY STAR label. At the time of publication, EPA had not made a determination on efficiency levels.

With EPAct, a new standard motor was introduced that just met the energy-efficient definition at the lowest possible initial cost. For the most part, these motors have used design and precision manufacturing to achieve these efficiencies, rather than active material. The more efficient and higher-quality materials are now being used in premium-efficiency motors.

Other countries are also considering the establishment of standard efficiency levels. The development of Canadian standards has paralleled that of the United States, and the national efficiency levels have been harmonized (CSA 1993). The European Union (EU) and Committee of European Manufacturers of Electrical Machines and Power Electronics (CEMEP), the European association of motor manufacturers, have developed a motor efficiency classification scheme covering motors in the range of 1.1–75 kW. Table 2-12 presents the two proposed higher-efficiency levels. A voluntary agreement associated with the classification scheme calls for the motor manufacturers to progressively reduce their output of motors not meeting these levels. If

Table 2-12

Proposed European Union–CEMEP Energy Efficiency Classification Scheme for Two- and Four-Pole Induction Motors

	Minimum Nominal Efficiency (as determined by Method IEC 34.2) (%)			
	Class 2		Class 1	
Motor Size (kW)	2-pole	4-pole	2-pole	4-pole
1.1	76.2	76.2	82.2	83.8
1.5	78.5	78.5	84.1	85.0
2.2	81.0	81.0	85.6	86.4
3.0	82.6	82.6	86.7	87.4
4.0	84.2	84.2	87.6	88.3
5.5	85.7	85.7	88.5	89.2
7.5	87.0	87.0	89.5	90.1
11.0	88.4	88.4	90.6	91.0
15.0	89.4	89.4	91.3	91.8
18.5	90.0	90.0	91.8	92.2
22.0	90.5	90.5	92.2	92.6
30.0	91.4	91.4	92.9	93.2
37.0	92.0	92.0	93.3	93.6
45.0	92.5	92.5	93.7	93.9
55.0	93.0	93.0	94.0	94.2
75.0	93.6	93.6	94.6	94.7

Source: European Union–CEMEP 1999

progress is not significant, mandatory minimum-efficiency standards may be applied (European Union–CEMEP 1999). Several other countries are also considering standards, including Australia (Standards Australia 1999), Brazil (Geller 2000), Thailand (ERM 1999), and China (Liu 2000).

Availability of Different Motor Efficiencies

Premium-efficiency three-phase induction motors are available from most manufacturers in T-frame ODP and TEFC enclosures; in speeds of 1,200, 1,800, and 3,600 rpm; and in sizes from 1 to 200 hp. Certain manufacturers make a premium-efficiency product as small as 0.5 hp and as large as 350 hp. Some make a premium-efficiency line that runs at 900 rpm. Because a majority of the design and tooling of the premium-efficiency motors can be used in many special and definite purpose motors that are not covered by EPAct, some manufacturers offer an efficient product in these classes as well. Single-phase motors also are often available in standard and efficient lines.

Standard single-phase motors have extremely poor efficiencies. For example, 0.25 hp motors from different manufacturers range in efficiencies from 52% to 60%, with power factors of 53% to 62%. Energy-efficient single-phase motors in this size can achieve efficiencies of 75%. Efficient single-phase motors are currently available through most distributors that stock small motors. They are also appearing as an option in some packaged equipment, including commercial refrigeration cases.

Benefits of Efficient Motors

The most obvious benefit of an efficient motor, whether it is an EPAct or premium product, is energy savings. Even in the largest motors, in which the efficiency improvement between standard- and high-efficiency models is small in percentage terms, a minor relative improvement can yield substantial energy savings. For example, a 1% improvement for a 500 hp motor operating at 75% load saves 2.8 kW. If the motor operates almost continuously, as many large motors do, that 1% improvement could yield annual energy savings of nearly $1,500 at $.06/kWh. Also, for utilities with a monthly demand charge of $6/kW, the demand savings will be $202/yr. In contrast, improving the efficiency of ten 1 hp motors by 10% (from 75 to 83%) saves 0.72 kW, which yields annual energy savings of $380 and demand savings of $52.

Efficient motors not only reduce energy consumption and contribute to reduced demand but also save energy in the cables and transformers that feed the motor. Most efficient motors have a higher power factor than standard-efficiency motors. Efficient motors are also likely to last longer because they run at cooler temperatures, resulting from 20–40% lower losses. The decrease in operating temperature is not as dramatic as one would expect, however, since efficient motors have downsized ventilation to decrease the ventilation losses. Manufacturers no doubt vary in how they make the tradeoff between ventilation (hence cooler, longer operation) and efficiency.

Along with temperature effects, the lubrication procedure is another critical factor that will affect the lifetime of a motor. Many motor failures are caused by bearing failures, the majority of which are in turn caused by underlubrication or overlubrication. Lubrication is a function of the maintenance practices of each plant, not motor design.

Efficient motors often use heavier-duty bearings that are presumably more resilient when faced with poor lubrication. Therefore, on the basis of cooler operation and better bearings, efficient motors should tend to last longer than standard-efficiency motors. How much longer is a matter of speculation, however, since limited data are available on this issue.

Because of their lower losses, efficient motors suffer less thermal stress than standard motors when they are started and operated at small overload intervals. This makes them attractive in some duty cycling applications since they can withstand a higher on-off cycle rate.

Efficient motors may possess a few drawbacks. Efficient motors tend to have a lower slip rate, as well as operating speeds that are slightly higher than in their standard-efficiency counterparts because of their lower losses. When an efficient motor is driving loads where the power increases with the cube of the speed (such as in many pumps and fans), the higher speed causes the power drawn by the motor to rise, and a portion of the savings associated with the efficient motor can be lost. This cube law phenomenon and ways to mitigate losses from faster rotation are discussed further in Chapter 5.

The smaller slip of efficient motors causes them to have a lower starting torque than standard-efficiency units. Thus, they should not be used in certain applications where starting torque is critical. Because the actual starting torque of many standard-efficiency NEMA Design A and B motors was higher than the minimum specified in MG 1, they were used in some applications for which a Design C motor was appropriate. The Design A and B motors were less expensive and more readily available than the Design C. When the motor is replaced with an efficient motor, the new model may be overloaded. In those cases, a Design C motor should be specified. In addition, some efficient motors also have lower power factors than standard-efficiency motors.

Later in this chapter we discuss the economics of efficient motors, but first we will address another important issue—the repair of failed motors.

Motor Failure

Motors don't fail because of age or operating hours, but rather from a form of stress. Overloading, power supply anomalies, improper lubrication, corrosion, or contamination can cause stress. If these stresses are minimized, motors can operate for hundreds of thousands of hours (Douglass 1999a). The most common motor failures result from either bearing or winding failure. Windings fail when their insulation degrades, usually because of some combination of overheating, insulation aging, and overvoltage transients. Overheating can result from overloading the motor, blocked ventilation, or voltage imbalance. For example, a common problem for three-phase motors is single-phasing, in which one or two of the phases are lost because of a distribution malfunction. Minor insulation failure can also lead to poor motor performance, shock, and fire hazard. Major insulation failure will trip the overcurrent protection devices.

Many winding failures result from mechanical failures, such as of bearings. When the bearing fails, the motor overheats. If a motor problem can be identified through a preventive maintenance program *before* an electrical failure occurs, significant costs can be avoided (Suozzo et al. 2000).

Motor Repair

Each year, more motors are repaired than new motors are sold. For every new motor bought, approximately 2.5 motors are repaired. It is estimated that motors are repaired on the average of every 5–7 years. Since they are frequently operated for 20–30 years, a motor may be repaired three to five times in its serviceable lifetime (Schueler, Leistner, and Douglass 1994).

The terms *repair* and *rewinding* are frequently used interchangeably. They are in fact two separate procedures. A major electrical failure of the motor requires replacement of the stator winding. This process, called rewinding, consists of stripping out the old windings and replacing them with new wire. When the damage is restricted to the bearings, only those parts need to be replaced. In addition, shops will recondition functional motors as a preventive maintenance measure. This procedure includes cleaning, inspection, and rebalancing. As a result, most shops prefer to call their businesses motor repair or motor service providers, rather than rewinders.

A rewound motor can use the same rotor, stator iron, and case, leading to considerable savings in raw materials. Rewinding is very common because it is economical in terms of initial cost. Commercial or industrial facilities that purchase motors at the trade price (the price paid by low-volume customers) rewind most models over 10 hp. Larger users that receive a volume discount on motor purchases may restrict rewinding to only those larger than 40 hp. It is generally cheaper to replace failed standard motors under 10 hp with efficient motors rather than to rewind them.

Users must consider two key economic criteria when deciding on a motor rewind. The first is the cost difference between buying a new efficient motor and rewinding (see Tables A-1 and A-2 in Appendix A).

The second consideration is that the motor might not be as efficient or reliable as the user expects when it returns from the repair shop; this could result either from some preexisting damage that is not detected and therefore not corrected during the repair, or because the repair itself damages the motor. The possibility of such performance degradation is often overlooked in an effort to minimize initial cost. An efficiency loss of only 1% in a large motor can cost $1,500/yr in energy bills, yet some rewound motors run several percentage points below the nameplate efficiency (Montgomery 1989).

Either the repair shop or the user typically identifies severely damaged motors as such when they fail prematurely after being repaired. Slightly damaged units that look fine but are running, say, 1–5% below nameplate efficiency can tally up to thousands of dollars in excess losses over the years. For instance, in the course of severe bearing failure, the rotor may hit the stator and damage the magnetic properties of the iron core. If the bearing is replaced but the magnetic damage is ignored, the repaired motor will appear to be as good as new, while in actuality it will be sustaining excess operating losses. The only way to quantify these losses is to test the motor. Such testing is discussed later. First we address the question of how poor rewind practices can damage motors.

Impact of Rewinding on Motor Losses

In theory, most motors can be restored to their original efficiency rating. In practice, however, motor efficiency is often degraded through normal rewind practices, making the initial low cost a potentially poor investment. An efficient rewind is defined more by what is not done than by what is done. Maintaining efficiency consists of attention to detail and quality control. Quality repair practices fall into two major categories: avoiding practices that degrade efficiency; and appropriate testing before and after repair to diagnose possible problems (Schueler, Leistner, and Douglass 1994).

Research and experience in Canada and by the Electric Apparatus Service Association (EASA), a North American trade group representing motor repair shops, have shown a strong relationship between maintained efficiency and a robust quality assurance program. Quality repair practices also deliver a more reliable motor. This finding is reasonable since increased losses are manifested as increases in motor operating temperatures, which shorten a motor's life. As a result of this finding, it has now become an accepted practice to use the terms *quality repair* and *efficient repair* interchangeably (Schueler, Leistner, and Douglass 1994).

Table 2-13

Empirical Studies of Efficiency Loss during Motor Repair

Study	Sample Size	Decrease in Full-Load Efficiency	Comments
McGovern (1984)	27	1.5–2.5%	Motors ranged from 3 to 150 hp; wide range of motor age and rewind histories (General Electric)
Colby & Flora (1990)	4	0.5–1.0%	Standard- and premium-efficiency 5–10 hp motors (North Carolina)
Zeller (1992)	10	0.5% @ rated load 0.7% @ load	Controlled test; identical 20 hp premium-efficiency motor shops (British Columbia)
Dederer (1991)	9	1.1% @ rated load 0.9% @ load	Controlled test; identical 20 hp standard-efficiency motor shops (Ontario)
Ontario Hydro (1992)	2	2.2% (40 hp) @ rated load 0.4% (100 hp) @ rated load	Motors rewound four times each

Source: WSEO 1994

No comprehensive studies of the impact of motor repair on efficiency are available. A review of the literature has identified five empirical studies covering 52 motors, all less than 150 hp (see Table 2-13). Across these five studies, following repairs, the full-load efficiency decreased 0.5–2.5%, with an average of 1%. It is likely that the impact of rewinding is somewhat lower for larger motors because they are usually repaired by larger shops that are more likely to have quality assurance programs in place (Schueler, Leistner, and Douglass 1994).

The initial repair studies focused on increased core losses (McGovern 1984; Seton, Johnson, and Odell Inc. 1987a). The insulation materials used in the past few decades in stator copper windings are solvent-resistant and very hard to remove. The conventional approach to softening the windings for removal is to bake the stator in an oven. If the stator gets too hot, however, its magnetic properties can be damaged, leading to an increase in core losses. These losses are primarily due to damaged insulation between the laminations in the core. Very high temperatures can distort the iron and the air gap, also resulting in increased losses (Schueler, Leistner, and Douglass 1994).

EASA published a study recommending that oven set-points not exceed 650°F (343°C) (EASA 1985). This recommendation takes into account that stator core temperatures sometimes reach up to 150°F (66°C) higher than the oven set-point, largely because of heat released by the combustion of insulation materials. Other analysts, citing EASA's own test data and reports from motor manufacturers, suggest that oven set-points should not exceed 500°F (260°C) (Lovins et al. 1989).

The EASA test results shown in Figure 2-24 indicate a general correlation between oven temperature and efficiency degradation. Although the sample size is small, the results vary widely among individual motors, and many motors in the sample inexplicably gained efficiency on the second rewind. Even if the EASA recommendations are correct, many rewind shops do not follow them. A study performed by the Bonneville Power Administration (BPA) on rewind practices in the Pacific Northwest found that about half of the motors are baked at oven set-point temperatures above 650°F (343°C) and thus

Figure 2-24

Test Results from Rewound Standard-Efficiency Motors That Were Stripped Using a Burnout Oven

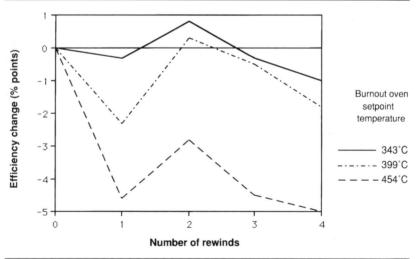

Note: Efficiency degradation becomes more severe as the temperature increases. Efficiency also generally decreases with subsequent rewinds.

Sources: EASA 1985; Lovins et al. 1989

Table 2-14

Major Sources of Decreased Efficiency during Motor Repair/Rewind

Action	Losses Affected
Change in type of bearings	Windage and friction
Change in type or size of fan	Windage and friction
Excessive burnout temperatures (over 650°F)	Core
Core lamination damage during winding removal or repair	Core
Winding with smaller wire	Stator I^2R
Change in winding configuration	Stator I^2R /stray load
Increased air gap	Stator I^2R
Rotor bars cracked or loose	Rotor I^2R
Degrade air gap symmetry (reduced rotor diameter, bent shaft)	Rotor I^2R

are likely subjected to efficiency-degrading core damage (Seton, Johnson, and Odell Inc. 1987a).

More recent studies (Colby and Flora 1990; Dederer 1991; Ontario Hydro 1992; Zeller 1992) have shown that core loss is not the only source of efficiency loss. Table 2-14 summarizes the sources of losses reported in the 1994 BPA repair study (Schueler, Leistner, and Douglass 1994).

The Zeller study segregated losses by type. Table 2-15 demonstrates the interaction between the different sources of loss.

Larger losses result in higher energy bills and reduced motor lifetimes. For example, consider a 50 hp continuous-duty standard motor that has losses increased by 50%, 100%, 150%, and 200% in the rewinding process. Table 2-16 shows the increase in losses as well as the extra cost of losses for an electricity price of $.06/kWh. If there is an increase of 100% in the motor core losses, the additional annual operating cost will be similar to the rewinding cost.

The increased losses also raise motor temperature, which decreases the insulation lifetime. Table 2-16 also shows the corresponding temperature increases and associated reduction in insulation lifetime when the core losses go up. A large increase in core losses dramatically affects the lifetime of the rewound motor. Higher motor temperatures also influence the lifetime of the lubricant in the bearings: to avoid premature failure, the user can perform more frequent regreasing, although this will incur additional costs. Thus, as the rewind quality worsens, the motor life shortens.

One alternative stripping method widely practiced in Europe (Dreisilker 1987; Lovins et al. 1989) applies moderate temperatures up to 300°F (150°C) to soften the insulation so that it can be safely removed

Table 2-15

Effect of Changes in Segregated Losses on Total Losses for Ten Repaired Motors

Motor	Core	Windage	Stray	Stator	Rotor	Total
A	C	II	C	C	C	C
C	II	C	I	C	C	C
D	C	III	III	C	C	III
E	C	III	II	C	C	II
F	C	III	D	C	C	C
G	II	C	I	I	C	C
H	C	I	C	C	C	C
I	III	C	C	III	C	I
J	D	III	C	D	C	C
K	I	II	II	C		C

Notes:
III = Relatively large increase after rewinding
II = Moderate increase after rewinding
I = Relatively small increase after rewinding
C = Insignificant change after rewinding
D = Decrease after rewinding

Source: Zeller 1992

Table 2-16

The Effect of Increased Core Loss on Motor Operating Cost and Insulation Life for a 50 hp, 3,600 rpm ODP Motor

Core Loss Increase		Increase in Annual Operating Cost		Temp. Rise °C	Approximate Decrease in Insulation Life %
%	watts	$	% of rewind cost		
50	515	271	28	7	62
100	1,030	542	55	14	38
150	1,545	813	83	21	24
200	2,060	1,084	110	29	14

Source: Montgomery 1989

with mechanical pulling. The low temperatures used in this process are much less likely to damage the core. Unfortunately, this technique is not well known or widely used in the United States. Its main champion in the United States is Dreisilker Electric Motors, Inc., located in Glen Ellyn, Illinois. (See Appendix D for sources on motor testing and repair equipment.)

Some rewind shops use chemicals to loosen the windings. However, these solvents pose health and environmental problems because of their toxicity and are not able to dissolve modern epoxy-based varnishes. High-pressure water jets can also be used for stripping, but the equipment is very costly.

In addition to the methods used to remove old windings, the materials and techniques utilized in reassembling a motor can affect its subsequent performance. Most motor rewind shops install materials in the motor equal to or better than those in the original motor. For example, most shops use wire with at least a Class F insulation, even on motors that were originally equipped with Class B insulation. In some cases, however, this attempt to improve the motor can backfire and result in a lower operating efficiency.

For example, data from the BPA study on rewind practices indicate that approximately 20% of rewound motors are U-frames, which were originally equipped with an older style of wiring containing bulky insulation. When modern, thinner, insulation is installed, most rewind shops will use more than in the original design. By using the same number of turns of thicker wire, motor efficiency will be improved since the extra copper will reduce resistance losses. However, if a larger number of turns are used, as is common, resistance losses will rise and magnetic losses will substantially increase, resulting in decreased operating efficiency.

Obtaining a Quality Repair

What is done before a motor is sent for repair has more to do with obtaining a quality repair than what is ultimately done at the shop. Three steps can help ensure a quality repair:

- Evaluate the prospective repair shop

- Do not pressure the shop for an unrealistic turnaround time

- Develop and clearly communicate your requirements to the provider

Planning leads to a more favorable outcome, as does having a good process for deciding on whether to repair or replace a motor.

Evaluate the Prospective Repair Shop

There are a number of ways to evaluate a repair shop. Since motor performance has been linked to the presence of a quality assurance program, an ISO-9000 certification is a strong indicator of a quality shop. EASA has developed a quality repair certification, EASA-Q, which builds on ISO-9000, adding repair-specific elements (EASA 1998). Advanced Energy (AE), a North Carolina research and educational organization, has implemented a quality assurance program for motor repair facilities, the Proven Excellence Verification Program, which involves an inspection of the facility's equipment and procedures, independent testing of several repaired motors, and an annual review of shop performance (AE 2000). These certifications are expensive and time-consuming to acquire; therefore, the number and geographic distribution of certified shops is limited. Also, certified shops tend to be the biggest shops, which work with large industrial customers that require their suppliers to be ISO-9000 certified. Most motor manufacturers have an evaluation program in place for selecting shops that provide warranteed repair service. This selection can be used to indicate a quality repair shop (CEE 1998).

Certification is not the only sign of a quality repair shop since many smaller shops are unwilling or unable to expend the resources to acquire certification. Customers can evaluate a service center themselves. Washington State University (WSU) has developed a *Service Center Evaluation Guide* (Douglass 1999b) that is available from DOE's *Industrial Best Practices: Motors* program, formerly the *Motor Challenge* program (see Appendix C and the Annotated Bibliography), as well as many regional and utility motor programs. The guide provides a questionnaire and checklist that can be used on a visit to the shop. Elements of the evaluation include

- Staff training and morale
- Presence of facilities and materials for handling the type and size of motors that may be repaired
- Presence and use of
  - Core loss tester or EASA loop test setup
  - Surge comparison tester
  - Voltage-regulated power supply for running at rated voltage
  - Vibration-testing equipment
- Review of record-keeping practices
- Method of insulation removal
- Overall cleanliness of the shop

While a low rating on any one element of the evaluation should not disqualify a shop, several questionable items may indicate that it is a poor choice (Douglass 1999b).

Do not Pressure the Shop for an Unrealistic Turnaround Time

The most expedient way to avoid needing fast turnaround is to implement a motor systems management plan that will ensure that spares are available to replace critical motors while they are being repaired. A management system also determines whether a motor should be repaired or replaced, describes a motor's repair history, and tracks routine maintenance. Assistance in setting up a motor systems management plan can be found in the *Energy Management Guide for Motor-Driven Systems* (McCoy and Douglass 1997) and in the *MotorMaster+®* software package (WSU 1999). Some examples of the assistance that is available from service centers are managing of the motor tracking system, guaranteeing ready spares for the specified motors, and preventive/predictive maintenance services (Douglass 1999a).

Develop and Clearly Communicate Your Requirements to the Provider

When maintenance must be performed on a motor, specifications should be provided that outline the requirements for before and after testing, the varnish application method, record keeping, and so on. A model for this system can be found in *The Model Repair Specifications for Low Voltage Induction Motors* (Douglass 1999c). A "medical history" of the motor should also be given. This history includes past repairs, results of predictive testing, lubrication and other maintenance activities, and operating characteristics such as method and frequency of starting and load and power source information (Douglass 1999b).

Economics of Energy-Efficient Motors

The economics of efficient motors must be evaluated separately for three distinct situations: (1) installing a premium-efficiency instead of an EPAct motor in a new application; (2) installing a new efficient motor instead of rewinding a failed motor; and (3) installing a new efficient motor as a retrofit for an existing operational motor.

Figure 2-25

Ranges for Full-Load Efficiency vs. Size, and Costs vs. Size

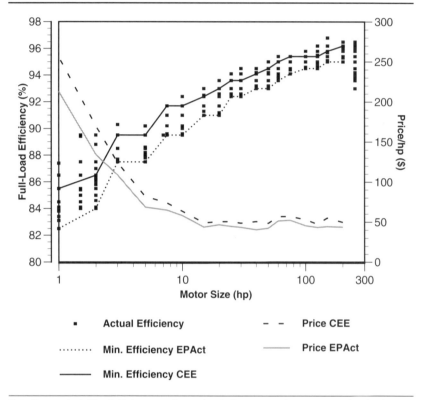

■ Actual Efficiency	– – Price CEE
········ Min. Efficiency EPAct	—— Price EPAct
—— Min. Efficiency CEE	

Note: Ranges for full-load efficiency vs. size, and costs (average per horsepower trade prices) vs. size for NEMA Design B standard and high-efficiency, 1,800 rpm, three-phase induction motors. Distribution of efficiency data points reflects variation among manufacturers. EPAct and CEE premium-efficiency and minimum-efficiency values are provided for reference. Price per horsepower values is based on average price for qualifying product in the *MotorMaster+* database.

Source: WSU 1999

Economics of Premium vs. EPAct Motors

The cost of efficient motors is typically 10–25% higher than for standard motors (Easton Consultants and XENERGY 1999a). This higher price is due to increases in the quantity and quality of materials, the cost of designing the motor, and the cost of tooling. Figure 2-25 illustrates the efficiency and trade price ($/hp) ranges of both EPAct and premium-efficiency motors. The market ("trade") price of electric motors can be substantially lower than the list price. A medium-size user can obtain a

25–50% discount off the list price, and a large user can receive a discount as high as 50–70% (Easton Consultants and XENERGY 1999a; Seton, Johnson, and Odell Inc. 1987b). Some high-volume dealers can sometimes provide discounts as high as 60% in bid situations (Easton Consultants and XENERGY 1999a; Stout 1990).

In many new installations, the extra cost of a premium-efficiency motor is justified by the energy and demand savings. Consider a new application of a 50 hp motor with the following specifications:

- 6,000 hours of annual use at 75% load
- Cost of electricity = $.06/kWh
- Demand charge = $70/kW-yr
- Efficiency of EPAct motor = 93.9% at 75% load
- Efficiency of premium-efficiency motor = 94.8% at 75% load
- Extra list cost of premium-efficiency motor = $470
- Price is 65% of list
- Actual extra cost = $305

The yearly savings afforded by the premium-efficiency motors are as follows:

- Demand savings = 50 hp × (1/0.939 − 1/0.948) × 0.75 × 0.746 kW/hp = 0.283 kW
- Energy savings = 0.283 kW × 6,000 hr/yr = 1,697 kWh
- Cost savings = $.06/kWh × 1,697 kWh + $70/kW-yr × 0.283 kW = $122/year
- Simple payback period = $305/$122 = 2.5 years

Economic calculation methods other than simple payback are discussed in Appendix A.

The payback decreases linearly with the number of operating hours. Therefore, for a new application, a premium-efficiency motor can be an attractive investment if it has high operating hours and/or it is used in areas where the electricity cost is high.

Figure 2-26 shows payback periods for TEFC motors in new applications. These values draw on Tables A-1 and A-2 (see Appendix A) for cost and performance data. The values assume an average price of about 60% of list price (Easton Consultants and XENERGY 1999a; Seton, Johnson, and Odell Inc. 1987b), which is the price actually paid by typical commercial and industrial customers. Very large customers can often purchase motors at a larger discount. There is a significant variation in the economics depending on motor size, which is most likely due to the recent implementation

Figure 2-26

Simple Payback Times for New, Premium-Efficiency TEFC Motors vs. New EPAct TEFC Motors as a Function of Motor Size and Annual Operating Hours

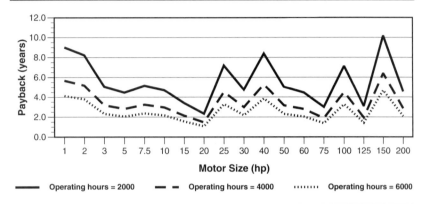

Note: The assumptions, efficiencies, and motor costs are listed in Table A-1 (see Appendix A) and are based on the *MotorMaster+*® database.

Source: WSU 1999

Figure 2-27

Simple Payback Times for New, Premium-Efficiency ODP Motors vs. New EPAct ODP Motors as a Function of Motor Size and Annual Operating Hours

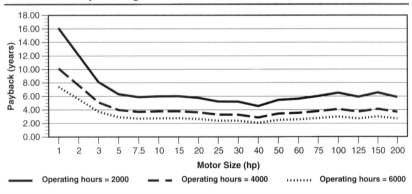

Note: The assumptions, efficiencies, and motor costs are listed in Table A-1 (see Appendix A) and are based on the *MotorMaster+*® database.

Source: WSU 1999

of EPAct. For other values of annual operating hours or electricity costs, the paybacks can be adjusted linearly. For example, a motor operating 3,000 hrs/yr would take twice as long to pay back as one operating 6,000 hrs/yr. A similar analysis for ODP motors (Figure 2-27) shows that the economics is slightly less attractive than for TEFC motors for most sizes of motors.

Life-cycle cost analysis will determine the threshold number of annual operating hours at which a premium-efficiency motor becomes cost-effective. The results of such an analysis are illustrated in Figure 2-28. As can be seen, the economics is very size-specific because of the variation in the price differential between EPAct and

---

Figure 2-28

Present-Value Savings from Premium-Efficiency Motors Compared with the Marginal Cost of Premium-Efficiency Motors Relative to Standard-Efficiency Motors as a Function of Motor Size and Annual Operating Hours

---

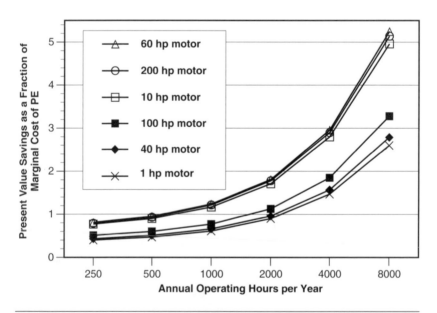

Note: Assumptions include 1,800 rpm TEFC motors; values for efficiency and cost from Table A-1 (see Appendix A); 75% load; 15-year life and 7% real discount rate; energy and power savings are valued at $0.06/kWh and $70/kW-yr; and demand coincidence factor of 0.75. High-efficiency motors are cost-effective from a life-cycle cost perspective whenever the present-value savings exceed the marginal cost of a high-efficiency motor (i.e., wherever present-value savings exceed the 100% line in the graph).

---

premium-efficiency motors. Note that in almost all cases premium-efficiency motors are cost-effective in applications in which they operate above 2,000 hrs, but some sizes are cost-effective with annual operating hours as low as 500 hrs.

Because no one manufacturer has the most efficient motor in every style and size, users should comparison-shop across brands. However, manufacturers change their designs from time to time, and it is therefore important to obtain current efficiency information on the specific motors under consideration. Washington State University, with funding from DOE's *Industrial Best Practices: Motors* program, has developed a selection and comparison tool called *MotorMaster+®*. This guide, complied from a database of over 25,000 motors sold in the United States, allows the comparison of different motors for particular applications. The database contains information provided by manufacturers to WSU on each motor, including full- and part-load efficiency, power factor, catalog number, list price, and full-load speed. It can also be used to build site-specific motor inventory databases that can be used to make future motor decisions (WSU 1999).

Economics of Efficient Motors vs. Rewinding

As discussed above, rewinding can reduce the efficiency of a motor. In such cases, the energy savings from installing a new, efficient, motor rather than rewinding an existing motor can be more attractive than nameplate comparisons would suggest.

Consider an application in which the economics of rewinding a motor is compared with the economics of purchasing a new EPAct motor. This application uses a 50 hp motor with the following specifications:

- 6,000 hours of annual use at 75% load

- Cost of electricity = $.06/kWh

- Demand charge = $70/kW-yr

- Nominal efficiency of standard motor when new = 90.6% at 75% load

- Increased losses due to rewind = 1 efficiency percentage point

- Efficiency of EPAct motor = 93.9% at 75% load

- Price is 65% of list

- Extra cost of new motor = $3,252 (EPAct) × 0.65 – $980 (rewind) = $1,133

Figure 2-29

Simple Payback Times for New, EPAct TEFC Motors vs. Repairing Standard TEFC Motors as a Function of Motor Size and Annual Operating Hours

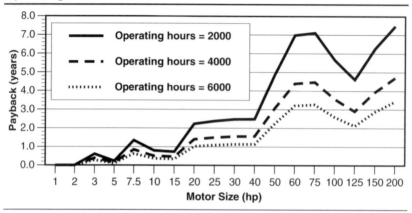

Note: The assumptions, efficiencies, and motor costs are listed in Table A-1 (see Appendix A) and are based on the *MotorMaster+®* database.

Source: WSU 1999

Figure 2-30

Simple Payback Times for New, EPAct ODP Motors vs. Repairing Standard ODP Motors as a Function of Motor Size and Annual Operating Hours

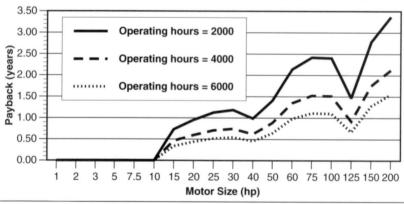

Note: The assumptions, efficiencies, and motor costs are listed in Table A-1 (see Appendix A) and are based on the *MotorMaster+®* database.

Source: WSU 1999

Figure 2-31

Simple Payback Times for New, Premium-Efficiency TEFC Motors vs. Repairing Standard TEFC Motors as a Function of Motor Size and Annual Operating Hours

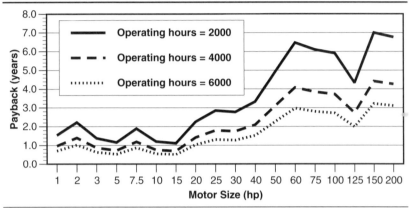

Note: The assumptions, efficiencies, and motor costs are listed in Table A-1 (see Appendix A) and are based on the *MotorMaster+*® database.

Source: WSU 1999

Figure 2-32

Simple Payback Times for New, Premium-Efficiency ODP Motors vs. Repairing Standard ODP Motors as a Function of Motor Size and Annual Operating Hours

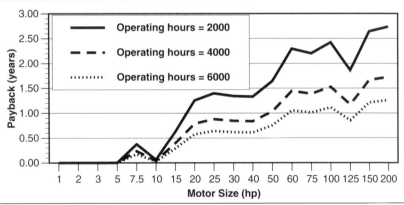

Note: The assumptions, efficiencies, and motor costs are listed in Table A-1 (see Appendix A) and are based on the *MotorMaster+*® database.

Source: WSU 1999

The yearly savings afforded by the EPAct motor are as follows:

- Demand savings = 50 hp × (1/0.896 – 1/0.939) × 0.75 × 0.746 kW/hp = 1.43 kW

- Energy savings = 1.43 kW × 6,000 hr/yr = 8,580 kWh

- Cost savings = $.06/kWh × 8,580 kWh + $70/kW-yr × 1.43 kW = $615

- Simple payback period = $1,133/$615 = 1.8 years

Note that the economics of replacing motors instead of rewinding them does not hinge on the 1% damage assumed from rewinds. Ignoring the 1% damage in the above example, the replacement motor still has a very attractive 2.4-year payback. Figure 2-29 shows the generally very favorable economics of replacing failed standard-efficiency TEFC motors with an EPAct motor. In almost all cases it is more cost-effective to replace rather than repair a motor of 40 hp or less. For ODP motors (see Figure 2-30), the economics of replacement is even more favorable. It is almost always more economic to replace rather than repair all but the largest motors, while for motors of 15 hp and less, replacement is actually less expensive on a first-cost basis alone.

While the payback for choosing to replace a failed standard motor with a premium motor will in most cases be slightly longer than for an EPAct motor, the additional incremental investment will often be attractive. As can be seen for TEFC motors (Figure 2-31), the payback for replacement rather than repair in motors operating only 2,000 hrs/yr up to 15 hp is less than 2 years, and in other cases up to at least 40 hp. As with TEFCs, the economics for replacing failed ODP motors with premium motors is more attractive in almost all cases (Figure 2-32).

Economics of Replacing Operating Motors with Efficient Motors

Efficient motors are clearly economic for most new applications and attractive compared to rewinding in most instances. But what about the economics of replacing operating motors with efficient motors? The incremental cost of the efficient motor in such instances is generally its full purchase plus installation cost, not the marginal difference between it and a standard motor or a rewind. Thus, the payback of such replacements is considerably longer than with new applications or rewinds. The right combination of conditions, however,

Figure 2-33

Simple Payback Times for Retrofitting New, EPAct TEFC Motors vs. Stock Standard TEFC Motors as a Function of Motor Size and Annual Operating Hours

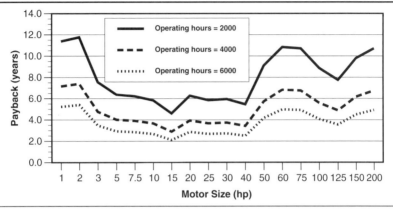

Note: The assumptions, efficiencies, and motor costs are listed in Table A-1 (see Appendix A) and are based on the *MotorMaster+*® database.

Source: WSU 1999

Figure 2-34

Simple Payback Times for Retrofitting New, Premium-Efficiency TEFC Motors vs. Stock Standard TEFC Motors as a Function of Motor Size and Annual Operating Hours

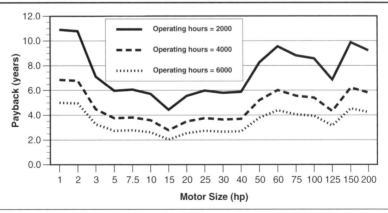

Note: The assumptions, efficiencies, and motor costs are listed in Table A-1 (see Appendix A) and are based on the *MotorMaster+*® database.

Source: WSU 1999

Figure 2-35

Simple Payback Times for Retrofitting New, EPAct ODP Motors vs. Stock Standard ODP Motors as a Function of Motor Size and Annual Operating Hours

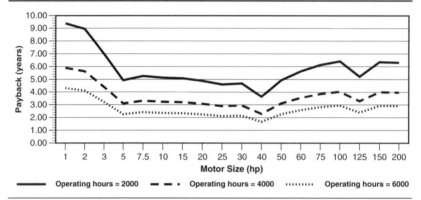

Note: The assumptions, efficiencies, and motor costs are listed in Table A-1 (see Appendix A) and are based on the *MotorMaster+®* database.

Source: WSU 1999

Figure 2-36

Simple Payback Times for Retrofitting New, Premium-Efficiency ODP Motors vs. Stock Standard ODP Motors as a Function of Motor Size and Annual Operating Hours

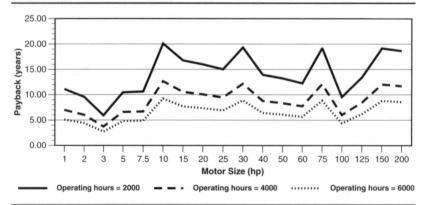

Note: The assumptions, efficiencies, and motor costs are listed in Table A-1 (see Appendix A) and are based on the *MotorMaster+®* database.

Source: WSU 1999

can make replacements cost-effective. Some of these conditions are listed below:

- The existing motor has an efficiency below its nameplate rating, a likely condition if the motor has been damaged in rewinding or if it is grossly oversized.

- The number of operating hours near full load is high.

- The replacement motors can be purchased in bulk at a large discount.

- The price of electricity is high.

A motor replacement program conducted at Stanford University several years ago offers an interesting example (Wilke & Ikuenobe 1987). Seventy-three standard-efficiency motors in HVAC applications were replaced in a group retrofit with efficient motors. Motor sizes ranged from 7.5 to 60 hp, with annual operating hours from 2,000 to continuous (8,760). Because of the size of the purchase, the university received a substantial cost break on the motors. Field tests showed that many of them were operating below their nominal efficiency ratings, possibly because of previous rewind damage. In addition, many of the motors were oversized for their loads: 32 of the replacements involved downsizing (see "Motor Oversizing" in Chapter 3). The overall payback period on the project, counting a utility rebate, was under 3 years. The program's payback without the rebates would have been under 5 years.

This case study is instructive on several counts. First, there is no reason to believe that this fleet of motors was specified or maintained with anything but typical skill. This suggests that many such motors in institutional and commercial buildings are operating well below their published efficiencies because of oversizing, rewind damage, or both. The attractive paybacks achieved in the Stanford program further suggest that group replacement of standard-efficiency motors might be cost-effective in many settings.

Even without credit for downsizing, replacement is cost-effective in many cases, particularly in TEFC motors between 5 and 40 hp sizes for applications that operate 4,000 hrs/yr or more. Figure 2-33 shows the paybacks for retrofitting operating standard-efficiency TEFC motors with EPAct motors. The analysis is based on the values in Table A-1 in Appendix A and assumes that the user pays 60% of list price. In practice, large users may be able to negotiate higher price breaks on group purchases, which would improve the economics. The payback for replacing operating standard-efficiency motors with premium-efficiency motors will be slightly longer but is generally attractive in applications that have high loads and operating hours (i.e., 4,000 hrs/yr or more (see Figure 2-34).

The economics of retrofitting ODP motors is slightly better than with TEFC motors because of the lower first cost of the motor (Figure 2-35), especially in the 5 to 100 hp range. As with TEFC motors, the payback for replacement with a premium motor is slightly longer (see Figure 2-36). For further discussion of the economics of motor efficiency options, see Appendix A.

Field Measurements of Motor Load and Efficiency

Measurement of load and efficiency in the field can be challenging because of variations in units and limitations of test equipment. A number of techniques are available for measurement of both determinations, with the most accurate techniques requiring the greatest expense in both equipment and effort. Some commercial equipment claims to be able to accurately measure both motor load and efficiency. However, for many applications, simpler approaches may be sufficient to characterize the most important loads in the building or plant.

Motor Load Determination

One of the simplest methods of load determination is the *slip* method. This method takes advantage of the nearly linear relationship between motor slip and load (see Figure 3-9). Although this method should be used with caution because it can produce values significantly in error, it may be about as good as anything else available. This is particularly true at low loads, which can challenge the range of most wattmeters used in the method discussed next (Douglass 2000). Slip also varies with motor voltage, possibly resulting in errors of over 5% because of voltage variation (Nailen 1987). Voltage compensation can reduce some of the error if the motor is not powered exactly at nameplate voltage.

The slip method requires a voltmeter and a tachometer. The method, expressed mathematically, is

$$\text{Percent load} = 100 \times \text{slip} / ((S_s - S_r) \times (V_r / V)^2)$$

where
    slip = measured slip
    S_s = synchronous speed
    S_r = nameplate speed
    V_r = nameplate voltage
    V = mean measured line-to-line voltage

The voltage error is likely to be smaller than the error associated with the nameplate rpm. NEMA tolerance on the slip requirement for nameplate labeling is currently 20%. Most manufacturers round nameplate rpm to the nearest 5 rpm (Douglass 2000).

A more accurate method for load determination is the watt method. A three-phase wattmeter is attached to the motor input leads, usually at the motor controller disconnect. The motor load is estimated by multiplying the motor input power by an approximate motor efficiency. The nameplate efficiency, manufacturer's data, or data from the *MotorMaster+®* database (WSU 1999), mentioned earlier in this chapter, can be used to find the value for motor efficiency. The watt method works well above 50% load since the efficiency curve for most motors is relatively flat between 50% and 100%, and at these loads watt measurements are reasonably accurate. The method has two main drawbacks, however. First, the results are dependent on the accuracy of the efficiency estimate. Second, they are sensitive to the voltage applied to the motor (E Source 1999). Although it is possible to correct for voltage error, correction requires knowledge of the effects of voltage variations on the efficiency of any specific motor design.

The *MotorMaster+®* software (WSU 1999) uses a variant of the watt method, applying an iterative approach. It starts with an efficiency value and divides it into the wattage to get output power and compute percent load. Then it looks up a new efficiency from a partial-load efficiency table and recomputes load. The software requires several iterations to converge on load (Douglass 2000).

Motor Efficiency Determination

No good methods exist for measuring motor efficiency *in situ*. All available options have major drawbacks: they tend to be time- and labor-intensive, require expensive test devices, or produce estimates of questionable accuracy. However, although these methods are not as accurate as the IEEE 112-B method, which requires a specially equipped laboratory, they can provide some useful estimates of the motor load and efficiency. These estimates can help identify low-efficiency motors, motors damaged in rewinding, and oversized motors. Some current methods are described below.

Three devices were tested at Oregon State University that produced reasonably accurate results when operated correctly (Douglass 2000). However, they too have flaws: they are costly and require uncoupling the motor from the load for a no-load test.

Baker Instrument Company of Colorado and PdMF of Florida have both recently brought out new models of their current-signature predictive maintenance testers that supposedly determine efficiency with good accuracy (Douglass 2000). More information on these companies can be found in Appendix D.

WSU has developed a spreadsheet that works quite accurately for measuring motor efficiency, as long as accurate readings of temperature, power, and speed under load; power at no-load; and cold winding resistance are available. The problem is that it is difficult to find a portable wattmeter that is accurate for measuring motor power at no load because the no-load power factor is extremely low (Douglass 2000).

A quicker, though less accurate, method for checking motor efficiency in the field builds on the slip method of load determination. A tachometer is used to measure the actual motor speed, which is used to determine the motor load (output power). The wattmeter is used to measure the motor input power. For example, a 10 hp motor was rated at 1,745 rpm at full load. The measured speed was found to be 1,778 rpm, and the measured power was 3.8 kW. Load is nearly proportional to slip, so the fraction of full load is approximately proportional to the fraction of the full-load slip:

$$\frac{1{,}800 \text{ rpm} - 1{,}778 \text{ rpm}}{1{,}800 \text{ rpm} - 1{,}745 \text{ rpm}} = 40\% \text{ of full load (or 4.0 hp output)}$$

The efficiency is thus approximately

$$\frac{4.0 \text{ hp} \times 0.746 \text{ kW/hp}}{3.8 \text{ kW}} = 79\%$$

If the application never runs at a higher load, a 5 hp premium-efficiency motor might make a good retrofit or replacement (from Table A-1, the efficiency of a 5 hp premium-efficiency motor is about 90.5%), depending on the operating hours, load profile, and cost of electricity.

None of these in-service methods is sufficiently accurate for actual motor evaluation, but they can be a good way to screen candidate motors before sending a motor out for a lab test (Douglass 2000). Then, the results of lab tests, together with the motor load profile and motor age and condition, can assist the user in evaluating whether to keep the motor, replace it with a more efficient version, or replace it with a different motor that is better matched to the load.

Summary

The three-phase squirrel-cage induction motor accounts for over 75% of U.S. drivepower input, followed by single-phase induction motors and all others (synchronous, wound-rotor induction, DC, and so forth). Premium-efficiency motors are available for many three-phase and single-phase induction motor applications. These motors offer reduced energy and peak-power costs as well as increased life compared with EPAct and standard-efficiency motors. Premium-efficiency motors are often cost-effective in new applications or as alternatives to rewinding, and they can be cost-effective in some retrofits, depending on the specifics of the application. It is important to obtain up-to-date information on the efficiency of the specific motor in question. Motor rewinds can seriously degrade efficiency, and specifying quality rewinds can improve the efficiency and reliability of repaired motors.

System Considerations

In this chapter, we discuss a number of important but often over-looked determinants of motor system efficiency, including power supply quality, the distribution network that feeds the motor, the match between the load and the motor, the transmission and mechanical components, and maintenance practices. We also present simple and inexpensive diagnostic techniques for identifying some common motor system problems. Unfortunately, the lack of field data makes it difficult to quantify the extent of energy losses and equipment damage from poor system optimization. In general, older facilities modified in pieces and loaded closer to capacity are more likely to have problems than newer facilities.

Power Supply Quality

AC motors, particularly induction motors, perform best when fed by symmetrical, sinusoidal waveforms of the design voltage and frequency values. Deviations from these ideal conditions can reduce the motor's efficiency and longevity. Such distortions in power quality include voltage unbalance, out-of-specification voltage and frequency, and harmonics.

Voltage Unbalance

In a balanced three-phase system, the voltages in the phases can be represented by three vectors of equal magnitude, each out of phase by 120°. A system that is not symmetrical is called an unbalanced system.

The following formula can be used for the approximate calculation of the voltage unbalance:

$$\text{Voltage unbalance (\%)} = \frac{\text{Maximum difference of the voltages in relation to the average voltage}}{\text{Average voltage}} \times 100$$

Suppose the measurements in the three phases give the following values:

$$V_a = 200 \text{ V}$$
$$V_b = 210 \text{ V}$$
$$V_c = 193 \text{ V}$$

The average voltage = 201 V

Maximum difference from the average = 210 V – 201 V = 9 V

Voltage unbalance = (9 V/201 V) × 100 = 4%

Voltage unbalance is problematic for several reasons. First, it wastes energy. As Figure 3-1 illustrates, voltage unbalance leads to high current unbalance, which in turn leads to high losses. A modest phase unbalance of 2% can increase losses by 25%. Second, prolonged operation under unbalanced voltage can damage or destroy a motor. The excess heat generated in a motor running on a 2% unbalance can reduce the insulation lifetime by a factor of eight (Andreas 1982). An unbalance of 5% or more can quickly destroy a motor. To address this problem, many designers include phase unbalance and phase failure protection in motor starters. Another negative impact of phase unbalance is a reduction in motor torque, particularly during start-up. Figure 3-2 shows the reduction of rated power as a function of the voltage unbalance.

While severe unbalance (over 5%) causes immediate, obvious problems, small unbalances in the 1–2% range are insidious because they can lead to significant increases in energy use without being detected for a long period of time, particularly if a motor is oversized. To avoid this situation, the voltages in a facility should be regularly monitored. NEMA recommends that the voltage be balanced to the best degree possible. An unbalance of over 1% should be remedied immediately, and motors should not be operated with unbalances of greater than 5% (NEMA 1999).

Other equipment, such as variable-frequency drives and transformers, is also sensitive to voltage unbalance. As with motors, unbalance can lead to increased phase currents that produce heat, which reduces efficiency and can damage the equipment.

While a voltage unbalance can occur from the electricity service to a facility, there is some debate about how common this problem is. The

Figure 3-1

Effect of Voltage Unbalance on Three-Phase Induction Motor Currents

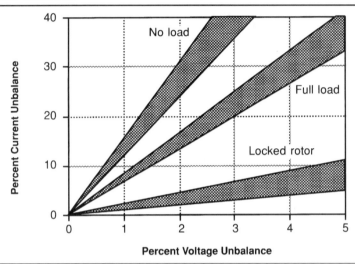

Source: Andreas 1982

Figure 3-2

Derating Factor Due to Unbalanced Voltage for Integral-Horsepower Motors

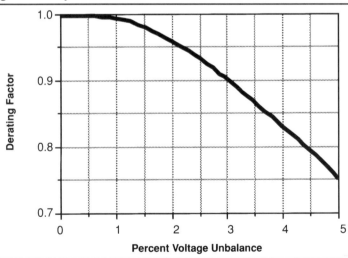

Source: NEMA 1999

American National Standards Institute's (ANSI 1995) report C84.1-1995, Appendix D, suggested that the no-load voltage unbalance at the meter should be less than 5%. The document indicated that 98% of utilities' customers have less than 3% unbalance, while 66% have less than 1%. It is unclear how this relates to the voltage unbalance under load. However, a study in the PG&E service territory indicated that the vast majority of customers had very limited voltage unbalances (E Source 1999). Discussions are underway among NEMA, electric utilities, and ANSI to clarify this confusion (Bonnett 1999).

Whether the service voltage is in balance is less important than whether the voltage at the motor control center is in balance. Many problems result from issues with facility distribution systems. There are several common causes of voltage (or phase) unbalance. The first is a nonsymmetrical distribution of single-phase loads in the facility. Most facilities contain a mixture of three-phase loads (such as motors) and single-phase loads (such as most lighting, electrical outlets, single-phase motors, and processes—for example, electric arc furnaces). Putting a disproportionate share of the single-phase loads on one of the three phases can cause voltage unbalance. To make the identification more challenging, some of these imbalance problems are transient, as when a bank of lights is turned on or off.

A second cause of voltage unbalance is an open circuit in one of the phases, often caused by a blown fuse. This problem often results from lightning strikes at or adjacent to the facility and will lead to motor failure if immediate steps are not taken. For large or critical motors, phase protection devices that alarm or trip in the event of serious unbalance are recommended (E Source 1999).

Finally, different-size cables carrying the phases of a three-phase load can lead to unbalanced conditions. This can happen in an older facility when a load is converted from single- to three-phase. Different cable sizes produce different voltage drops, which in turn lead to the unbalanced voltages.

The diagnosis of voltage unbalance is a simple operation requiring the measurement of the voltages in the three phases. It is prudent to check these voltages over a full cycle of facility operation since the unbalance may occur only during certain situations. For example, voltage balance may be fine except when an intermittent single-phase load is in service.

Voltage and Frequency

When an induction motor is operated at a voltage or frequency other than its rated value, its performance changes. Motors are designed to operate successfully at full load with a ±10% voltage fluctuation (NEMA 1999). However, a 10% change will increase the operating temperature for a given load, which will accelerate deterioration of the insulation. Voltage increases will usually reduce the power factor while voltage decreases will usually increase the power factor. A 10% increase in voltage will also affect slip, decreasing it by 21%, while a 10% decrease in voltage will increase slip by 21%. Higher-than-rated frequencies usually improve the power factor but decrease locked rotor torque while increasing speed and friction and windage losses. Operation at lower frequencies decreases speed and power factor while increasing locked rotor torque. If variations in both voltage and frequency occur, the effects are superimposed (NEMA 1999).

Voltage fluctuations normally result from improperly adjusted transformers, undersized cables (leading to large voltage drops due to higher resistance in the small cables), or a poor power factor in the distribution network.

When a motor is underloaded, reducing voltage can improve the power factor and efficiency, mainly by reducing the reactive current. This practice works for both standard and efficient motors, although efficient motors are less affected by voltage fluctuations.

Since torque is proportional to the square of the voltage, motors operating at undervoltage might have a hard time starting or driving a high-torque load. For instance, if the voltage slips to 80% of the rated value, the available starting torque is only about 60% of its rated value.

The diagnosis of voltage level problems requires monitoring and recording voltages, preferably for a whole cycle of the facility's operation. Patterns in fluctuation over time sometimes help to reveal the cause(s). Measuring voltage is normally easiest at the motor starter terminals; to estimate the voltage at the motor terminals, calculate the voltage drop in the cable connecting the starter to the motor. NEMA-rated voltages for three-phase, 60 Hz induction motors appear on the motor nameplate and typically allow for a voltage drop of about 4% in the motor feeder cables.

Harmonics and Transients

Under ideal conditions, utilities supply pure sinusoidal waveforms of one frequency (60 Hz in North America and 50 Hz in Europe), similar to those shown in Figure 2-3. Resistive loads, such as incandescent lights, use all of the energy in that waveform. Other loads (including ASDs and other power electronic devices, arc furnaces, and overloaded transformers) cannot absorb all of the energy in the cycle.

Figure 3-3

Example of a Distorted (Nonsinusoidal) Wave

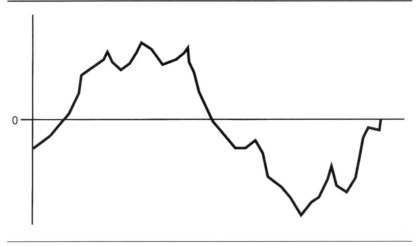

Source: Eaton Corporation 1988

In effect, they use energy from only part of the sine wave and thus distort it (see Figure 3-3).

The resulting distorted waveform contains a series of sine waves with frequencies that are multiples of the fundamental 60 Hz frequency. These distortions are called harmonics. The 180 Hz (or 3 × 60 Hz) component is the third harmonic; the 300 Hz (or 5 × 60 Hz) component is the fifth harmonic; and so on.

Harmonics can increase motor losses, reduce torque, and cause torque pulsation and overheating. Vibration and heat in turn can shorten motor life by damaging bearings and insulation. Harmonics may also cause malfunctions in electronic equipment, including computers; induce errors in electric meters; produce radio frequency static; and destroy power system components.

Electronic ASDs, discussed further in Chapter 4, can both generate and be damaged by harmonics from other sources. It is thus very important that they be installed properly and, in some cases, be isolated from other equipment by separated feeders, transformers, and harmonic filters. Serious harmonics problems from properly installed ASDs are rare. Problems are most likely to occur with large drives and in situations where ASDs control a large fraction of the total load.

Standard-efficiency motors must sometimes be derated by 10–15% when supplied by an ASD that produces substantial

Figure 3-4

Derating Curves for Motors Operating on Adjustable-Speed Drives

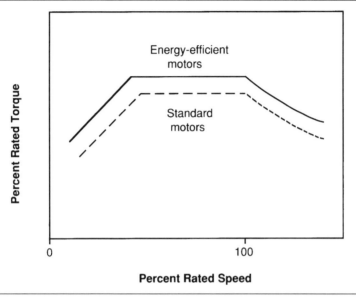

Source: NEMA 1999

harmonics. Derating is more likely to be needed for constant-horse-power installations than for variable-torque loads. For example, a motor with a nominal rating of 100 hp that was derated by 15% would only be able to drive an 85 hp load. Efficient motors may be better able to cope with harmonics due to these motors' higher thermal margins and lower losses, and these motors therefore are seldom derated. One notable exception is when the motor runs below 30–40% of rated speeds. Under these conditions, high losses caused by harmonics combined with the lower ventilation available at reduced fan speeds often require efficient motors to be derated. As Figure 3-4 shows, the allowable torque of both types of motors falls off sharply at low and very high speeds when powered by an ASD. Under all circumstances, the motor manufacturer should be consulted before using a motor with an ASD. Particular care should also be taken in grounding the motor because high-frequency harmonics can increase leakage currents, which can result in an unsafe installation (NEMA 1999).

Harmonics can substantially disrupt conventional electric meters. One study sponsored by EPRI found errors ranging from +5.9% to –0.8%

in meters subjected to harmonics from ASDs (EPRI 1982). During periods when ASDs operated at very light load, errors exceeded 10% and caused severe waveform distortion. Harmonics generated by ASDs or other equipment may cause metering errors and overbilling, providing customers with another reason to correct or suppress harmonics at the source. Customers with medium-size or large ASDs on their premises might even consider installing solid-state watt-hour meters that generally give accurate readings even in the presence of harmonics (Peddie 1988).

In addition to variable-frequency drives producing harmonics, these drives can also be sensitive to harmonics. Power supply harmonics can lead to nuisance trips, and in extreme situations even drive failure.

Besides harmonics, the voltage waveform may also contain another form of undesirable distortion called transients. These are brief events, usually microseconds in length, and appear either as voltage spikes or voltage notches in the sinusoid. Fast transients result from the commutation of power electronic devices and circuit breakers, as well as from lightning.

If transients occur rarely, they have little impact on energy consumption, but if they occur repeatedly and at frequent intervals, they can behave like harmonics and thus increase losses in a motor. Very large transients, as in the case of a lightning strike, can damage or destroy equipment. Generally, transients are a problem only in facilities where large loads are cycled, producing distortion in the voltage waveforms. For example, a facility with a large induction furnace where power is applied intermittently might have a problem with transients.

Diagnostics and Mitigation of Harmonics and Transients

Equipment to accurately monitor transients and harmonic distortion is readily available (see Appendix D). Several less expensive tools can roughly assess the level of harmonics and indicate whether more precise measurements are warranted.

For instance, an oscilloscope can be used to generate a picture of the voltage waveform, which can be inspected for distortions that signal the presence of meaningful harmonics, which should then be measured with a harmonic analyzer. Another technique is to compare the voltage readings from two AC digital voltmeters, one with true root mean square (RMS) capabilities and the other without. The true RMS meter gives the correct voltage even if there is harmonic distortion, whereas a normal meter only gives the correct value if there are no harmonics. If the two meters are calibrated, readings differing by more than a few percentage points indicate a significant harmonic content.

Harmonics should be reduced to an acceptable level (less than 5% of the fundamental current in medium-voltage systems and less than 1.5% in high-voltage systems) as close as possible to the source (IEEE 1981). Mitigation at the source is normally most effective, as it prevents the losses from harmonics propagation in the network.

Surge suppressors are available and effective for suppressing transients that may interfere with the operation of computing and communications equipment. Claims that these devices save energy, however, are unfounded.

In ASDs, harmonics are most commonly controlled by installing filters at the ASD input circuit to provide a shunt path for the harmonics and to perform power-factor compensation. IEEE Standard 519 (IEEE 1981) contains guidelines for harmonic control and reactive-power compensation of power converters. The cost of the harmonic filter to meet this standard is typically about 5% of the cost of the ASD.

The installation of inductors on drives can also be used to address harmonics issues. Inductors are particularly effective in eliminating drive trips due to problems in the power supply to the drive.

The Federal Communications Commission (FCC) has produced a set of regulations regarding the electromagnetic interference (EMI) produced by computing devices. These regulations, which are also becoming widely accepted in the ASD market, set permissible radiation and conduction levels. FCC standards define two classes of products: Class A systems used in commercial and industrial environments and Class B systems used in residences. The Class B standards are stricter to avoid noticeable interference with radio and television use in the home. Although ASDs are expected to meet only Class A standards, some manufacturers offer ASD equipment that performs within Class B requirements. Radiated EMI can be brought down to FCC standards by proper layout and by shielding the enclosure.

System Oversizing

Motor systems become oversized when designers adopt successive safety factors or when the requirements of the motor-driven equipment are reduced due to system changes. Designers often project growth in a system's peak requirements, and they assume the extra capacity cost is a small premium to pay to ensure that the system will be able to cope with maximum demand. For example, a designer might choose a pump with a 30% safety margin (a certain amount of the margin is for increase in the process requirements and the rest is for scale build-up in the pipes) and then will round up when choosing among standard motor sizes, thereby specifying a model with 20% extra horsepower. Such oversizing may be warranted in some cases but many times it can lead to costly waste and other problems.

Motor Oversizing

As Figures 3-5 and 3-6 illustrate, the efficiency and power factor of a motor vary with the load. The efficiency of most motors peaks at approximately 75% load and drops off sharply below 40% load, although this range varies among different designs.

One study found that standard-efficiency motors peaked near 100% load, and the high-efficiency models peaked nearer 75% load (Colby and Flora 1990). Power factor drops steadily with the load. Even at 60% load, the power factor often needs compensation, and it drops even more sharply below 60% load. Figure 3-7 shows that low-

---

Figure 3-5

Typical Efficiency vs. Load Curves for 1,800 rpm, Three-Phase, 60 Hz, Design B Squirrel-Cage Induction Motors

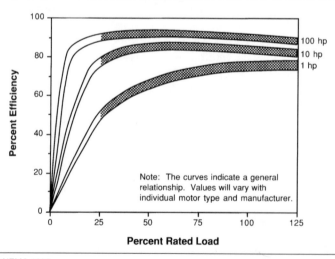

Note: The curves indicate a general relationship. Values will vary with individual motor type and manufacturer.

Source: NEMA 1999

Figure 3-6

Typical Power Factor vs. Load Curves for 1,800 rpm, Three-Phase, 60 Hz, Design B Squirrel-Cage Induction Motors

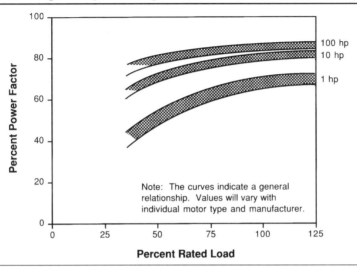

Note: The curves indicate a general relationship. Values will vary with individual motor type and manufacturer.

Percent Rated Load

Source: NEMA 1999

Figure 3-7

Typical Full-Load Power Factor vs. Horsepower Rating Curves for Three-Phase, 60 Hz, Design B Squirrel-Cage Induction Motors

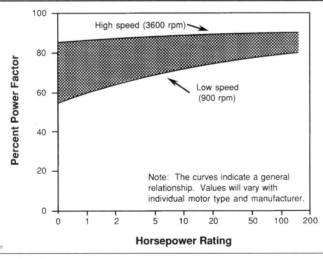

Note: The curves indicate a general relationship. Values will vary with individual motor type and manufacturer.

Horsepower Rating

Source: NEMA 1999

speed motors have substantially lower power factors than high-speed motors of the same size.

To stay within its optimal operating limits, a motor should be sized to run at 50% or more of its rated load a majority of the time. A grossly oversized motor (generally below 40% loading) will run at decreased efficiency and low power factor, thus increasing energy costs and requiring either costly power-factor compensation or added utility charges to pay for the reactive current. In addition, larger motors cost more to buy and install, as well as requiring larger and more expensive starters.

Although oversized motors present these drawbacks, they also can accommodate unanticipated high loads and are likely to start and operate more readily with undervoltage conditions. A modest sizing margin, however, can generally provide these advantages.

The question of motor sizing is not limited to new installations. Often it is economic to replace existing motors with smaller, high-efficiency units. Users must carefully evaluate which motors to downsize. Those that run many hours a year at light loading are obvious candidates because they are running far below their optimal efficiencies. Motors that operate in the 50–100% load region are not likely candidates because they are operating at close to peak efficiency. For example, a 150 hp motor driving a 120 hp load is normally more efficient than a 125 hp motor driving the same load because the efficiency of many efficient motors peaks at approximately 75% load, and larger motors generally have higher efficiencies.

One must also be careful to address the differences in operating speed that can result from downsizing the motor. As discussed in Chapter 4, small changes in the speed of centrifugal loads can result in large changes in energy consumption. And as analyzed in Chapter 2, the speed of the motor varies with load. As a result, the resized motor may not operate at the correct speed for the application. Usually these problems can be easily and inexpensively addressed, but one must be cognizant of this possibility when downsizing motors.

There is no definitive rule about downsizing because the relationship between efficiency and load varies among different sizes and types of motors. Larger motors generally maintain efficiency at low loading better than smaller motors do. Similarly, high-efficiency motors have a flatter efficiency curve than standard-efficiency models. For example, efficiency might drop rapidly at 48% of full load in a standard motor, but for an efficient motor, efficiency might not drop until the motor reaches 42% of its rated load. In general, motors that always run below 40% load are strong candidates for downsizing.

Figure 3-8 shows the change in efficiency over a range of 25% to

Figure 3-8

Two Representative Efficiency Curves for 5 hp Motors and 10 hp Motors

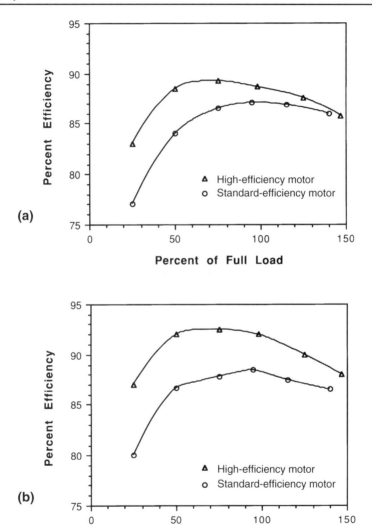

Note: Two representative efficiency curves for 5 hp motors are shown in (a). The energy-efficient motor maintains its high-efficiency level over a wider load range than does the standard motor. A similar comparison for 10 hp motors is presented in (b) (Colby and Flora 1990). Current premium-efficiency motors and most EPAct motors have an efficiency/load similar to those of the energy-efficient motors pictured here.

150% load for seven new 5 hp motors (three efficient motors and four standard units) and eight new 10 hp motors (five efficient motors and three standard units) tested by the North Carolina Alternative Energy Corporation (Colby and Flora 1990). These figures represent one of the few published measurements of efficiency below 50% loading. Manufacturers typically list efficiencies at 100%, 75%, and 50% loading only.

Two points are notable. First, the high-efficiency motors maintained their efficiencies better across the full range of loading. Second, the efficiency of all motors fell off sharply as loads fell from 50% to 25%; the efficiency fell six to seven points in the standard-efficiency units and about four points in efficient motors.

Evaluating whether to downsize a motor requires knowing both its typical and maximum loads. Consider, for example, two 100 hp motors running different fans in a facility. Both motors are metered, and neither requires more than 35 hp during the metering period. The first fan's specifications reveal it will never need more than 40 hp. Since the motor will never exceed about 40% of its rated 100 hp load, it is appropriate to replace it with a high-efficiency 50 hp unit. The second fan's specifications indicate that it occasionally must use up to 80 hp, even though it typically draws only 35 hp. Thus, while the motor often runs at only 35% of its rated 100 hp capacity, it should not be downsized—or at least not by much—unless the system's maximum-power requirements are reduced. Such measures are discussed in Chapter 5.

Another option in such a situation is to use a smaller motor with a high service factor so that it can withstand overloading on rare occasions. However, this determination should be made by an engineer familiar with the process.

Diagnostics of Motor Oversizing

The load of a motor can be roughly estimated by comparing a wattmeter reading of the power input with the motor's rated power. A clamp-on ammeter does not give a good estimate of the motor load since the power factor drops sharply at low loads, and amperage readings are greatly affected by power factor. It is preferable to use one of the load determination methods or devices discussed in Chapter 2. In general, the watt method is preferable for motors that are loaded at more than 50% of rated capacity, while the slip method is preferable for motors operating at low load (Douglass 2000).

If the motor load changes over time, a simple wattmeter will not suffice. Instead, a data logger should be used to monitor the absorbed power under different operating conditions.

Figure 3-9

Variation of Slip and Current (Amperes) with Motor Load

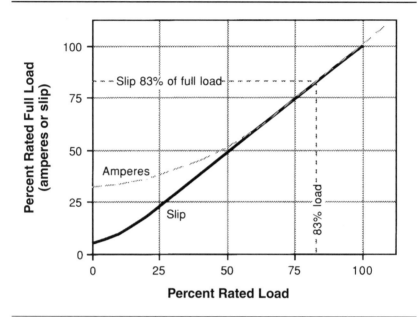

Note: This relation holds only at the rated motor voltage; slip varies inversely with the square of the actual motor voltage.

Source: Nailen 1987

Extent of Oversizing

As data that will be presented in Chapter 6 suggest, approximately two-fifths of motors 5 hp and larger are running at or below 40% of rated load (XENERGY 1998). As discussed above, there is very little published information on efficiency loss at loading below 50%. The data shown in Figure 3-9 suggest that loading between 25% and 40% leads to a drop of two to eight percentage points below the performance at 50% load. Loading below 25% will lead to an even larger drop in efficiency. Based on admittedly sketchy data, we assume an average efficiency loss of five percentage points in those motors running below 40% loading.

While some oversized motors should be downsized, many should instead be equipped with controls that enable them to operate more efficiently at partial load. Various means of doing this are discussed in Chapter 4.

Distribution Network Losses

In-facility distribution losses can be reduced by selecting and properly operating efficient transformers and by correctly sizing distribution cable.

Distribution transformers normally operate above 95% efficiency unless they are old or very lightly loaded. Transformers over 30 years old should be replaced by new models that are more efficient. It is usually more effective to run one transformer at moderate to full load than to operate two of them in parallel lightly loaded.

Many large customers are primary metered (on the high-voltage side of the distribution transformer) and own their own transformers. Some utilities are encouraging their primary-metered customers to install high-efficiency transformers in new facilities or for facility expansions. While all transformers are relatively efficient, they last a long time; small improvements in efficiencies can result in large lifetime savings. Efficiency in transformers comes from many of the same measures we see taken in motors: the use of better steels for the core, and more (i.e., larger-diameter wire) and better (e.g., copper in place of aluminum) winding materials.

Transformers can be purchased in three basic types, and with a range of efficiencies for each. The most common small transformers are dry-type, which are cooled by convection. In liquid-immersed transformers, the core and windings are placed in a synthetic oil bath that transfers the heat. For aggressive environments, an encapsulated transformer can be specified. In general, the liquid-immersed versions tend to have lower losses and are more compact, though concerns about oil leakage may limit where they can be placed. Much of this concern stems from past use of polychlorinated biphenyl (PCB) oils, which were flammable and toxic. In modern designs, these have been replaced with synthetic oils that address both concerns. The dry-type transformers are readily available in a wide range of sizes but tend to have lower efficiencies. They are also larger than the liquid-immersed transformers. The encapsulated transformers are significantly more expensive than either of the other designs and are restricted to special applications requiring their features.

Since transformers can be bought in a wide range of efficiencies and costs, it is best to perform a total-cost-of-ownership calculation. NEMA Standard TP-1 (1996) and IEEE Standard PC57.12.33 (1998) describe how to carry out these calculations, which require information regarding the transformers under consideration, projected loading, annual duty hours, and cost of electricity. For those applications where it is impractical to do a cost-of-ownership calculation, TP-1, Table 4-2 (see

Table 3-1 below) specifies designs for dry-type transformers that are considered energy efficient based on average national data. While some people have used low temperature rise as a proxy for energy efficiency, it is not a good indicator of efficiency.

Cable Sizing

Cabling represents another opportunity for saving energy. Table 3-2 compares losses (watts per foot) and costs of using several cable sizes to supply a 30 amp load, assuming 8,000 hours of annual operation and an electricity cost of $.06/kWh. The larger cables offer very attractive payback times. Note that the payback is very sensitive to operating hours: at 4,000 hours, the paybacks in Table 3-2 would double.

Table 3-1

NEMA Class I Efficiency Levels for Dry-Type Distribution Transformers

Reference Condition		Temperature		% of Nameplate Load	
Low Voltage		75°C		35%	
Medium Voltage		75°C		50%	

Single-Phase Efficiency			Three-Phase Efficiency		
Low KVA	Medium Voltage	Voltage	Low KVA	Medium Voltage	Voltage
15	97.7	97.6	15	97.0	96.8
25	98.0	97.9	30	97.5	97.3
37.5	98.2	98.1	45	97.7	97.6
50	98.3	98.2	75	98.0	97.9
75	98.5	98.4	112.5	98.2	98.1
100	98.6	98.5	150	98.3	98.2
167	98.7	98.7	225	98.5	98.4
250	98.8	98.8	300	98.6	98.5
333	98.9	98.9	500	98.7	98.7
500	—	99.0	750	98.8	98.8
667	—	99.0	1,000	98.9	98.9
833	—	99.1	1,500	—	99.0
			2,000	—	99.0
			2,500	—	99.1

Source: NEMA 1996

Table 3-2

Savings from Lower-Loss Distribution Wiring

Marginal Savings and Costs of Lower-Loss Distribution Wiring
(Compared to #8 Wire) Assuming 100% Load

				Installed 1989 $/Ft First Cost				
Wire Size[a]	I^2R/Ft Loss (W/Ft)	Loss ($/Ft-Yr)	Saving ($/Ft-Yr)	Conduit	Wire	Total	Marg. Cost (+$/Ft)	Simple Payback (Yrs)
#8	2.01	$0.96		$3.16	$1.21	$4.37		
#6	1.28	$0.61	$0.35	$3.16	$1.59	$4.75	$0.38	1.1
#4	0.81	$0.39	$0.57	$3.86	$2.20	$6.06	$1.69	3.0
#3	0.65	$0.31	$0.65	$3.86	$2.67	$6.53	$2.16	3.3

Marginal Savings and Costs of Lower-Loss Distribution Wiring
(Compared to #8 Wire) Assuming 75% Load

				Installed 1989 $/Ft First Cost				
Wire Size[a]	I^2R/Ft Loss (W/Ft)	Loss ($/Ft-Yr)	Saving ($/Ft-Yr)	Conduit	Wire	Total	Marg. Cost (+$/Ft)	Simple Payback (Yrs)
#8	1.18	$0.56		$3.16	$1.21	$4.37		
#6	0.75	$0.36	$0.20	$3.16	$1.59	$4.75	$0.38	1.9
#4	0.47	$0.22	$0.34	$3.86	$2.20	$6.06	$1.69	5.0
#3	0.38	$0.18	$0.38	$3.86	$2.67	$6.53	$2.16	5.7

[a] Sizes in American Wire Gauge. For diameters, see Appendix A.
Assumptions: 30 amp load, 8,000 hrs/yr, 6¢/kWh.

Source: Lovins et al. 1989

Small feeders typically use the minimum-size conduits through which an electrician can easily pull the wire. In general, the size of small feeders can be increased without extending the size of the conduit, making the use of oversized feeders cost-effective. Conduit for feeders of larger motor sizes is determined by the diameter of the wire, so the conduit size will often need to be enlarged if the wire size is increased. As a result, the use of oversized wires for larger circuits must be evaluated on a case-by-case basis.

Energy savings are not the only reason to install larger distribution cables. The added distribution capacity provides room to expand loads in the future without having to remove and replace the old wiring. It also provides lower voltage drops, which improve the motor's starting and operating performances.

The Southwire Company's Wire-Sizing Policy

The Southwire Company, a billion-dollar industrial firm with an aggressive energy management program, wires all new loads under 100 A with a conductor one size above code, as well as using larger than normal wire for larger loads when doing so is cost-effective. Jim Clarkson, corporate energy manager, said that because his staff does not have time to evaluate every new wiring job, uneconomic oversizing may occur in some installations, but overall the policy saves the firm money, energy, and time (Clarkson 1990).

Power-Factor Compensation

As discussed in Chapter 2, low power factor has undesirable and costly effects that are often worth mitigating. Figure 3-10 shows the extent to which improving power factor can reduce losses. As examples, increasing power factor from 0.75 to 0.90 would reduce cable and transformer copper losses by 32%, while improving power factor from 0.60 to 0.90 would reduce losses by 57%.

In smaller motors, power factor generally is lower and drops more rapidly as the load decreases. As a result, a facility with a large number of small motors without power-factor correction will typically have a lower power factor than a facility with predominantly large motors. Poor power factor can be caused not only by lightly loaded motors but also by other loads such as fluorescent lighting ballasts and certain types of ASDs.

Improving the power factor can save energy and dollars by reducing losses in the customer's distribution system. Greater savings can often be achieved by reducing power-factor penalty charges (if these charges are imposed by the utility). Such charges are normally substantial enough to make it cost-effective for the customer to improve the power factor to 0.90. For example, a utility might increase the demand charge by 1% for every 1% the power factor drops below 90%. If a facility has a peak demand of 1,000 kW and a power factor of 81%, the facility will be charged for peak demand as follows:

Adjusted peak = (1,000 kW) × [1 + (0.90 – 0.81)] = 1,090 kW

If the utility demand charge is $70/kW-yr, the power-factor penalty will be:

Power-factor penalty = (90 kW) × $70 = $6,300/yr

In contrast, the energy savings for the same load, assuming it is fed at 480 V, three-phase through 500 feet of cable, would be about 21,000 kWh, resulting in a cost reduction of $1,300/yr at $.06/kWh.

The consumer can correct power factor either in a distributed manner (capacitors connected to the motor terminals) or in a centralized manner (a capacitor bank at a central location in the facility). Some large facilities may have an intermediate scheme with several capacitor banks, each serving several motors. The distributed option reduces the losses between the motors and the central capacitor bank. This procedure also costs less to install.

The centralized scheme requires controlled switching of the capacitor bank. Switching avoids overcompensation of the power factor

---

Figure 3-10

Reducing Losses in Electrical Distribution Systems through Power-Factor Improvement

---

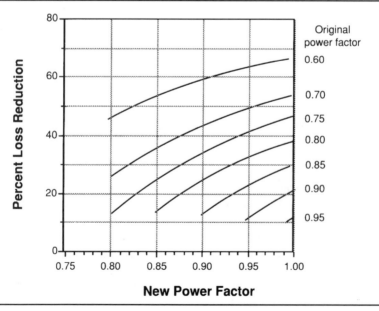

Figure 3-11

Correcting Power Factor with Capacitors

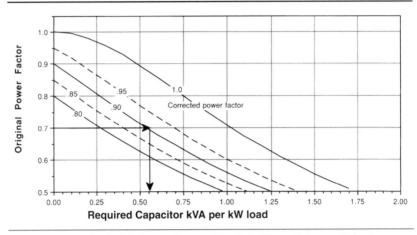

Note: Given the original (existing) power factor and the desired (corrected) power factor, the required capacitor kVA/kW of load can be determined. For example (shown with arrows), if the existing power factor is 70% and the desired value is 90%, about 0.54 kVA of capacitance/kW of load is required.

when only a limited number of motors are running. Overcompensation causes the same undesirable effects as a low power factor.

The installed cost for capacitors ranges from $20–30 per kilovolt-ampere-reactive (kVAR) of reactive power for dispersed units to $50–75/kVAR for central capacitors. In the above example, 240 kVAR are required (see Figure 3-11). Assuming $25/kVAR, the cost would be $6,000. The resulting payback is less than 1 year just from penalty reduction, 5 years from energy savings alone, and approximately 9 months from the combined savings.

Motor manufacturers generally recommend a maximum capacitor size at the motor terminals (see Figure 2-13). Figure 3-11 shows the capacitor kilovolt-ampere-reactive required to improve the power factor by various amounts. Thus, improving the power factor of a 1,000 kW load from 70% to 90% requires capacitance of just over 500 kVA.

The concern over electrical equipment (primarily transformers and capacitors) containing PCBs has resulted in the removal and disposal of many power-factor correction capacitors in customers' distribution systems. If these capacitors aren't replaced, power factor will decrease. Since the switchgear and mounting are already in place, new capacitors might be installed cheaply enough to obtain a reasonable payback on energy savings alone, depending on the specifics of each application.

As mentioned in Chapter 2, another potentially lucrative benefit of power-factor improvement is increased capacity of the distribution system. This advantage is especially relevant in new construction and whenever load approaches the systems capacity (as when a facility is expanded or equipment is added). If installing power-factor compensation eliminates or even postpones the need to replace the transformer, switchgear, feeders, or other equipment, it can be very cost-effective, again depending on the specifics of the situation.

Diagnostics of Power-Factor Compensation

Medium-size and large customers generally know if they have low power factor because the utility charges them for the reactive power they draw. An industrial facility where most of the load is for motors that do not have power-factor correction will typically exhibit a power factor of 70–80%. Other equipment, such as rectifiers or arc furnaces, can have power factors as low as 45%. To avoid a large reactive power bill, most facilities have already installed equipment for power-factor correction. Small customers, such as residential consumers and small commercial buildings below 50–100 kVA that typically pay small or no power-factor charges, generally do not install corrective equipment. Most small consumers have a fairly good power factor because they have a high fraction of resistive loads or compensated loads such as lighting ballasts with internal power-factor correction.

The measurement of the power factor requires the use of a wattmeter to measure power, a voltmeter, and a clamp-on ammeter. There are also meters that read power factor directly, either as dedicated power-factor meters or as part of more elaborate analyzers. Also on the market is equipment that can measure watts, volts, and amps and can register these continuously on paper or in computer storage (see Appendix D). The power factor in a symmetrical three-phase system is given by

$$\text{Power factor} = P/(3\ V \times I)$$

where
  P is the three-phase power
  V is the phase-to-neutral RMS voltage
  I is the RMS current in each phase

Load Management and Cycling: General Considerations

Most energy-saving measures described previously reduce demand except in instances of variable loads where the peak demand does not coincide with the utility's peak. For example, installing an ASD on a variable load saves energy at partial load but does not reduce demand at full load. Consumers, especially those operating large motor systems, should take into account the economic benefits of demand reduction when evaluating the cost-effectiveness of energy conservation investments.

Motor cycling and scheduling can reduce power demand during peak periods. Loads that can be suspended periodically with no serious cost or inconvenience are likely candidates for cycling. Examples include refrigeration equipment, air conditioners, and heat pumps. Loads that frequently idle for extended periods also are good candidates for shutdown or cycling to lower power during idle periods.

Whether equipment cycling lowers energy use as well as demand depends on the application, as described in the following examples.

A large retail store with a constant-volume HVAC system installs an energy management system that turns off one of the four 30 hp fans in a staggered 15-minute rotation every hour. Cycling is an acceptable control method in this building since it creates only minor temperature swings that are well within the limits of comfort for both shoppers and employees. In this case, along with a fall in demand, energy use also declines by 25% for each fan. If the building comfort level can be maintained using only 75% of the HVAC system's ventilation capacity, another option is to slow the fans by means of ASDs or resheaving (changing the pulleys that help connect the motor to the fan).

In another building, several small air conditioning units are each turned off for 15 minutes every hour. However, the set-points on the thermostats in the building do not change. In this case, while there is a substantial decrease in demand, there is only a small reduction in energy use because the air conditioners have to work harder when they are operating.

Potential Cycling Problems

Starting a motor causes an inrush of current that generates a great deal of heat. During start-up, ventilation fans are turning slowly, so they remove only a small portion of this heat. If this heat buildup is excessive, it can reduce the lifetime of the insulation and bearings and possibly lead to rapid failure. There are also mechanical stresses from electromagnetic forces associated with large starting currents. In particular, the ends of the windings can suffer fatigue and cracking.

These thermal and mechanical stresses limit the frequency at which a motor can be cycled. Additionally, the electrical equipment that feeds the motor and the mechanical equipment driven by the motor are stressed each time it is started. These types of drawbacks can be miti-gated with the use of starting controls, which are discussed below.

Allowable Cycling

The heating that occurs when a motor starts is a function of the current and the time used to accelerate the load. The longer the starting time, the more the motor heats up. The time it takes to accelerate a load from start to the rated speed depends on several factors, including the load's inertia, which depends on both the load's mass and its radius. A fan with a large radius has a much greater moment of inertia than a smaller-radius pump of similar shaft power requirement. The higher the load's rated speed, the longer it takes to start. Kinetic energy is pro-portional to the square of the speed. Torque is also important: the higher the torque required by the load relative to the torque available from the motor, the longer it takes to accelerate to rated speed.

NEMA Standards MG 1-1988, Section 12.55 (NEMA 1999), and MG 10-1994 (NEMA 1994) provide guidance on the number of suc-cessive starts that can be made each hour without causing motor damage. Table 3-3 presents the allowable number of starts per hour and the minimum time between starts, considering the effects of motor horsepower, number of poles (rated speed), and inertia of the load. If a motor operates at close to the upper bounds derived from Table 3-3, some reduction in motor lifetime should be expected.

Starting Controls

Three-phase motors use starters that apply all three phases to the motor simultaneously. These starters generally include a motor con-tactor (a relay to control the flow of electricity to all three phases) as well as devices that protect the motor and wiring from either a pro-longed small overload or a sudden severe overload.

Because the switching mechanism is a contactor, the conventional three-phase motor starter will apply the full voltage to a motor as soon as the contactor receives power. Since the motor is starting from a dead stop, extra current is required to produce the magnetic field that drives the motor and to supply the initial energy to move the motor and load. As a result, a motor will use between five and seven times the current when starting as it will when operating at full load. This current surge typically lasts for approximately 30 seconds but may range from only a few seconds to several minutes in the case of heavy loads.

These large starting currents may also produce large voltage drops in the feeders, making starting difficult and causing comput-ers to malfunction, lights to dim, and other motors to stall. These

Table 3-3

Allowable Number of Starts and Minimum Time between Starts for Designs A and B

hp	2-pole A	2-pole B	2-pole C	4-pole A	4-pole B	4-pole C	6-pole A	6-pole B	6-pole C
1	15	1.2	75	30	5.8	38	34	15	33
1.5	12.9	1.8	76	25.7	8.6	38	29.1	23	34
2	11.5	2.4	77	23	11	39	26.1	30	35
3	9.9	3.5	80	19.8	17	40	22.4	44	36
5	8.1	5.7	83	16.3	27	42	18.4	71	37
7.5	7	8.3	88	13.9	39	44	15.8	104	39
10	6.2	11	92	12.5	51	46	14.2	137	41
15	5.4	16	100	10.7	75	50	12.1	200	44
20	4.8	21	110	9.6	99	55	10.9	262	48
25	4.4	26	115	8.8	122	58	10	324	51
30	4.1	31	120	8.2	144	60	9.3	384	53
40	3.7	40	130	7.4	189	65	8.4	503	57
50	3.4	49	145	6.8	232	72	7.7	620	64
60	3.2	58	170	6.3	275	85	7.2	735	75
75	2.9	71	180	5.8	330	90	6.6	904	79
100	2.6	92	220	5.2	441	110	5.9	1,181	97
125	2.4	113	275	4.8	542	140	5.4	1,452	120
150	2.2	133	320	4.5	640	160	5.1	1,719	140
200	2	172	600	4	831	300	4.5	2,238	265
250	1.8	210	1,000	3.7	1,017	500	4.2	2,744	440

A = Maximum number of starts per hour
B = Maximum product of starts per hour times load work^2
C = Minimum rest or off time in seconds

Allowable starts per hour is the lesser of A or B divided by the load work^2, i.e.,

$$\text{Starts per hour} \le A \le B/\text{Load work}^2$$

Note: Table is based on the following conditions:
1. Applied voltage and frequency in accordance with MG 1-1998, Section 12.45 (NEMA 1999).
2. During the accelerating period, the connected load torque is equal to or less than a torque that varies as the square of the speed and is equal to 100% of rated torque at rated speed.
3. External load work^2 is equal to or less than the values listed in MG 1-1998, Section 12.50 (NEMA 1999).
4. For other conditions, the manufacturer should be consulted.

Source: NEMA 1994, Table 2-3

problems deserve special attention with large motors and those with long feeders or feeders with small cross-sections.

Certain types of electronic controls can ramp up the power during starts instead of forcing the motor to go to full speed from a dead stop.

This system, known as a soft start, reduces the inrush of starting current and thus decreases equipment wear. A soft-start feature is incorporated in most inverter-type ASD controls.

Transmission

The transmission subsystem, or drivetrain, transfers the mechanical power from the motor to the driven equipment. The efficiency of drivetrains (output power × 100/input power) ranges from below 50% to over 95%. As a result, the type of drivetrain used for a given application can have a greater effect on overall system efficiency than the efficiency of the motor itself.

The choice of transmission type depends upon many factors, including the desired speed ratio, horsepower, layout of the shafts, and type of mechanical load. The major varieties include direct shaft couplings, gearboxes, chains, and belts. There is no large-scale survey of the distribution of the different transmission types in the field. Lovins et al. (1989) estimated the distribution for commercial and industrial motors to be as follows:

- 30–50% shaft couplings
- 10–30% gears
- 34% belt drives
- 6% chains

These data were compiled from a small number of sources and may differ from the proportions in any given geographic area.

Shaft Couplings

Shaft couplings have low losses if precisely aligned. Misalignment of the shafts will not only increase losses but also accelerate wear on the bearings. The use of couplings is constrained by space and shaft location and is limited to applications where load speed does not vary with respect to motor shaft speed.

Gears

Gears or gear reducers are the primary drive elements for loads that must run slowly (generally below 1,200 rpm) and require high torque that might cause a belt to slip. Gears are also frequently used for loads exceeding 3,600 rpm. The ratings for gear drives depend on the gear ratio (or the ratio of the input shaft speed to the speed of the output shaft) and on the torque required to drive the load. Several types of gears can be used in motor transmissions, including helical, spur, bevel, and worm gears (see Figure 3-12 and "Gears" in Appendix C).

The losses in gears result from friction between them as well as in the bearings and seals, from windage, and from lubricant churning. A large number of gear combinations can be used for a given speed ratio.

Helical and bevel gears are the most widely used and are quite efficient, reaching 98% efficiency per stage (each step of reduction or increase in shaft speed). With helical gears, the input and output shafts are parallel; with bevel gears they are at right angles. Spur gears are used for the same purpose as helical gears but are less efficient and therefore should not be used in new applications.

Worm gears allow a large reduction ratio (5:1–70:1) to be achieved in a single stage. Their efficiency ranges from 55% to 94% and drops quickly as the reduction ratio increases due to the rise in friction between the gears. For this reason, worm gears should only

---

Figure 3-12

A Worm Gear Set (a) and a Three-Stage Helical Gear Set (b)

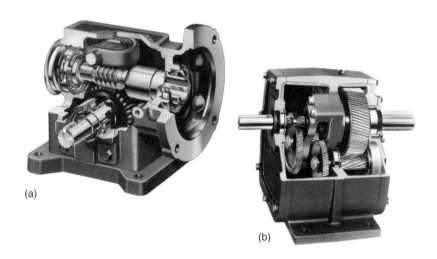

(a)

(b)

Note: In (a), the worm is shown mounted on the upper shaft; the worm is a helical gear (i.e., the teeth trace out helical paths, in this case much like a screw thread). The worm wheel is a gear wheel with a concave face to mesh with a worm. The worm normally drives the wheel, providing a gear set with high reduction ratio connecting shafts with nonintersecting axes at right angles. In (b), gears with helical teeth are used to transmit power between parallel shafts. For bevel and spur gears, see "Gears" in Appendix C.

Source: Reprinted with permission from Reliance Electric

---

be used in drives below 10 hp where operating costs are low. A large reduction ratio is more efficiently achieved by several stages of helical or bevel gears.

Worm gears cost less than helical gears for applications up to 10–15 hp, but helical gears are less expensive above this rating and are becoming the standard for larger drives. The different efficiencies of these two types of gears also affect cost. For example, in low-horsepower ranges, the efficiency of worm gears at full load is typically 70–80%, compared to approximately 90–96% for a helical gear. Worm gears' lower efficiency will often force the user to increase the size of the motor, and this added cost must be taken into account when comparing gears; a helical gear with a smaller motor may have a lower initial cost than a worm gear with a larger motor, even for applications below 10 hp.

Figure 3-13 shows the comparative efficiencies of several types of gearboxes as a function of the speed ratio.

Gear drives are similar to motors in that their efficiency drops markedly below 50% of full load (see Figure 3-14) because some of

---

Figure 3-13

Typical Range of Gearbox Efficiencies Based on 1,750 rpm Input

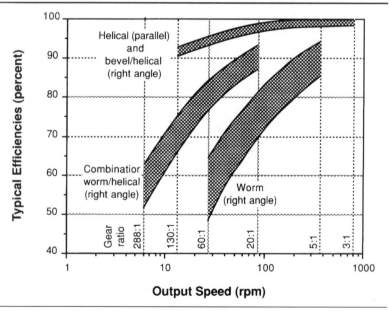

Source: Reprinted with permission from Reliance Electric

Figure 3-14

Helical Reducer Efficiency vs. Load and Speed for a Typical Single-Reduction Gear Unit

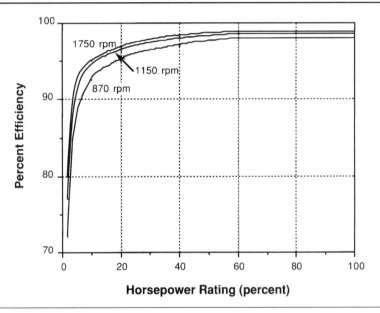

Source: Reprinted with permission from Falk Corporation

the losses are not direct functions of load. For a large gearbox, these fixed losses represent about half the total losses at full load.

In large gearboxes, reducing loss is even more important because lubrication effectiveness and lifetime are diminished by high temperatures. Using low-friction bearings, gears with a high-quality finish, and improved lubricants can bring the efficiency of a single-stage helical gear to over 99%.

Because gear reducers come in an assortment of in-line and right-angle configurations and sizes, more efficient reducers are difficult to retrofit without major changes to the equipment, as the new reducers are likely to have different dimensions, configurations, or both.

Belt Drives

About one-third of motor transmissions use belts (E Source 1999). Belts allow flexibility in the positioning of the motor relative to the load and, using pulleys (sheaves) of suitable diameters, belts can increase or decrease speeds. There are several types of belts: V-belts, cogged V-belts, synchronous belts, and flat belts (see Figure 3-15).

V-belts are the most common type and have an efficiency of

Figure 3-15

Belt Drives, Including (a) V-Belt Cross-Section, (b) Cogged (or "Toothed") Belt Drive, and (c) Synchronous Belt Drive

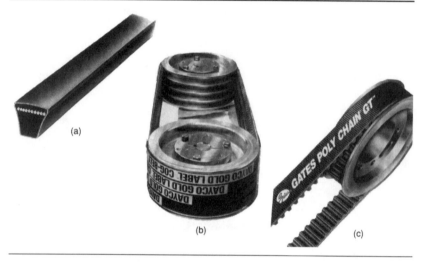

Note: The cogged belt drive uses the conventional V-belt (smooth) pulleys, while the synchronous belt has meshing teeth on the belt and pulleys (or "sprockets"), preventing slip. Flat belts are similar to synchronous belts (wide and thin) but are smooth on both sides and ride on smooth flat pulleys.

Source: Reprinted with permission from Gates Rubber Company (a and c), and Dayco Products (b)

Figure 3-16

A Belt Drive, Showing the Four Flexing Points, Two at Each Pulley (A, B, C, and D)

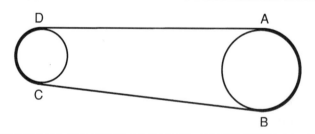

Source: Nailen 1987

90–96%. V-belt losses stem from flexing (see Figure 3-16), slippage, and (to a lesser extent) windage. The bending and unbending of the belt material when it enters and leaves the pulley causes the flexing losses.

A belt's tension critically determines its performance. Too much tension can stress the belt, bearings, and shafts; too little tension causes slip, high losses, and premature failure of the belt. With wear, V-belts stretch and need retensioning. They also smooth with wear, and thus become more vulnerable to slip. This slip, if the V-belt is not properly maintained, will increase and therefore efficiency will be lower, possibly to below 90%.

Cogged V-belts have lower flexing losses since less stress is required to bend the belt, and therefore they deliver 1–3% better efficiency than standard V-belts. Cogged V-belts can easily be retrofitted on the same pulleys when V-belts wear out. They cost 20–30% more than V-belts but the extra expense is recovered over a few thousand operating hours. In addition, they typically last twice as long. Efficiencies with cogged V-belts are greatest when they are used with the smallest appropriate pulley (de Almeida and Greenberg 1994).

The only caveat to the use of cogged V-belts is that some engineers use their inherent slip characteristics to provide overload protection. If other protection, such as sheer pins, has not been installed, the use of cogged V-belts in those systems can lead to equipment damage in an overload condition.

A review by E Source of five studies in which V-belts were replaced with cogged V-belts reported savings of 0.4% and 10%, with a median savings of 4.1%. At the 4.1% savings level, the payback from energy savings alone ranges from 1 to 5 months (E Source 1999). Similar savings are reported in a Ford Motor Company case study (Elliott 1995).

The most efficient belts are the synchronous and flat belt designs, which can be 97–99% efficient because they have low flexing losses and little or no slippage. Synchronous belts are applicable to low- and medium-speed applications, while flat belts are appropriate for medium- and high-speed applications.

Figure 3-17 shows the relative performance of synchronous belts in comparison with conventional V-belts. Synchronous belts have no slip because their teeth engage in the teeth of the sprocket pulleys. V-belts rely on friction between the belt and the pulley grooves to transmit the torque, and that friction can be affected by liquids, dust, wear, and other factors. Synchronous belts are designed for minimum friction between the belt and the pulley and can withstand much harsher conditions. The efficiency curve of synchronous belts is not only higher but also flatter than that of V-belts, with larger percentage savings as the load decreases.

Figure 3-17

Efficiency vs. Torque for V-Belts and Synchronous Belts in a Typical Application

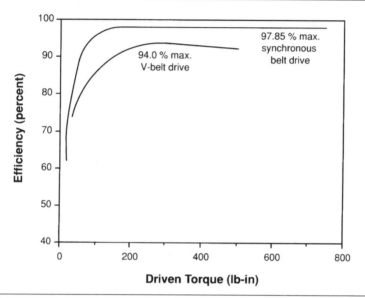

Source: Reprinted with permission from Uniroyal

Due to their construction, synchronous belts stretch very little and do not require periodic retensioning. They typically last over four times longer than V-belts, and the savings in labor and materials for replacements in most cases more than offset the extra cost of the belts. Retrofitting synchronous belts requires installing sprocket pulleys that cost several times the price of the belt. In cases where pulley replacement is not practical or cost-effective, cogged V-belts should be considered.

Synchronous belts are available in sizes from fractional-horsepower applications to over 1,000 hp. Due to their positive transmission, they are suitable for applications requiring accurate speed control. They are not, however, suited for shock loads where abrupt torque changes can shear sprocket teeth. To alleviate this problem, some manufacturers have doubled the belt's resistance to shock loads by using polyurethane compounds instead of neoprene rubber. Another drawback of synchronous belts is that they do not slip if a machine jams, and can thus pose a possible safety threat. Possible solutions, other than installing a different belt type, include using a clutch or a shear pin that breaks and disengages the equipment in the event of a jam.

The meshing of the belt teeth in the sprocket makes synchronous belts noisier than V-belts, but a sound-reducing shield can mitigate this problem. It can also protect personnel from the belt and other moving parts and guard the equipment from debris.

Several practical considerations may limit the benefits of synchronous belts. Although their alignment specifications are the same as for V-belts, they demand closer adherence to specification. Also, the limited number of available sprocket diameters makes speed-matching more difficult. For centrifugal applications, one report indicates that the greater efficiency of power transmission is outweighed by improper speed, resulting in an average 2% increase in energy consumption (Greenberg 1996).

Flat belts are thin belts constructed of aramide fibers and high-friction surface compounds. These high-performance belts, which are common in Europe, feature low stretching and flexing losses, achieving efficiencies similar to those of synchronous belts but offer additional advantages. These belts can accommodate some slip when there is a surge in the torque, yet they still maintain an efficiency level close to that of synchronous belts under normal conditions. The cost of the pulleys for flat belts is lower than for synchronous belts, and due to the absence of teeth, flat belts do not have the noise problems associated with synchronous belts.

Table 3-4 summarizes the characteristics of different belt types and can be used as a guide to selecting the appropriate model for different applications.

Table 3-4

Comparison of Belt Drive Characteristics

	Typical Efficiency Range (%)	Suitable for Shock Loads	Periodic Maintenance Required	Change of Pulleys Required	Special Features
V-Belts	90–98	Yes	Yes	No	Low initial cost.
Cogged V-Belts	95–98	Yes	Yes	No	Easy to retrofit. Reduced slip.
Flat Belts	97–99	Yes	No	Yes, but low cost	Medium- to high-speed applications. Low noise. Low slip.
Synchronous Belts	97–99	No	No	Yes, with higher cost	Low- to medium-speed applications. No slip. Noisy. May have problems matching speed.

Source: de Almeida and Greenberg 1994

Chains

Chains, like synchronous belts, do not slip. Traditionally, belts have been applied in relatively high-speed, low-torque applications, whereas chains have been used in low-speed, high-torque applications.

Chains also feature high load capacity, the ability to withstand high temperatures and shock loads, long life if properly lubricated, and virtually unlimited length. Chain drives of several thousand horsepower have been built. The efficiency of well-maintained chain-and-sprocket combinations can reach 98%, but wear lowers their efficiency a few percentage points.

There are several types of chains, including standard roller (both single strand and double strand), double pitch, and silent chains.

With the exception of the silent kind, chains are noisier than belts. Compared to roller chains, silent chains offer slightly higher efficiency (up to 99%) but are 50% more expensive in the low-horsepower range and 25% more expensive in the high-horsepower range.

Although the steel in the chain stretches only minimally when tensed, the chain sags and needs readjustments as links and sprockets wear. Inadequate lubrication increases wear. Keeping high-speed chains well lubricated is difficult because centrifugal forces eject the lubricant; enclosing the chains and providing constant relubrication, however, as is done in camshaft drives in many auto engines, can solve this problem.

Lubricants can also quickly lose their effectiveness in environments contaminated with dust or liquids. Under these conditions, the use of synchronous belts may prove more attractive. Another drawback is that, as the chain wears, the sprockets normally need to be replaced, which increases maintenance costs.

Maintenance

Regular maintenance of the motor system, including inspection, cleaning, and lubrication, is essential for peak performance of the mechanical parts and to extend their operating lifetime.

Lubrication

Lubrication is required to reduce the friction and rapid wear of metal parts moving against one another. Most lubricants fall into two categories: oils or greases. Oils are liquid lubricants, traditionally based on animal, vegetable, or mineral sources, with a wide variety of composition, viscosity, and other properties. The value of a liquid lubricant depends primarily on it ability to form and maintain a film between contact surfaces. Greases are gels made of a mixture of a lubricating oil and

Figure 3-18

Grease Life vs. Bearing Temperature

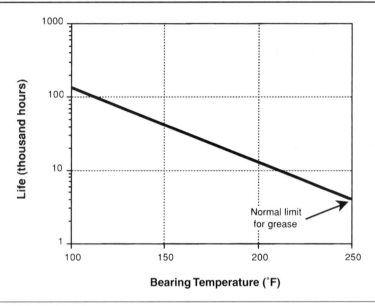

Source: Montgomery 1989

soap. They are primarily used to lubricate rolling bearings and gears. The soap provides no lubrication directly but instead releases the oil when heated or agitated to coat the contact surfaces (E Source 1999).

Both underlubrication and overlubrication can cause higher friction losses in the bearings and shorten their lifetime. Underlubrication may occur because either an insufficient amount of lubricant was applied during routine maintenance or routine maintenance was not done frequently enough. In either case, the friction of the bearings will rise, and the energy used by the motor will increase to overcome the increased resistance. Consequently, the motor will run hotter, further decreasing its efficiency, and the higher temperature will lower the lubricity and lifetime of the lubricant (see Figure 3-18).

Most maintenance staff will try to avoid undergreasing by applying "plenty of grease," which, unfortunately, often leads to overgreasing of motor bearings. Bearing grease must be highly viscous so as to properly lubricate the moving parts when the motor gets hot. If applied in excess, grease develops internal friction that impedes the bearings and increases the force necessary to turn the shafts. Tests have shown that overgreasing can raise bearing losses up to 25%, thereby dropping the overall motor efficiency by perhaps 0.2–0.5 percentage point (Katz 1990). In addition, overgreasing may damage the seals and increase churning losses, which leads to overheating and early failure of the bearing. Overgreasing can also cause the accumulation of grease and dirt on the motor windings, causing overheating and premature failure.

Old grease should be removed before greasing, and the bearing chamber should generally be filled not more than one-third full of grease. Contamination of the lubricant, especially with water, can also substantially degrade the lubricant performance and lifetime.

Oils and greases are available in a variety of special formulations, with additives to decrease friction and wear and increase lubricant life. Additives may be put in natural (usually mineral) oils, or the oil may be entirely synthetic to meet specific lubrication needs. The most common friction-reducing additives include molybdenum disulfide (MoS_2) and polytetrafluoroethylene (Teflon™). The energy-savings potential of such lubricants with these additives in gearboxes and motors is discussed in Chapter 7. Other additives and synthetic formulations are used to improve lubricants' ability to resist degradation due to high temperatures. While the primary benefit of this improved high-temperature stability is longer lubricant life, indirect energy savings can result from the ensuing constancy of desired lubricant properties. This is especially important where lubricant maintenance is neglected (Lovins et al. 1989).

Recently, synthetic, engineered lubricants have entered the marketplace. These products, which are optimized for a specific application, can replace conventional petroleum-based oils and greases, reducing energy consumption and equipment wear by reducing friction. While friction is relatively small in motors themselves, it can represent a large loss in mechanical equipment like compressors, pumps, and gear drives. Synthetic lubricants have been demonstrated to reduce energy consumption from 2% to 30% in these applications. Though synthetics cost 1.5 to 3 times more than conventional lubricants, they do have a longer life, mitigating the initial cost. In many cases, the additional cost can be more than justified based on the longer lubricant life alone (Nadel et al. 1998).

Periodic Checks

Temperature (the first and quickest indicator of trouble) and the electrical and mechanical condition of a motor should be checked periodically. In general, most facilities with a good maintenance program will grease and inspect a motor every 6 months. However, recent cutbacks in maintenance staffs have led to increased incidences of underlubrication.

Bearing wear may be signaled by overheating, increased noise, or vibration; a cracked rotor cage can produce the same effects. The condition of the motor windings should be checked by measuring the resistance of the windings and of the insulation between the windings and the ground. Maintenance is key to efficient operation of belts as well. The motor drivetrain should also be checked in order for the belt's tension to be adjusted or worn belts replaced. Higher-efficiency belt operation leads to a lower belt temperature. As with motor insulation, belt life is reduced by half if the operating temperature increases by 10°C. Gear reducers should be checked as well to see if they are properly lubricated.

If a motor is left idle for a considerable number of hours and is located in a humid place, a heating resistor should be placed inside the motor to avoid condensation. Moisture will decrease the insulation resistance between the windings and the ground. Motors with abnormal conditions should be repaired or replaced.

Cleaning and Ambient Conditions

As noted in Chapter 2, the cooler a motor operates, the higher its efficiency and the longer its lifetime. Higher temperature increases the

Figure 3-19

The Effect of Ambient Air Temperature on a Motor's Load-Carrying Ability

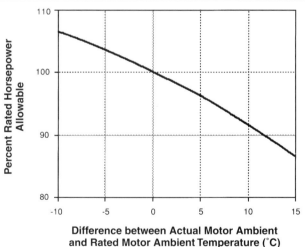

Source: Nailen 1987

windings' resistivity and, therefore, their losses. Cleaning the motor casing and ventilation filters as well as the apertures of open drip-proof motors is important because the operating temperature increases as dust and dirt accumulates. In extreme cases, failure may occur if thick layers of dust accumulate. Adding paint to the casing is not recommended since the paint acts as insulation and decreases the ability of the motor to dissipate heat. Figure 3-19 shows the effect of ambient temperature on the allowable horsepower.

Summary

We have discussed many factors that determine the reliability, longevity, and efficiency of a motor-driven system. Key goals include high-quality power supply; proper equipment sizing; careful attention to harmonics, transients, power factor, and distribution loss; good load management practice; optimized transmission systems; and careful maintenance of the entire drivepower system. In Chapter 4 we turn to one particularly important set of system components: the motor controls.

Motor Control Technologies

Introduction

Motor speed control offers the potential for the single largest amount of energy savings in drivepower systems. Most motors are fixed-speed, AC models. However, adjusting the speed to match the requirements of the loads, which generally vary over time, can enhance the efficiency of motor-driven equipment. The potential benefits of speed variation include increased productivity and product quality, less wear in mechanical equipment, and energy savings of 50% or more for certain applications.

Speed controls can save the most energy in centrifugal machines, which include most pumps, fans, and blowers and some compressors. Speed control is also effective in mills (such as the rolling mills that produce sheet metal in a steel plant), traction drives (such as subway cars), conveyors, machine tools, and robotics.

The available options for motor speed controls include multi-speed and DC motors, shaft-applied drives (including mechanical drives, hydraulic couplings, and eddy-current drives), and electronic adjustable-speed drives (ASDs). In this chapter these are discussed and compared in light of their typical applications, advantages, limitations, and costs. Electronic ASDs, since they have become the dominant technology, are covered in detail later in this chapter, although much of this information is summarized in Table 4-1. (Readers not concerned with the technical details of how electronic ASDs work should skip the section "Characteristics of Electronic Adjustable-Speed Drives.")

A speed-control technology should match the characteristics of the

Table 4-1

Adjustable-Speed Motor Drive Technologies

	Technology		Applicability (R = retrofit; N = new)	Cost	Comments
Motors	Multispeed (incl. PAM[1]) Motors		Fractional—500 hp PAM: Fractional—2,000+ hp R, N	1.5–2 times the price of single-speed motors	Larger and less efficient than single-speed motors. PAM more promising than multi-winding. Limited number of speeds.
	Direct-Current Motors		Fractional—10,000 hp N	Higher than AC induction motors	Easy speed control. More maintenance required.
Shaft-Applied Drives (on motor output)	Mechanical	Variable-Ratio Belts	5–125 hp N	$350–$50[2]/hp (for 5–125 hp)	High efficiency at part load. 3:1 speed range limitation. Requires good maintenance for long life.
		Friction Dry Disks	Up to 5 hp N	$500–$300/hp	10:1 speed range. Maintenance required.
	Eddy-Current Drive		Fractional—2,000+ hp N	$900–$63/hp (for 1–150 hp)	Reliable in clean areas. Relatively long life. Low efficiency below 50% speed.
	Hydraulic Drive		5–10,000 hp N	Large variation	5:1 speed range. Low efficiency below 50% speed.
Wiring-Applied Drives (on motor input)	Electronic Adjustable-Speed Drives	Voltage-Source Inverter	Fractional—1,000 hp R, N	$1,500–$80/hp (for 1–300 hp)	Multi-motor capability. Can frequently use existing motor. PWM[3] appears most promising.
		Current-Source	100—100,000 hp R, N	$200–$30/hp (for 100–20,000 hp)	Larger and heavier than VSI.[4] Industrial applications including large synchronous motors.
		Others	Fractional—100,000 hp R, N	Large variation	Includes cycloconverters, wound rotor, and variable voltage. Generally for special industrial applications.

[1]PAM means "pole amplitude modulation." [2]The prices are listed from high to low to correspond with the power rating, which is listed from low to high. Thus, the lower the power rating, the higher the cost per horsepower. [3]PWM means "pulse width modulation." [4]VSI means voltage-source inverters.

load. These characteristics include the load profile (number of hours per year at each level of load from minimum to maximum), horsepower range, speed range, price of energy, overall energy efficiency of the motor and control systems, reliability and maintenance requirements, physical size limitations, control and protection requirements, equipment lifetime, and first cost of the drive system. The ideal drive for any given application should be capable of varying both speed and torque to match the requirements of the load. Adjustable-speed loads can be classified into the following three groups according to relationship between torque and speed: variable-torque loads; constant-torque loads; and constant-power loads.

Variable-Torque Loads

In this case, the torque increases with the square of the speed. Examples can be found in centrifugal pumps, fans, and compressors common in heating, ventilating, and large air conditioning systems. The design of centrifugal equipment is such that, at low speed, the equipment can match the low pressure and flow requirements of most systems. In pump systems, static head will increase the pressure and power required at lower speeds.

Constant-Torque Loads

A classic example of constant-torque loads is the conveyor belt. The torque required to move a conveyor depends on the load on the belt, not its speed. Since the load is independent of the speed, the drive may need to produce maximum torque at any speed.

Figure 4-1

Types of Motor Loads

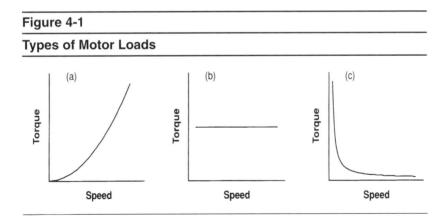

Note: (a) torque increasing with speed (e.g., in centrifugal fans and pumps); (b) constant torque (e.g., in positive displacement pumps and compressors); and (c) constant power (e.g., in vehicle drives).

Constant-Power Loads

With constant-power loads, torque decreases with increasing speed but power (the product of speed times torque) remains constant. The most familiar example is a lathe or grinding machine.

Figure 4-1 shows the typical torque-to-speed characteristics for the three classes of loads. Variable-torque loads are by far the most common, accounting for 50–60% of the total motor energy used in the commercial and industrial sectors (see Table 6-18).

Speed-Control Technologies Other than Electronic ASDs

Multispeed Motors

Some motors are designed to operate at two, three, or four speeds; two-speed motors are the most common. As explained in Chapter 2, the speed of an induction motor depends on the number of pole pairs in the motor. Multispeed motors are available up to 500 hp and are very reliable but have the following drawbacks:

- The stator slots must be bigger than those of single-speed motors in order to accommodate two or more windings. As a result, the motors are bulkier and cannot be easily retrofitted.

- The current-carrying capacity of the copper is poorly used since only one set of windings is active at any one time.

- Fundamental aspects of their design lead to a lower efficiency level than for comparably sized single-speed motors.

- The available speed ratios are limited.

- The motor starters typically cost up to twice as much as single-speed motor starters.

- Multispeed motors cost 50–100% more than single-speed motors.

Two-speed motors can be used to save energy in such applications as air volume control in facilities that have large differences in their day-to-night or weekday-to-weekend airflow requirements. A 1,800/1,200 rpm motor, for instance, can reduce fan energy requirements at night and on weekends by 70%. All that is required is a two-speed motor with a starter, a timer, and a relay.

The pole-amplitude-modulation (PAM) motor is a single-winding, two-speed, squirrel-cage induction motor that avoids some of the drawbacks of conventional two-speed designs. PAM motors are available in a wider range of speed ratios than standard multispeed motors but they are limited to ratios based on synchronous speeds. They include 900/720, 1,200/720, 1,200/900, 1,800/720, 1,800/1,200, 3,600/720, and 3,600/900 rpm versions. PAM motors are more compact than other multispeed motors. In fact, they have the same frame size as single-speed designs.

The lower speed can be used for soft starting, resulting in a smaller inrush of current and less heating. In applications for which a two-speed duty cycle is appropriate, PAM motors are especially well suited for driving large fans or pumps with ratings from a few horsepower to thousands of horsepower. In the case of a retrofit, using an existing throttling device (valve or damper) allows for fine-tuning the flow once the main adjustment is made through speed selection while reducing the heavy losses of the throttle-only control.

Like multispeed motors, PAM motors are available for variable-torque, constant-torque, or constant-horsepower applications. They and their starters cost about the same as standard multispeed motors and have similar efficiencies.

Pony Motors

An increasingly popular technique for motor drives with two distinct operating conditions is to use two separate motors for a single application. The second, smaller motor is called a pony motor. For shaft-driven equipment, two motors can drive pulleys for the same shaft with controls so that only one motor can operate at a time. In pumping applications, two pumps with different capacities and speeds will often be installed in parallel.

Pony motors are becoming a common option for cooling towers, municipal water systems, and air handlers. They produce energy savings with the use of standard motors and starters, and are easy to maintain and repair. In addition, since they use two different drive belts, they offer greater flexibility in speed selection than the limited ratios available in multispeed and PAM motors.

Direct-Current Drives

Although expensive and of limited reliability (see Chapter 2), DC motors can produce high starting torques. Their speed can be controlled with great precision—down to 1% of the nominal speed of the motor—typically by varying the voltage level. They are used in applications up to about 10,000 hp. Figure 4-2 shows the torque-to-horsepower characteristics for DC motors.

The basic operating theory of DC motors is covered in Chapter 2. Their weakness is the commutation subsystem: both the brushes and

---

Figure 4-2

Speed Control of Direct-Current Motors, Showing the Torque-Horsepower Characteristics

---

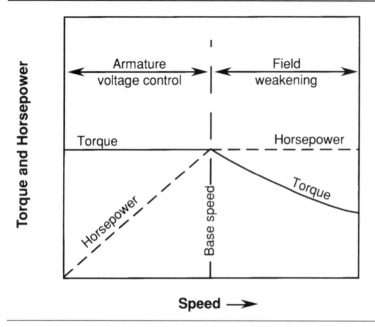

Speed ⟶

Note: Below the base speed the torque is almost constant; speed control of the motor is achieved by varying rotor ("armature") voltage. Speeds above the base speed can be obtained by decreasing the stator ("field") current, which causes the motor to operate in a constant-horsepower mode.

---

the commutator wear because of friction and arcing. For this reason, DC motors require periodic maintenance and are not suitable for use in explosive or corrosive environments. In addition, due to the complexity of the rotor, DC motors are substantially more expensive, bulkier (since they have "dead volume" required by the commutator and brushes), and less efficient than AC induction motors. High-horsepower DC motors also have lower speed limits than their AC counterparts because of the centrifugal stresses on their larger, heavier rotors. Large AC drives can make use of higher voltages than DC motors, which are subject to a voltage limit due to arcing in the commutator. Higher voltages are desirable since they result in proportionately lower currents for the same power consumption. Lower currents lead to lower power losses in the electrical distribution system and allow smaller, cheaper wire to be used.

DC motors have traditionally been used in applications where high starting torque is required (such as traction devices and cranes) or where accurate speed control is needed (such as rolling mills, lathes, paper machines, and winders). Using DC motors for speed control requires converting the available AC power to DC. Historically this was done using either a motor-generator set or a rectifier. In a motor-generator set, an AC motor is used to operate a DC generator, which in turn powers the DC motor. Because each piece of equipment has inherent energy losses, the overall efficiency of this system can be 50% or lower. For example, it might require 100 kW of electrical input at the AC motor to drive a machine with a DC motor that needs 67 hp (or 50 kW) of power applied to the shaft.

The AC-to-DC conversion efficiency can be higher (up to 98%) with solid-state rectifiers than with motor-generator sets or mercury rectifiers. However, many facilities that use rectifiers were initially designed with a central DC power supply and one large rectifier. As new DC tools with their own rectifiers are added to such facilities, the load on the central rectifier decreases, lowering the overall efficiency of the central system.

In recent years, more efficient solid-state controllers for DC motors have appeared on the market. These units, which have some features in common with AC ASDs, provide DC power at relatively high efficiencies for many existing speed control applications.

However, due to the drawbacks with DC motors mentioned above and the availability of better alternatives discussed later in this chapter, DC motors are now seldom used in new applications and their production is rapidly dwindling. As discussed in Chapter 5, some applications using DC drives should be replaced with AC motors and ASDs. Examples include high-performance drives in steel and paper mills, as well as electric transportation.

Shaft-Applied Speed Control: Mechanical, Hydraulic, and Eddy-Current Drives

Mechanical, hydraulic, and eddy-current (induction clutch) drives are grouped together because they are all installed between the constant-speed motor shaft and the driven equipment. Usually these drives are bulky and not very efficient and require regular maintenance.

Shaft-applied drives are not normally used in retrofits due to their space requirements. In new applications, they are generally installed only in low-horsepower applications where they may be less expensive per horsepower than electronic ASDs. However, when ongoing maintenance and energy costs are included in the analysis, it is often more cost-effective to use an electronic ASD.

In addition, because of the relatively low efficiency of many of these drives, particularly when operating at low loads, it is sometimes cost-effective to retrofit a shaft-applied drive with an electronic ASD based on the value of the energy savings.

Mechanical Drives

Mechanical devices for controlling speed include variable gearboxes, adjustable pulleys (sheaves), and friction dry discs. Variable gearboxes usually employ conical drums and can be applied only to small and medium-size drives, generally under 100 hp. Belt-slipping problems and maintenance requirements are making them less and less attractive relative to other drive options.

Adjustable pulleys (see Figure 4-3) are simple devices that allow speed to be varied typically over a 3:3 range by adjusting the gap between flanges of the pulley sheaves. This adjustment can be performed either pneumatically or by a small servomotor. These devices are very efficient (in the 95% range) and fairly inexpensive (from $50/hp for a 100 hp drive to $300/hp for a 5 hp drive). Due to belt-slipping problems, they are not suitable for shock loads and are available only below 125 hp. Adjustable pulleys have been used to control the speed of small and medium-size fans.

Friction dry discs (see Figure 4-4) allow a wide range of speed ratios (up to 10:1) but are limited to small loads (up to a few horsepower) and are costly ($300–500/hp). They are expensive because they require precision parts, and are only used with small motors (most drives are less expensive per horsepower when used with a large motor). Speed is varied by manually turning a crank, which changes the transmission ratio (Payton 1988). These drives are typically 95% efficient. However, the high level of maintenance they require, their inability to be automatically controlled, and their low power-handling capability make friction dry discs inappropriate for many applications.

In general, mechanical drives have a limited horsepower range. They require regular maintenance because they have movable parts, some of which rely on friction for transmitting power. Developments in electronic ASDs provide more reliable, flexible, less bulky, and increasingly cost-effective alternatives.

Hydraulic Couplings

The output speed of a hydraulic (or fluid) coupling is controlled by the amount of slip between the input and output shafts. Thus, the

Figure 4-3

Typical Adjustable Pulley Drive

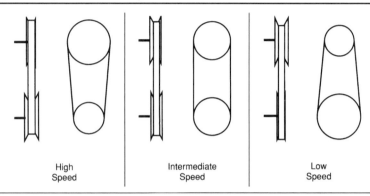

| High
Speed | Intermediate
Speed | Low
Speed |

Note: Changing the gap between the flanges of the pulley sheaves changes the effective pulley diameters, thus varying the speed. The top shaft is connected to the motor, the bottom to the driven load.

Figure 4-4

Operation of Friction Disc Speed Control

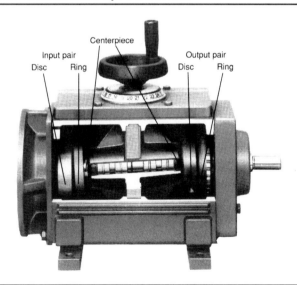

Note: Speed variation is achieved by manually turning a crank, which moves the point of contact between one or more friction disc/ring pairs. Moving the contact point in turn changes the effective diameter of the friction discs, thus changing the transmission ratio.

Source: Reprinted with permission from Reliance Electric

Figure 4-5

Operation of Hydraulic Drive Speed Control

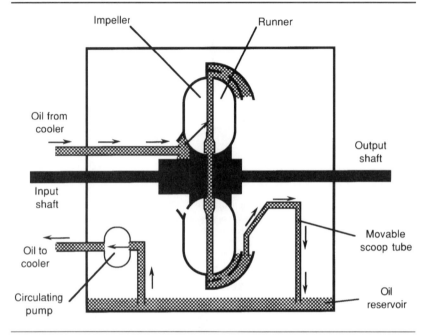

Note: The input shaft drives a vaned impeller, and a vaned "runner" drives the load through the output shaft. The input and output shafts are not connected except through the hydraulic circuit. Speed variation is achieved by changing the amount of oil in the working circuit (between the impeller and the runner) through a movable scoop tube. Since the output speed is controlled by the amount of slip between the impeller and the runner, the output shaft speed cannot exceed the input shaft speed. The available speed range is typically 5:1.

Source: Andreas 1982

output shaft speed cannot exceed the input shaft speed while the motor is driving the load. The torque converter in automobiles with automatic transmissions is a type of hydraulic coupling. In the fluid coupling, the input shaft drives a vaned impeller, and a vaned runner drives the load.

Figure 4-5 shows the structure of a hydraulic drive. Speed is controlled by varying the amount of oil in the working circuit, achieving a typical speed range of 5:1. This speed ratio is changed by deliberately introducing losses in the system. As a result, greater speed reduction results in lower system efficiency. For an output speed of 50%, the overall efficiency of the hydraulic coupling is typically 40%.

Figure 4-6

Power Flow and Control of an Eddy-Current Drive System

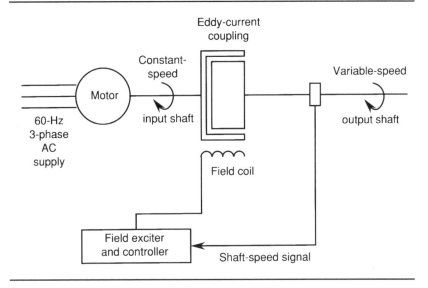

Although hydraulic couplings can be used in applications from a few horsepower to tens of thousands of horsepower, their use is acceptable only when most of the duty cycle is in the upper speed range. At lower speeds the losses are too high. In addition, because the couplings are bulky, retrofits, which generally require repositioning of heavy equipment and construction of new foundations, can be prohibitively expensive.

Eddy-Current Drives

The eddy-current drive couples an eddy-current clutch to an AC induction motor (see Figure 4-6). A rotating drum connected to the induction motor surrounds a cylinder attached to the output shaft. The concentric cylinder and drum are coupled by a magnetic field, and its strength determines the amount of slip. A low-power solid-state controller varies the current in the winding that produces the magnetic field, thereby varying the speed. This field excitation typically consumes 2% of the drive's rated power (Magnusson 1984).

The eddy-current drive is a slip device like the hydraulic coupling, albeit with slightly better efficiency. Waste heat, generated by the motion of the drum and cylinder relative to the magnetic field, is the main source of power loss and is removed either by air or water cooling. Air-cooled drives are available with ratings from 114 hp through 200 hp. Water cooling is also used for some drives ranging from 200 hp to over 2,000 hp.

Eddy couplings operate reliably in a clean environment. They are bulky, typically occupying twice the space of the induction motor itself. Typical prices range from $200/hp for a 5 hp drive to $150/hp for a 15 hp drive to less than $100/hp for a 100 hp drive. Prior to recent decreases in the cost of electronic ASDs, eddy-current drives were often specified for speed control in HVAC systems and wastewater treatment plants. Although today's electronic ASDs have higher efficiencies and are competitive in cost, eddy-current drives have the advantage of not producing significant harmonics or voltage transients. Eddy-current drives may still be an acceptable choice in installations where the load operates at 70% or more of the rated speed most of the time.

Characteristics of Electronic Adjustable-Speed Drives

Solid-state electronic ASDs were developed about 40 years ago. Early versions were complex, expensive, and only moderately reliable. Advances in semiconductor technology for power devices and especially for microelectronics have been dramatic in the past two decades. ASDs' costs have decreased substantially, and their performance and reliability have improved dramatically. Therefore, electronic ASDs are becoming the preferred motor speed control technology.

This section provides the reader with a technical overview adequate for understanding application issues and costs. Readers interested in a somewhat more technical discussion of ASD design and operation should consult the *ASD Master User's Guide* (Jacobs Engineering 1996).

Most electronic ASDs control motor speed by synthesizing electrical power of the desired frequency since the speed of AC motors is proportional to the frequency of the power supply. This makes it possible to control the speed over a wide range—from 0% to 300% of rated speed.

Because ASDs are more compact than mechanical or hydraulic adjustable-speed controls, and also because they do not have to be mechanically coupled to the motor, they can be more readily retrofitted. The main ASD components do not have moving parts, and therefore require little periodic maintenance. When properly applied, ASDs can be extremely reliable. They are available in a power range that covers fractional horsepower (typical of home appliances) to a few hundred horsepower (as in commercial building HVAC systems) to the tens of thousands of horsepower used by the pumps and fans of large electric power plants.

Types of ASDs

Electronic ASDs are characterized by the type of electronic input they require and the way they control a motor's speed. There are four basic types of ASDs: inverter-based; cycloconverters; wound-rotor slip recovery; and voltage-level controls.

Inverter-Based ASDs

Inverter-based ASDs are the most common systems for induction motors, and can be used with synchronous motors as well. They account for well over 90% of the ASDs currently sold (PEAC 1987).

The general diagram for an inverter-based ASD is shown in Figure 4-7. Some ASDs operate on single-phase power (which is found in most residences and many small commercial buildings) and drive single-phase motors; others operate three-phase motors.

Figure 4-7 shows that, in the first stage, the input AC power supply is converted to DC using a solid-state rectifier. The DC link, which carries the DC power from the first stage to the second, includes a filter to smooth the electrical waveform.

In the second stage, the inverter uses this DC supply to synthesize an adjustable-frequency, adjustable-voltage AC waveform by releasing short steps or pulses of power. The speed of the motor will then

Figure 4-7

General Inverter Power Circuit with Motor Load

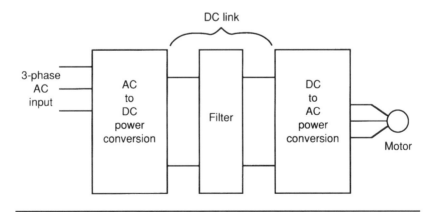

Figure 4-8

Pulse-Width Modulation

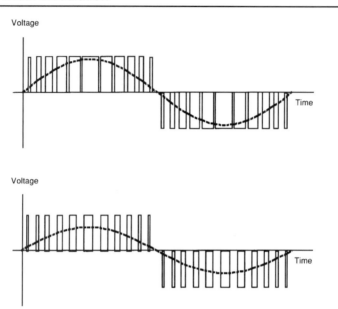

Note: Changing the width of the voltage pulses varies output voltage. Changing the length of the cycle varies output frequency.

change in proportion to the frequency. Usually the output voltage waveforms can be synthesized over the frequency range of 0–120 Hz, but they are available up to 180 Hz.

There are three main types of inverter-based ASDs: voltage-source inverters (VSIs); pulse-width-modulation (PWM) inverters, and current-source inverters (CSIs). Each has its own advantages and disadvantages as well as its own niche in the market.

VSIs and PWM inverters generate variable-frequency, variable-voltage waveforms. The former synthesize a square wave; the latter create a pulse-width-modulated output made of a series of short duration pulses, as shown in Figure 4-8. In both cases, the output has the frequency that will produce the desired speed, but the shape of the output is not as smooth as the sinusoidal AC waveform of a conventional power distribution system.

VSIs (also known as single-step or square-wave inverters) are used in low- to medium-power applications, typically up to several hundred horsepower, and can operate several motors at once. Multimotor operation is desirable when several motors are in operation at the same adjustable speed, as is often the case in the textile industry. Moreover, it is much cheaper to use one 200 hp ASD to drive ten 20 hp motors than to buy ten ASDs to drive the same motors. One drawback to multimotor operation is that external overload protection must be provided to each motor.

PWM inverters have fewer problems than square-wave inverters although their efficiency is a bit lower due to the higher switching losses. Because of their better performance at low speed, lower harmonics, and ability to maintain good efficiency in the high-frequency range (by switching from pulse-width-modulated to square wave output), PWM inverters have become predominant in applications below 200 hp and are available up to approximately 500 hp.

Neither square-wave nor PWM inverters have regeneration capabilities. Regeneration is the ability to return energy to the supply system when a motor is slowing down, essentially operating the motor as a generator. This feature saves energy in drives with a high start-stop duty cycle (such as electric traction in urban rapid transit systems) as the braking energy is pumped back into the AC supply.

The third type is CSIs, also called current-fed inverters, which behave like a constant current generator, producing an almost square wave of current. CSIs are used instead of VSIs for large drives (above 200 hp) because of their simplicity, regeneration capabilities, reliability, and lower cost. Although more rugged and reliable than VSIs, CSIs have a poor power factor at low speeds and are not suitable for multimotor operation. A special type of CSI, the load-commutated inverter, can be used with synchronous motors, typically in applications above 1,000 hp.

Cycloconverters

Cycloconverters convert AC power of one frequency to AC power of a different frequency without using an intermediate DC link. The output frequency can range from 0% to 50% of the input frequency. Cycloconverters feature regeneration capabilities and are used in large drives (above a few hundred horsepower) for low-speed applications. There are no common applications for cycloconverter ASDs in residential or commercial buildings, but typical industrial applications include ball mills and rotary kiln drives in the cement industry, where cycloconverters' low-speed capability eliminates the need for gears.

Wound-Rotor Slip Recovery ASDs

As noted in Chapter 2, inserting an external resistor in the circuit can alter the speed of a wound-rotor induction motor. However, controlling the speed in this manner is very inefficient. A wound-rotor slip recovery ASD recovers and reuses some of the power wasted when an external resistor controls the speed of a wound-rotor motor. Its use is limited to very large motors (typically over 500 hp).

Two types of wound-rotor ASDs are available. The static Kramer drive is commonly used in applications requiring 50–100% of the synchronous speed, such as large pumps and compressors (Bose 1986). The more expensive static Seherbius drive is used in large pumps and fans where higher-than-synchronous speeds or regenerative braking is important (Leonard 1984).

Voltage-Level Controls

Unlike other electronic ASDs, variable-voltage controls do not vary the frequency of power supplied to the motor. Instead, the effective AC supply voltage applied to the stator windings is varied. When the applied voltage level decreases, the motor slows down. Although simple, this control method is not widely used due to its low efficiency and the high level of harmonics generated (Mohan 1981). Essentially the same technology is used in the so-called power-factor controllers, described below.

Applications of ASDs

ASDs are used for the following basic reasons: (1) to provide accurate process control or (2) to match the speed of a motor-driven device to varying load requirements. The most dramatic energy savings from speed control occur with loads that have losses that fall at reduced speeds. This is true of centrifugal machinery, including most pumps, fans, and some compressors. Their energy use is often proportional to the cube of the flow rate, so small reductions in flow can yield disproportionately large energy savings. For instance, a 20% reduction in flow, under the conditions spelled out in Chapter 5, can reduce energy requirements by nearly 50%.

ASDs are ideally suited for modifying the speed of centrifugal machines to provide the exact flow required by the system. This is in contrast to the conventional practice in fan and pump systems of running the motor at full speed and controlling flow via throttling devices, like inlet vanes or outlet dampers on fans, and valves on pumps. Such flow constriction is analogous to controlling the speed of a car with the brake while the accelerator is pushed to the floor—a very wasteful practice. Fans and pumps represent such a large proportion of drive-power energy use and are such attractive candidates for speed control with ASDs that we devote much of Chapter 5 specifically to these applications.

In other equipment, lowering the speed produces less dramatic savings. For example, conveyors are sometimes equipped with speed controls for process reasons. The energy needed to drive a conveyor depends primarily on the load carried by the belt and secondarily on its speed. Therefore, the energy savings that can be achieved with speed controls depend on the load profile of the conveyor.

Features of this technology other than the energy efficiency advantages of speed control may be even more important to the user, particularly in industry. For instance, response time and process control may be improved. Moreover, by eliminating control valves, ASDs reduce the number of parts exposed to fluid, which may be important in some applications where there is a possibility of contamination problems.

General Considerations for Selecting ASDs

Although pumps and fans provide the best applications for ASD retrofits, speed controls are not necessarily cost-effective for all pumps and fans. The load profile (time variation of the pressure and flow requirements) is very important for determining the cost-effectiveness of an application. For example, if a system must operate at full flow at all times, then a flow control scheme is never a good choice. However, the typical system will have flow requirements that vary considerably over time. The greater the amount of operation at relatively low flows, the more cost-effective it is to efficiently provide flow control.

The best way to determine the cost-effectiveness of a proposed ASD installation is to look at the power that would be needed at each operating condition with and without an ASD. The energy savings can then be calculated by taking the reduction in power at each condition and estimating the savings based on the actual (or expected) operating time at that condition. Sample calculations appear in Appendix A.

In general, the following are good applications for variable-speed flow control, and in particular ASD control. The applications

- Are fixed at a flow rate higher than that required by the load.

- Are variable-flow, where the variation is provided by throttling (by valves or dampers) and where the majority of the operation is below the design flow.

- Use flow diversion or bypassing (typically via a pressure-reducing valve).

- Are greatly oversized for the flow required. This situation can occur if successive safety factors were added to the design, a process was changed so that the equipment now serves a load less than in the original design, or a system was overdesigned for possible future expansion.

- Have long distribution networks.

- Have flow control by on-off cycling. Such systems are usually less cost-effective retrofit candidates than are those that use a throttling control.

- Have a single large pump or fan rather than a series of staged pumps or fans that come on sequentially as the process needs increase.

- Can reduce the pressure at the outlet of the fan or pump at lower flows. For example, a pump that discharges water into a long pipeline that can move the water at a lower pressure when the

Figure 4-9

Typical Efficiency Curves for an AC Inverter Drive

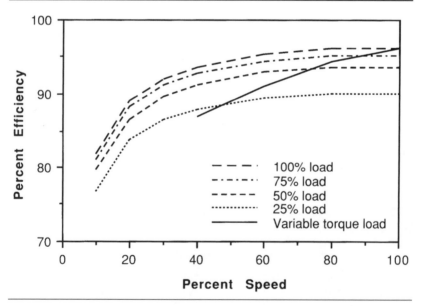

Note: The curves labeled with values at percent load are for constant-torque loads.

Source: Eaton Corporation 1988

flows are low (due to the decreased frictional losses in the pipes) would be a good candidate for an ASD. A pump supplying a fire protection system where the piping is oversized and a constant pressure is needed regardless of flow would probably not save enough for an ASD to be cost-effective.

Once a good ASD application is identified, the question arises of what type of drive to use. This choice involves selecting among the major categories of ASDs discussed above, as well as choosing a version suited for variable-torque loads (including pumps, fans, and compressors) or constant-torque loads (such as conveyors, some machine tools, and winders). Most manufacturers make two lines of ASDs, each suited to either variable- or constant-torque loads.

Constant-torque ASDs are typically 10–20% more expensive than variable-torque equipment because the electronics must be designed and built to withstand the high currents that occur when a constant-torque machine starts with a full load. Figure 4-9 shows the efficiency of variable-torque and constant-torque inverters as a function of both speed and load.

Process Controls and the Integration of ASDs

To work effectively and save energy, an ASD must be integrated into some type of control system. Most ASDs have a provision for controlling the speed by adjusting a setting on the local ASD panel. This type of relatively unsophisticated control is used where there are no major changes in the process that correspond to a desired speed of the equipment. For example, an ASD is often used to control the speed of a conveyor that carries a product through a freezing tunnel or a drying oven. The slower the belt, the longer the product is held in the tunnel. In these situations, the operator sets the belt speed based on some characteristics of the product and only changes the speed when the product changes. This is called open-loop control, as the system output is not monitored to regulate the performance. Typical applications, in addition to conveyors, include some types of ventilation equipment, and pumps and fans that run at constant speed where relatively large flow fluctuations due to external disturbances can be tolerated.

In more demanding applications, a control system, of which the motor and the ASD are a subsystem, must be designed to satisfy the process requirements. Generally the control system will have one or more sensors to monitor the state of the process variables. The sensor(s) provides data to a controller or computer programmed with the control strategy, or algorithm. The controller compares the actual level of the process variable with a preset desired level. Based on this comparison and the programmed algorithm, the controller or computer will send a signal to change the system operation so that the actual value of the variable correlates with the desired value. A common example is a building HVAC control system in which the speed of a chilled water pump might be controlled by the pressure of the water in the circulating loop of the building.

ASDs feature several types of control inputs that allow them to be easily controlled by an external signal. The value applied at the control inputs determines the speed of the motor. Most ASDs are equipped with low-voltage or low-current control inputs, or a low-pressure pneumatic control input. Additionally, modern ASDs also feature a digital interface that allows the plant's computer to communicate directly with the ASD. This situation is common in the process industries such as pulp and paper, chemicals, and refineries.

Potential Drawbacks of ASDs

Electronic ASDs hold great promise, but improper selection and use of the technology can lead to a number of problems.

An existing motor can be retrofitted with an ASD. However, caution must be exercised when doing this. Since the electrical wave generated by an ASD is slightly irregular, a motor will heat up slightly more when used with an ASD than when run off the standard line power. Inverter-duty motors use an upgraded insulation and other design features to address this problem. It is also important that the drive and motor be electrically compatible. If the electrical characteristics of the drive are not correctly matched, standing waves can be created in the motor, resulting in premature failure of both the motor and the drive. Most ASDs can be ordered with an option of driving a motor at up to twice its rated speed. At times it may be necessary to replace a motor if the system requires operation in these high speeds because the existing motor's rotor and bearings cannot handle the speed, or because the load demands more power at a higher speed. It is also important to be aware of any critical harmonic frequencies for the motor, the driven equipment (e.g., a fan), or a combination of those. The drive should be programmed to avoid operation at any multiples of these frequencies to prevent mechanical resonance problems that can destroy the equipment.

An ASD can save a great deal of energy by slowing a motor to match light loads. Care must be taken, however, because motors that are run at small fractions of their rated speed can overheat or suffer an irregular rotation known as cogging. Most manufacturers do not recommend operating a motor at less than 10–15% of rated speed.

ASDs have some losses in the circuitry. As with most electric equipment, these losses result in increased heating of the drive. Most drives are equipped with cooling fins so that the heat does not build up and trip the electronic circuitry; many are also equipped with fans. In general, the ASD cooling system will work adequately when the drive is located in a room with normal temperature conditions. Some large drives and those located near hot spots in a plant may need to be placed in an air-conditioned room.

A standard ASD houses the circuitry in a box that closes but does not seal tightly. Most manufacturers can package ASDs in housings that are impervious to dust, water, or explosive vapors. These special enclosures can add 10% to the cost of small ASDs and 5% to larger models.

Many add-on features are available that can increase the cost of an ASD installation. These can be required in any of the following situations: when special control interfaces are needed; when the machine has special requirements for acceleration, deceleration, or direction reversal; when manual or automatic bypass of the ASD is needed; or when equipment is required to protect against overload, voltage fluctuation, short circuits, loss of phase, harmonics, and electromagnetic interference.

The early generation of ASDs often had poor power factors. Today, most small ASDs (below 300 hp) use an input circuit with a high power factor over the entire speed range. In fact, these types of ASDs have a better power factor over the entire operating range than motors connected directly to the line power (which have lower power factors at low loads). These units are generally identified in catalogs as having high input power factor (0.95 or above) over the entire speed range or as a PWM-type input circuit. For small drive units with general applications, it is almost always possible to find an ASD with good power factor.

Larger current-source ASDs may have poor power factors at low speeds (see Table 4-2). The effects of poor power factor are partially mitigated by the fact that less power is used at low speeds than at full speed, especially in the case of variable-

Table 4-2

Displacement Power Factor vs. Speed for Typical Large Electronic Adjustable-Speed Drives

Percent Speed	Power Factor
100	.94
90	.95
80	.76
70	.67
60	.58
50	.50
40	.41
30	.32
20	.23
10	.14

Source: Eaton Corporation 1988

torque loads. With constant-torque loads, the reactive power consumed at low speeds can reach unacceptably high values. It is possible to correct the low power factor caused by ASDs with capacitors and filters (for displacement power factor and harmonics, respectively). However, an ASD differs from other motor loads in that the displacement power factor should be corrected at a central location with a switched capacitor bank instead of at the individual piece of equipment.

ASDs larger than 200 hp, such as cycloconverters and current-source inverters, can produce harmonics and electromagnetic interference that can disrupt power line signals, computers, and other electronics and communications equipment. Smaller pulse-width-modulated ASDs damp harmonics better and generally do not pose problems unless there are many of them in the plant. The problems caused by harmonics and EMI, and their diagnosis and mitigation, are discussed in Chapter 3.

Trends and Developments in ASD Technology

There has been continuous progress over the past decades in the technologies used in electronic ASDs, including the microelectronics used in control circuits; sensors that provide input to the controls; and the power electronics used to condition the input current to the motors. These technology trends and the associated cost reductions have helped accelerate ASD market penetration.

Increasingly powerful microprocessors and large-scale integration devices have allowed complex control functions and algorithms to be incorporated in compact and inexpensive ASD packages. Corresponding advances in power electronics technology have also been achieved.

The integration of power electronic devices and microelectronics into single packages known as power-integrated circuits (PICs), or smart power devices, will lead to further miniaturization. PICs have the potential to slash the number of ASD components, reduce costs, and improve reliability. One manifestation of this trend is the move toward integrated motor/drive packages, especially in the smaller motor sizes.

Another interesting development is the integration of sensors in a silicon chip for all kinds of variables, including temperature, pressure, light, force, acceleration, and vibration. The integration of PICs with sensors is helping to decrease the cost not only of the ASD, but also of the overall control system of which the ASD is a part. Electronic ASDs are now being routinely packaged with motors, especially for low-horsepower devices such as home air conditioners, heat pumps, washing machines, and other appliances. The Japanese are doing this in many heat pumps and air conditioners, and most U.S. manufacturers (including Trane and Carrier) have integrated ASDs in the fans and compressors for their high-end furnaces, air conditioners, and heat pumps. Also, advanced motors with speed control capability, such as switched reluctance, are beginning to be incorporated into appliances like clothes washers.

The general trends in electronic ASDs point to increased compactness, efficiency, and reliability, as well as more flexibility (e.g., added control and protection features), less power line pollution, and decreasing cost per horsepower. With these trends, a staggering growth in the electronic ASD market can be expected, particularly as the cost per horsepower decreases.

Economics of Speed Controls

The economics of motor speed control technologies are briefly discussed below. For further details, see Appendix A.

Mechanical ASDs, such as adjustable pulleys, are fairly inexpensive, with prices ranging from $50/hp (equipment only) for a 125 hp drive to $350/hp for a 5 hp drive. Typical equipment prices for eddy-current drives range from $200/hp for a 5 hp drive to $150/hp for a 15 hp drive, to less than $100/hp for a 100 hp drive. While the equipment prices are typically low, these mechanical ASDs are seldom used for retrofits due to the difficulty of repositioning the motor or the driven load. Electronic ASDs are generally much better suited to retrofit applications because they are connected to the motorized system only through the wiring.

The price of electronic ASDs, in terms of dollars per horsepower, is a function of the horsepower range, the type of AC motor used, and the additional control and protection facilities offered by the electronic ASD. In 2000, cost for low-voltage applications varied from about $160/hp at 50 hp to $100 at 500 hp. Typical costs per horsepower for ASDs used with induction motors are shown in Figure 4-10. These data are based on comparison of solicited bids and studies of actual implementations and assume variable-torque equipment. Most commercial applications use a standard installation with a NEMA 1 enclosure (indoor application). Installation costs typically run no more than 15% of equipment costs. Many industrial applications will require a higher-cost installation (such as a NEMA 12 "wash-down" enclosure or special ventilation of the cabinet) due to harsh environments or the need to isolate the drive from adjacent equipment. Installation can vary depending on the application and can cost substantially more than the standard installation because

Figure 4-10

ASD Costs per Horsepower for Low and Intermediate Voltages

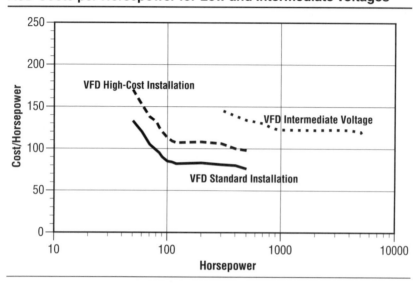

Note: Standard installations assume variable-torque load and a NEMA Type 1 enclosure for indoor application. High-cost installations assume use of a NEMA Type 12 wash-down enclosure and power conditioning equipment.

Source: Easton 2000

sometimes special controls and sensors are needed to integrate the ASD into the control system and/or abatement of harmonics is necessary (Easton 2000).

Drives for synchronous motors above 1,000 hp are about twice as expensive per horsepower as drives for induction motors of the same size. As noted earlier, the integration of ASDs into mass-produced appliances lowers costs dramatically. ASDs built into Japanese variable-speed heat pumps reportedly added only $25/hp to the manufacturer's cost (Abbate 1988).

Typical energy savings from ASDs range from 15 to 50%, and simple paybacks of 1–8 years are common, based on energy savings alone. The payback is, of course, sensitive to the price of electricity, labor costs, the size of the drive, the load profile, whether the application is new or a retrofit, and other factors.

In addition, there are costs and benefits that are difficult to quantify, including maintenance requirements, reliability, reduced wear on the equipment, less operating noise, regeneration capability, improved control, soft-start, and automatic protection features. In most instances, these difficult-to-quantify factors, like improved process control, motivate a user to install ASDs.

ASD Case Studies

The following descriptions of documented ASD applications profile the engineering background, cost and energy savings, and other benefits of the applications. Only projects that have submetered data have been used. In the past decade, since the publication of the first edition of this book, significant experience has been gained with ASD applications. While most of these applications have resulted in significant nonenergy benefits, these advantages are not as easily documented as the energy savings. The energy benefits are so great and readily available that, considering them alone, the projects prove economically attractive. While many of the applications are unique to a specific site, especially in industry, the examples presented below are reflective of the opportunities available from appropriate applications of ASDs.

Boiler Feed Pump at Fort Churchill Power Plant

Like many small to medium-size oil- and gas-fired power plants, Sierra Pacific Power Company's 110-megawatt (MW) Fort Churchill plant acts as spinning reserve, operating at minimum power until needed. Its minimum load, 16 MW, was provided at a relatively large cost due primarily to high fuel costs. An added problem, and one that limited the extent to which the plant's output could be reduced, was the large pressure drops in the throttling valves, which increased maintenance costs as well. The boiler feed pump provided over 2,700 pounds per square inch (psi) of pressure and all but 250 psi were wasted through restrictive operation of the feedwater control and turbine stop valves (EPRI 1985). In other words, the throttling valves dissipated most of the energy delivered to the water.

An analysis of the plant's operation, including the load profile (number of hours per year at each fraction of full load) and the heat rate (amount of fuel required per kilowatt-hour output at each fraction of full load), showed that the plant could be turned to an even lower load and valve wear reduced if the induction motor drive of the boiler feed pump was retrofitted with an ASD. The retrofit on the 2,000 hp pump was performed in 1984 as an EPRI demonstration project (Oliver and Samotyj 1989).

The following results were achieved:

- Minimum power was lowered to 12 MW, resulting in large fuel savings.

- Pump input power at minimum plant power was reduced from 815 kW to 293 kW.

- Annual savings from fuel and pump electricity totaled $1,600,000, which paid off the $480,000 installation in about 4 months.

- Other benefits (including reduced maintenance and reduced stack emissions) were significant but have not been quantified.

Coolant Pumps at Ford's Dearborn Engine Plant

The Dearborn Engine Plant's cooling system includes five 75 hp pumps for circulating cooling fluid to cutting tools. The pre-retrofit operation incorporated three pumps, operating in parallel at 64 psi and 1,325 gallons/minute each, for 5,700 hrs/yr (the other two pumps were maintained as spares).

An analysis of the system showed that the pumps were operating most of the time at excessively high pressure in order to meet occasional peak loads (the required pressure is 50 psi). The solution was to install an ASD on one of the pumps, along with a control system, not only to maintain pressure levels but also to shut off coolant flow to machines not operating. Thus, the pump staging, along with ASD control, was able to meet the system requirements exactly.

The results of the retrofit were as follows:

- Reducing the required flow and meeting that requirement more efficiently through reduced pressure reduced energy use by 48%.

- The $75,000 cost of the retrofit was paid back in 1.4 years from the annual $55,000 savings.

- Other benefits (not quantified) included reduced misting (from the coolant nozzles); improved indoor air quality; reduced pump wear; and improved direction control of coolant (due to constant pressure).

- The above benefits contributed to improved machining quality, improved coolant filter performance, and decreased filter maintenance due to reduced flows. The control system also allows pump wear to be monitored, assisting in planned maintenance (Strohs 1987).

Boiler Fans at Ford's Lorain Assembly Plant

The assembly plant at Lorain, Ohio, is served with 125 psi steam from three coal-fired boilers installed in 1957. In 1976, environmental regulations required replacement of the induced draft fans with models equipped with outlet dampers. This controls combustion better, thereby lowering emission levels. With this change, particulate emissions were reduced to acceptable levels at high boiler loads. However, this goal could not be met at loads below about 40% of the rated maximum capacity of each boiler (80,000 lb/hr of steam). In the summer, steam load dropped as low as 15,000 lb/hr, and large amounts of steam were vented to the atmosphere to prevent excessive emission levels.

After the damper-equipped fans were installed, tightened environmental requirements forced even further modification. In 1986, forced and induced draft fans were equipped with ASDs, which improved control of the boilers and met the particulate emission requirements even at boiler outputs below 25% of maximum.

In addition to avoiding fines for excessive emissions, the controls

- Saved $53,000 in coal costs by greatly reducing steam venting

- Saved $41,000 in electricity costs for the six fans

- Paid back the $90,000 retrofit cost in slightly less than 1 year

On the downside, the controls created a low power factor and potential electromagnetic interference, which may require ASD modification or additional power-factor correction equipment and filters (Futryk and Kaman 1987).

Ventilation Fans in a New Jersey Office Building

ASDs were installed on two supply fan and two return fan motors in variable-air-volume (VAV) ventilation systems in a 130,000-square-foot commercial office building. Inlet vanes on the fans had previously controlled the air volume.

As a test, two control schemes were compared. The first used the pre-retrofit duct static pressure control with the existing setting of 2.5 inches of water. The second reset the duct pressure to 1.5 inches when system loads were reduced.

The retrofit had the following results:

- With the same control strategy as before the retrofit, energy savings were 35%, amounting to $5,200 annually, which would pay off the $40,000 installation cost in 7.7 years.

- With the modified (pressure reset) strategy, savings increased to 52%, or $8,700, annually, with a simple payback period of 4.6 years.

- Due to the inefficiencies of the inlet vanes, resetting the duct pressure with inlet vane control did not result in significant savings compared to the base case (Englander and Norford 1988).

Gas Removed from Steel Making at Burns Harbor

Bethlehem Steel Corporation's (BSC) Burns Harbor Facility in northwest Indiana is one of the premier integrated steel production plants in the United States. Seeking opportunities to increase productivity while reducing energy costs, BSC retained General Conservation Corporation (GCC) to identify and implement energy efficiency measures on a shared savings basis. One opportunity they targeted was the application of an ASD on the primary induced draft (ID) fan of the #3 basic oxygen furnace (BOF). This fan, driven by a 7,000 hp motor with a design capacity of 40,000 cubic feet/minute (cfm) at 1,200 rpm, removes waste gasses from the BOF and draws them through several gas cleanup steps before reaching the fan. The fan exhaust passes through a silencer before reaching the stack. It operates continuously at 1,200 rpm, and gas flow is regulated by a set of upstream dampers. The gas volume required varies from negligible during idle periods to a peak of 25,000–30,000 cfm. This peak flow is only necessary for a third of the typical 45-minute heat cycle. As a result, not only does the gas flow vary significantly, but it shows that the fan is oversized for its maximum requirements.

When GCC proposed the installation of the ASD in 1987, BSC was uncertain about committing to it for three reasons: (1) the existing system already ran smoothly; (2) the ID fan was critical to plant operation; and (3) they were hesitant to spend the necessary capital. GCC agreed to fund the project in return for 50% of the savings, and BSC was given the option to purchase the equipment in 7 years at fair market value.

The need to maintain a critical pressure drop through the scrubber complicated the project design. The existing fan was deemed unable to withstand the cycle stress, so it was replaced with a new unit that was 5% more efficient. During process operation, the fan speed was set to vary between 960 and 1,075 rpm and to reduce to 560 rpm when the fan was idle. An added benefit was that these speeds could be adjusted to even lower levels immediately after system cleaning. With the modified system, average monthly energy consumption was reduced by almost 50%, from 2,602 to 1,310 megawatt-hours (MWh). In addition, the modifications reduced noise levels in the furnace area significantly and extended system component lives. The reduced fan speeds also improved the system's tolerance to slight fan imbalances, which in turn reduced operation and maintenance (O&M) costs. The resulting savings were $310,000 for each partner. This annual income for GCC allowed them to recover their $1,225,000 capital investment in about 4 years. BSC was so pleased with the project results that, at the end of the initial contract period in 1994, they extended the shared savings agreement (OIT 1998).

Improved Ventilation in a California Textile Mill

A Japanese-owned mill, Nasshinbo California, Inc., produces cotton fabric. This Fresno, California, facility is the only spinning and weaving plant in the western part of the United States. A ventilation system is needed to keep the plant's temperature at 85–95°F, with a relative humidity of 50–60%. This ensures reliable operation of the processing equipment and maintains produce quality. A system of nine supply and nine recirculation fans mixes outside air that has been cooled and humidified with air washers with plant air that has been filtered to remove suspended particles and fibers. Seasonal variations and different products cause differences in airflow requirements. Manually operated variable inlet guide vanes and outlet dampers control the flows. However, this control system is labor intensive and subject to corrosion due to the high humidity.

Pacific Gas & Electric Co. (PG&E), as part of their PowerSaving Partner (PSP) program, contracted with an energy services company, Tamal Energy, to finance the design and modification of the system with ASDs. ADI Control Techniques Drives implemented the project. They determined that all but one of the recirculation and two of the supply fans were oversized. The remaining 15 fans were fitted with ASDs. The flow was modulated using the ASDs, which produced better air quality and resulted in annual labor savings of 48 hrs/yr. Airborne lint in the plant was decreased, which reduced the number of equipment breakdowns while improving the product quality. In addition, the ASDs improved the plant's power factor, thereby cutting penalty costs.

The fan system modifications lowered system demand from 322 kW to 133 kW, and reduced electricity consumption by 59%, from 2,700 to 1,100 MWh annually. This produced annual cost savings of slightly more than $100,000. Nasshinbo realized these savings because the PG&E program covered the $130,000 cost of the project (OIT 1997).

Other Controls

The electronic ASD is only one of the new control technologies to have emerged from the electronics revolution of the past 10 years. Electronics process controls, sensors, fast controllers for compressors, power-factor controllers, and energy management systems (EMS) for controlling mechanical and lighting systems in buildings are all becoming more sophisticated and effective.

In many cases, the new control systems are used to improve the product being processed, to increase the yield of the product, or to improve the comfort in a building. Energy conservation is only some-times a primary goal; at other times, it is a side benefit when control systems have been installed for process reasons.

Improved controls and sensors can save energy by monitoring more system variables than was previously possible and by responding more rapidly and accurately than ever before. Because of their slow response time, process controllers have historically used set-points with built-in safety margins to ensure that the process meets minimum performance requirements. A new generation of intelligent controllers is currently available that looks at both the historic pattern of control and the recent system changes and automatically retunes the process control parame-ters to optimize response time for each situation.

These controllers yield energy savings for two reasons. First, the rapid response ensures that the process is operating within desired limits. Energy wasted when the process is slightly out of spec is now conserved. In addition, the increased accuracy allows the operator to set the process control variable to the exact set-point without incorpo-rating a safety factor.

Building and industrial process control loops in the past were typi-cally designed to control one or perhaps two variables. With the advent of the microprocessor, control systems can now monitor and respond to many parameters. For example, the temperature of heated or chilled supply air or water in building HVAC systems has traditionally been fixed at one preset level, or perhaps was varied with outside air temper-ature, which is only one indication of cooling or heating requirements in a large building. With this approach, the HVAC system tends to provide excessive cooling or heating much of the time. Newer energy manage-ment systems allow the temperatures of the cooling and heating medium to be reset based on the actual demand for heating and cooling in different zones of the building. Significant energy savings can result from this type of control strategy since only the minimum amount of heating and cooling is used to meet the needs of the building.

In the past, use of industrial process controllers was limited because sensors were not available for specific applications, or because they tended to drift and needed frequent recalibration. As a result, many processes were controlled manually with fairly crude adjustments. The electronics revolution has produced sensors that can detect small con-centrations of specific ions, sense humidity without fouling and drifting, and measure other process variables. These new sensors, combined with the advent of central control (which reduces the cost for each control point), have expanded the range of processes that can be automated. For example, humidity sensors can now be purchased that survive in the harsh environment of a lumber kiln and remain in calibration for long

Figure 4-11

**Schematic of a Vector-Control Drive
(Also Known as a Field-Oriented Control)**

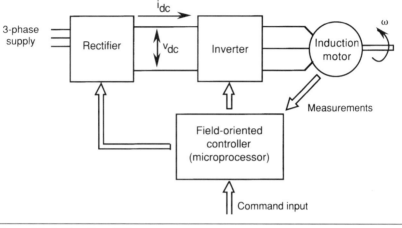

Source: Adapted from Bose 1986

periods of time. This innovation allows humidity to drive the operation of equipment.

Vector Control

DC motors have traditionally been used in high-performance applications, such as servodrives, rolling mills, robotics, and web winders, where accurate torque and speed control are necessary. The development of inexpensive microprocessors and ASDs now allows the more reliable AC-induction motor to be used in such tasks. In this process, known as vector control, the motor current, voltage, and position are continuously monitored. These values are then plugged into mathematical formulas called algorithms, which precisely control torque, speed, position, and other critical parameters. Vector control can be coupled with many types of ASDs and has been successfully used in a wide range of applications. The general approach to closed-loop, microprocessor-based vector control is shown in Figure 4-11. For more on vector control, see Leonard 1986 or Bose 1986.

Power-Factor Controllers

Many applications require a constant speed, even as the motor load varies. For example, a motor driving a saw blade must maintain constant speed regardless of whether the blade is cutting a 0.5-inch board or a 2-inch board, a hardwood or a softwood. As a result, the motor runs with a light load when cutting wood below its maximum capacity or when idling between cuts.

For applications below 15 hp, in which the motor runs much of the time at full speed but at light or zero loading capacity, electronic variable-voltage controls, also known as power-factor controllers (PFCs), can increase both power factor and efficiency. These devices eliminate part of the sine wave fed to the motor. As a result, the average current (represented by the area under the sine wave) is lower. Since the power is the product of the voltage times the current, PFCs reduce the power used at low loads and thus improve motor efficiency by reducing magnetic and I^2R losses.

Energy savings range from 10–50% at light loading to zero at full load. Overall energy savings of 10% typify the limited number of attractive installations (including saws, grinders, granulators, escalators, punch presses, lathes, drills, and other machine tools) that idle for extended periods. The potential savings are greater for single-phase motors than for three-phase motors because the former have much larger no-load losses. Similarly, the higher and flatter efficiency curves of large motors make them unattractive candidates for power-factor controllers. PFCs also can incorporate soft-start capabilities at little extra cost. In fact, many controllers marketed as soft-start devices include the power-factor control capability.

PFCs also improve power factor because of the way power consumption is sensed and controlled. In general, as the load on a motor decreases, the power factor deteriorates as a result of reactive current, which shifts the voltage sine wave out of phase with the current sine wave. When the controller eliminates part of the sine wave, it not only reduces the average magnitude of that waveform but also shifts the center of the waveform so that the voltage and current are closer to being in phase. In this way it improves the power factor.

PFCs can generate significant harmonics, which need to be suppressed. They also have internal losses typically equal to a few percentage points of rated power.

List prices of 10 hp PFCs are $30–60/hp. Their cost-effectiveness can be evaluated by estimating the load profile and the net efficiency gains at each point of part-load operation.

Summary

Speed control for motors, particularly when applied to fans and pumps, is an extremely effective way to produce energy savings in motor drives. While speed can be controlled using multiple motors, multispeed motors, and an assortment of mechanical devices, most retrofits and many new installations use electronic ASDs because these devices are easy to install on existing equipment and their costs and reliability are increasingly making them attractive to the user.

An ASD produces energy savings most effectively when integrated into a larger control system. Modern control systems not only control motor speed for adjustable-speed applications but also reduce energy use by improving process control.

Motor Applications

Previous chapters have discussed components of the drivepower system—motors, wiring, controls, and transmission hardware—upstream from the load. In this chapter we take a different perspective and focus on several principal loads, most notably fans, pumps, and compressors. We emphasize these topics for two reasons. First, these key loads account for more than half the drivepower energy used. Second, understanding the theory of fan, pump, and compressor applications is necessary if one is to design efficient, reliable systems to drive them.

Fans and Pumps

The fans, pumps, and compressors that move and compress air, water, and other gases and liquids (collectively known as fluids) in industry and commercial buildings consume approximately 50% of the electricity used by U.S. motor-driven systems. (See Chapter 6 for further discussion of end-use estimates.) Most of the electricity used by this group is for fans and pumps, the initial focus of this chapter.

Fans and pumps are used in many applications. Equipment ranges in size from fractional-horsepower units in residential appliances to tens of thousands of horsepower used in utility power plants. Despite the range of size and usage, nearly all fan and pump applications have time-varying flow requirements, and most of the flow variation is done inefficiently, if at all.

Fluid-Flow Fundamentals

All fan and pump applications share certain characteristics. One is the nature of fluid flow. To produce a flow of fluid through a

159

Figure 5-1

System Head Loss Curve, Showing Squared Relation between Pressure ("Head") and Flow

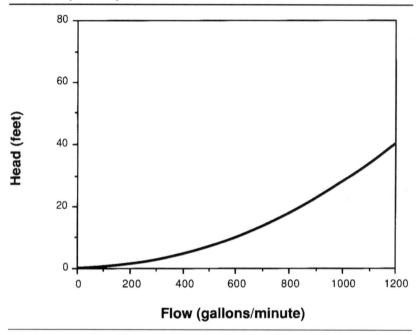

Flow (gallons/minute)

Note: This system has no minimum pressure requirement.

pipe, duct, damper, or valve, a pressure difference must be created across the component. A common example is the garden hose. To force water through it, the pressure must be greater at the faucet than at the far end of the hose. The greater the pressure difference, the greater the flow.

For a given system, a curve can be drawn to show the pressure difference required at any given flow. An example is shown in Figure 5-1, where the pressure difference (in pump jargon, "head") is shown as a function of flow (gallons per minute). As the figure shows, the relation is quadratic (i.e., the pressure difference is proportional to the square of the flow rate). Or, expressed in mathematical terms,

$$\Delta P \propto Q^2$$

where

ΔP is the pressure difference in pounds per square inch or inches or feet of water

160

Q is the flow rate in cubic feet per minute or gallons per minute (gpm)

$\propto$ means "is proportional to"

Thus, if the flow doubles, the pressure drop quadruples. This squared relationship holds true for systems of all fluid types. The slope of the curve is determined by the system components' resistance to flow. For example, a water distribution system constructed from 2-inch-diameter pipe will have much larger resistance than one constructed from 4-inch pipe, which would thus result in a greater pressure drop than for the 4-inch system for any given flow. In other words, the more restrictive the system, the steeper the system curve.

In the United States, units for measuring pressure and flow differ between fan and pump applications. For fans, the pressure difference is given in inches of water column (1 inch corresponds to 0.0361 psi; thus, 27.7 inches of water equals one psi). For pumps, the pressure (or head) is given in feet of water column (one foot corresponds to 0.434 psi; thus, 2.31 feet of water equals one psi). The flow rate for fans is given in cubic feet per minute; for pumps, in gallons per minute. (To convert U.S. units to the Système Internationale units of pascals, multiply inches of water by 249, or multiply psi by 6,894.)

The power required to create a given flow relates directly to the shaft power required by the fan or pump from the drive motor (which in turn relates to the required electrical input power). A basic relationship that follows from the physics of fluid flow is that the theoretical power required to create the pressure difference needed to produce a given flow is proportional to the product of the pressure and flow. That is,

$$\text{Power} \propto \Delta P \times Q$$

Thus, there is a unique theoretical power required for any given combination of pressure and flow. The terms *water-horsepower* (for water pumps) and *air-horsepower* (for air fans) are often used to denote the theoretical power required in these systems. The relationship between the theoretical and the actual power requirements is discussed in the following section, "Fan and Pump Characteristics."

A set of relations, known as affinity laws, exists for fans and pumps. One law states that for a given fan or pump installed in a given (unchanging) system, the flow rate is directly proportional to the speed of the fan or pump:

$$Q \propto N \quad \text{where } N \text{ is speed}$$

For example, if the speed of a fan is doubled, the flow through the fan and system attached to it is also doubled.

Another affinity law is that the theoretical power required by a fan or pump increases with the cube of its speed:

$$\text{Power} \propto N^3$$

For example, when a fan's speed is doubled, the power requirement grows eightfold (two to the third power). This cubic relation follows directly from the concepts previously discussed. The power required is proportional to the product of the pressure and the flow, and the pressure in a given system is proportional to the square of the flow. The power required is proportional to the cube of the flow. Since the flow is proportional to the speed, the power is proportional to the cube of the speed.

The "cube law" has a great significance for the energy used by motors in fluid-flow applications. For example, reducing the flow (by reducing the speed of the fan) in an oversized ventilation system by

Figure 5-2

System Curve with Minimum Pressure Requirement of 20 Feet

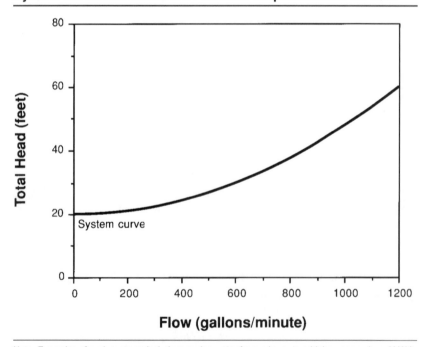

Note: Examples of such systems include pumping water from a lower to a higher reservoir and VAV building ventilation.

only 20% halves the power required by the fan. In many systems, the flow can be varied continuously to meet a constantly fluctuating demand. Methods of flow variation are discussed below, under "System Control and Optimization Techniques."

The affinity laws can be very useful in fan and pump applications, but the user must be sure that the laws hold in every specific case: they only hold if all other variables remain constant. For the cube law to apply, the system curve must be of the form in Figure 5-1: at zero flow, the pressure difference must be zero. Examples of such systems include residential and other ventilation systems that were originally designed to operate at constant air flow, and water circulation systems that do not use pressure controls or other means to create flow-independent pressure differences. The affinity laws also assume that the efficiency of the fan or pump remains constant at varying speeds.

Another common type of system has a fixed, or static, pressure requirement, even at zero flow, as shown in Figure 5-2. Examples include pumping between two reservoirs where there is an elevation increase, and most variable-air-volume (VAV) building ventilation systems that are designed to maintain constant pressure in the ductwork upstream from the dampers that serve each area. Since the cube law does not hold where the pressure does not drop to zero at zero flow, a more tedious analysis must be performed to determine the operating conditions of such systems.

Fan and Pump Characteristics

Just as each system has a characteristic curve for the pressure differences required by different rates of flow, each fan or pump also possesses a performance curve. More precisely, each fan or pump has a family of curves that, like system curves, are plotted on a graph with pressure on the vertical axis and flow on the horizontal axis. These curves describe where the energy goes (to some combination of pressure and flow) when a certain amount of energy is added to the fluid. Figure 5-3 shows a typical fan curve, and Figure 5-4 shows a typical pump curve.

We deal with only centrifugal fans and pumps in detail here because they collectively consume more energy than the other types (see Table 6-11). However, most of the analytical methods apply to other models as well, especially propeller-type (axial) fans. Centrifugal fans are used in many air-moving applications: residential furnaces; commercial and industrial HVAC equipment; and large blowers in utility power plants. Similarly, centrifugal pumps are used in applications ranging from fractional-horsepower residential units to industrial pumps of thousands of horsepower.

Figure 5-3

Typical Centrifugal Fan Curve

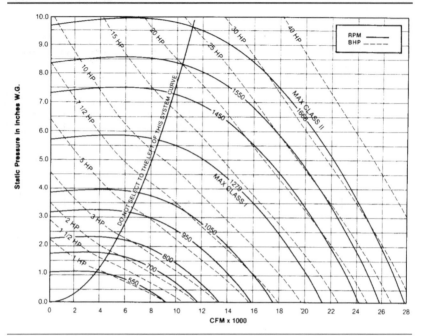

Note: The vertical axis is the pressure of fan operation, expressed in inches of water column. The horizontal axis is the flow rate, in thousands of cubic feet per minute of air. The solid curves labeled with different fan speeds (in revolutions per minute) show the pressure and flow relation of the fan. For example, if the fan is operating at 1,450 rpm and the system imposes 2.0 inches water column (w.c.) of pressure on the fan, there will be 22,000 cfm of flow through the system. The dashed curves represent the shaft power required at the fan to operate at any given point. In this example, the 22,000 cfm at 2.0 inches will require nearly 20 hp.

Source: Greenheck Fan Corporation 1986

The curves for centrifugal fans and pumps have similar shapes. They start at a pressure at zero flow; as flow increases, pressure remains constant or slightly increases; they then decrease in pressure as flow increases further. In Figure 5-3, the different curves represent the same fan operated at different speeds. The speeds are shown on the solid curves. In Figure 5-4, the different curves are for the same pump with different-diameter impellers (the bladed, wheel-shaped devices attached to the rotating shaft).

The reason that different speeds are shown for fans is that fan speed is usually easy to set at any given point by using different belts and pulleys that connect the fan to the motor. For pumps, the most

Figure 5-4

Typical Centrifugal Pump Curve

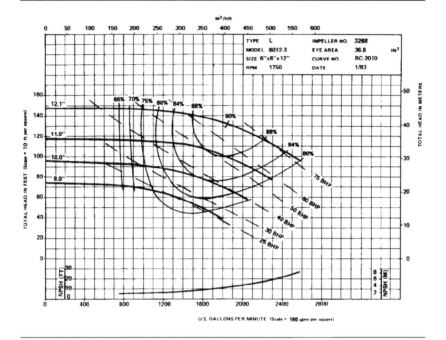

Note: The vertical axis is the pressure of pump operation, in feet of water. The horizontal axis is the flow rate, in gallons per minute. The solid curves labeled in inches show the pump characteristics for different impeller diameters. (All curves in this case are for 1,750 rpm pump speed.) For example, this pump with an 11-inch impeller will flow 1,200 gpm at 114 feet of head. The dashed curves labeled in BHP ("brake horsepower") show the horsepower required at the pump shaft for any given operating point. In the example above, about 43 hp would be required (interpolating between the 40 and 50 hp curves). The solid curves labeled with percentages are the pump efficiency. In this case, the efficiency is about 82%. The solid curve at the bottom (NPSH, or net positive suction head) refers to the minimum pressure required at the pump inlet to avoid pump damage through cavitation. The head and flow are also given in metric units.

Source: Paco Pumps 1983

common arrangement is for the motor to be directly coupled to the pump, forcing both to operate at the same speed. Machining the pump impeller to a smaller diameter provides different performance characteristics for the same pump. Of course, operating pumps at different speeds is possible, as discussed in the following section. With fans, on the other hand, it is generally not possible to reduce the impeller diameter to obtain different characteristics, so flow restriction or speed control must be used.

Table 5-1
Typical Centrifugal Fan Table

CFM	OV	¼" RPM	¼" BHP	½" RPM	½" BHP	¾" RPM	¾" BHP	1" RPM	1" BHP	1¼" RPM	1¼" BHP	1½" RPM	1½" BHP	1¾" RPM	1¾" BHP	2" RPM	2" BHP	2¼" RPM	2¼" BHP	2½" RPM	2½" BHP	2¾" RPM	2¾" BHP	3" RPM	3" BHP
4500	870	353	0.30	426	0.51	492	0.74	551	0.98	609	1.26	661	1.55												
5000	967	375	0.36	444	0.58	505	0.82	564	1.09	616	1.37	668	1.68	716	2.00										
5500	1064	398	0.42	463	0.67	521	0.93	576	1.20	628	1.50	675	1.80	723	2.14	767	2.49	809	2.85						
6000	1161	423	0.50	483	0.76	539	1.04	590	1.33	640	1.64	687	1.96	730	2.29	774	2.65	816	3.03	856	3.42	893	3.82		
6500	1257	448	0.59	505	0.87	558	1.16	607	1.47	653	1.79	700	2.13	743	2.48	783	2.83	823	3.22	863	3.62	900	4.04	936	4.46
7000	1354	474	0.69	527	0.99	577	1.30	625	1.62	670	1.96	712	2.30	755	2.67	795	3.05	833	3.43	870	3.83	907	4.26	943	4.70
7500	1451	500	0.81	550	1.12	598	1.45	644	1.79	687	2.14	728	2.50	768	2.88	808	3.27	845	3.67	881	4.08	915	4.50	950	4.95
8000	1547	527	0.94	574	1.26	620	1.61	663	1.97	705	2.34	745	2.72	783	3.10	820	3.51	858	3.93	893	4.36	927	4.79	960	5.23
8500	1644	554	1.09	599	1.42	642	1.79	683	2.16	724	2.55	762	2.95	800	3.35	835	3.76	871	4.19	906	4.64	940	5.09	972	5.55
9000	1741	581	1.25	624	1.60	665	1.98	705	2.37	744	2.78	781	3.19	817	3.62	852	4.05	886	4.48	919	4.94	952	5.41	985	5.88
9500	1838	609	1.43	649	1.79	689	2.19	727	2.60	764	3.02	801	3.46	836	3.89	869	4.34	902	4.80	934	5.26	965	5.73	997	6.23
10000	1934	636	1.63	675	2.01	713	2.41	750	2.84	786	3.28	820	3.74	855	4.19	887	4.66	919	5.13	951	5.61	981	6.09	1010	6.59
10500	2031	664	1.84	702	2.24	738	2.66	773	3.10	808	3.56	841	4.03	874	4.51	906	4.99	937	5.48	968	5.98	997	6.48	1026	6.99
11000	2128	692	2.08	728	2.49	763	2.92	797	3.38	830	3.86	863	4.35	894	4.84	926	5.34	956	5.85	985	6.36	1014	6.88	1043	7.41
11500	2224	720	2.34	755	2.76	788	3.21	821	3.69	853	4.18	885	4.68	915	5.19	945	5.71	975	6.24	1004	6.77	1032	7.31	1060	7.85
12000	2321	749	2.62	782	3.06	814	3.52	846	4.01	877	4.52	907	5.04	937	5.57	966	6.11	995	6.65	1023	7.20	1051	7.76	1077	8.32
12500	2418	777	2.92	809	3.38	840	3.86	870	4.35	901	4.88	930	5.41	959	5.96	987	6.52	1015	7.08	1043	7.65	1070	8.22	1096	8.80
13000	2515	806	3.25	837	3.72	867	4.21	896	4.73	925	5.26	953	5.81	981	6.38	1009	6.95	1036	7.54	1063	8.12	1089	8.71	1116	9.31
13500	2611	835	3.61	864	4.09	893	4.59	922	5.13	950	5.67	977	6.24	1004	6.82	1031	7.41	1058	8.01	1083	8.62	1109	9.23	1135	9.84
14000	2708	864	3.99	892	4.49	920	5.00	948	5.55	974	6.11	1002	6.69	1028	7.29	1054	7.89	1080	8.51	1105	9.14	1129	9.77	1154	10.40
14500	2805	893	4.40	920	4.91	947	5.44	974	6.00	1000	6.57	1026	7.17	1052	7.78	1077	8.40	1102	9.04	1127	9.68	1151	10.33	1174	10.98
15000	2901	922	4.84	948	5.36	974	5.90	1000	6.48	1026	7.07	1051	7.67	1076	8.30	1100	8.94	1124	9.59	1149	10.25	1173	10.91	1196	11.59
15500	2998	951	5.31	976	5.83	1001	6.40	1027	6.98	1052	7.59	1076	8.21	1100	8.85	1124	9.51	1148	10.17	1171	10.84	1195	11.53	1217	12.22
16000	3095	981	5.81	1004	6.34	1029	6.92	1053	7.52	1078	8.14	1101	8.77	1125	9.43	1149	10.10	1171	10.78	1194	11.47	1217	12.17	1239	12.88
16500	3191	1010	6.34	1032	6.88	1057	7.48	1080	8.08	1104	8.72	1127	9.37	1150	10.03	1173	10.72	1195	11.42	1217	12.13	1239	12.84	1262	13.57

(Column headings are grouped under STATIC PRESSURE IN INCHES W.G.)

Note: This is an alternative to the fan curve method (e.g., Figure 5-3) of showing the fan characteristics at various pressures and flows. Here the flow (in cubic feet per minute) is given in the first column; the second column gives the outlet velocity in feet per minute. The column headings across the top are pressure, expressed in inches of static pressure (i.e., inches of water column). For each pressure, the speed and shaft power required are shown for each flow. For example, if 15,000 cfm are required in a system that imposes 2.5 inches of static pressure on the fan, a fan speed of about 1,149 rpm is required and 10.3 hp will be required at the fan. Intermediate values are determined by interpolation. This table is for the same fan as in Figure 5-3.

Source: Greenheck Fan Corporation 1986

The cubic relation between speed and power requirement in certain fan and pump loads can affect the efficiencies gained by efficient motors, which have lower slip, often with up to 1% higher operating speed than their standard-efficiency counterparts. A 1% increase in motor speed can increase power-draw by 3%, negating much of the benefit gained from switching to an efficient motor. To ensure that such gains are not lost, pump impellers can be trimmed and fan sheave and belt systems modified to slow the fan slightly. When sheaves and belts are altered for this reason, the efficiency of the transmission system can also be improved by substituting cogged V-belts or synchronous belts for conventional V-belts.

The curves also depict efficiency (generally only for pumps in curves of constant efficiency in percent) and the shaft horsepower required to operate at any given combination of flow and pressure (shown in the figures as dashed lines sloping downward from left to right). The efficiency of a pump or fan is the ratio (usually expressed in percent) of the theoretical power required to the actual power needed and can range from well below 50% to above 90%. The information for fans is often provided not as a graph (as in Figure 5-3) but as a table containing the same information. For example, Table 5-1 lists flow rates in the first column; thus, the rows of data represent constant flow. Pressures are listed across the top, so the columns of data represent constant pressure. The intersection of flow and pressure pinpoints the necessary speed and power required.

A fundamental concept for analyzing fluid-flow applications is the so-called operating point—the combination of pressure and flow at which a given system and fan (or pump) operate. It is determined by plotting the system curve and the fan (or pump) curve on the same graph of pressure versus flow. The operating point is simply the intersection of the two curves; it represents the equilibrium flow point where the pressure drop through the system equals the pressure added to the fluid. For example, in Figure 5-3, assume the curve labeled "do not select to the left of this system curve" is the applicable system curve. If the flow requirement is 10,000 cfm, then the pressure required will be about 7.7 inches of water, the fan speed must be about 1,490 rpm, and the shaft power required is about 18 hp.

Once a desired operating point is determined, the system designer must choose a fan or pump to meet this condition. This choice is important in determining the energy and power requirements of the system; unfortunately, choosing the best equipment for the application is difficult. This is true even with pumps, where the efficiency is explicitly stated once the pump and operating point are known. Given an operating point, a designer wishing to get the

Figure 5-5

Typical Pump Selection Chart, Showing One Type of Pump of One Manufacturer at One Speed

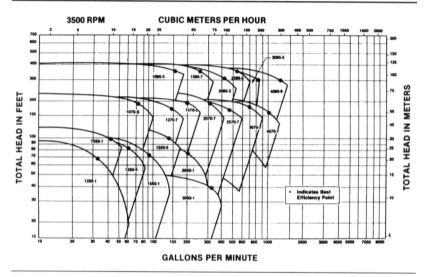

Note: Once the operating point is known (from the system design), the user can determine which pump is best suited to the application. For example, if the system demands 200 gpm at 175 feet, the best pump from this manufacturer at this speed is probably number 1570-5. The efficiency of an adjacent pump may be checked at the same operating point.

Source: Paco Pumps 1985

most efficient pump from a specific manufacturer would use an equipment selection chart such as the one in Figure 5-5.

The specific pump curve is then inspected (see Figure 5-4) to establish the efficiency at the operating point. These last two steps are repeated for all pumps available from all manufacturers until the best pump is found. For fans, where the efficiency is not directly available, the process is analogous except that the designer would search for the lowest shaft power requirement for any given operating point. This difficult and time-consuming process is rarely performed exhaustively, even in new systems. In existing systems that are modified (and thus have a different system curve), the fan or pump is seldom analyzed in an effort to reoptimize the system. The result is that many systems suffer from unnecessarily large energy consumption due to pumps and fans operating far from their points of maximum efficiency.

More recently, software has become available to aid in the selection of pumps and fans. While some pump and fan manufacturers provide software about their own products in their electronic catalogs, commercial software packages often detail equipment from multiple manufacturers; therefore, these commercial options provide the consumer with a more complete picture of the available pump and fan technology. The Annotated Bibliography lists examples of several software packages.

Another common cause of energy waste is the oversizing of the driven equipment, the motor, or both. As discussed in Chapter 3, systems are usually designed to have excess capacity. Pumps or fans are often oversized because exact determination of the system curve is difficult. In addition, the system may change over time. For example, pumps will require more capacity as scale and corrosion build up inside pipes, and fans may require additional capacity as the leakage and dirt buildup in the ductwork increases. To prevent excessive flows in oversized systems, some means of flow control are often necessary.

System Control and Optimization Techniques

In order to minimize the energy and peak power requirements of a fluid-flow system, a designer must "get the big picture" of why the flow is needed in the first place. For example, if a flow of chilled water is needed to cool a building, the following options could reduce the required flow: reducing the cooling load; increasing the size of the cooling coils; reducing the temperature of the chilled water; increasing the air flow through the cooling coils; or some combination of all these measures. Of course, some of these options involve tradeoffs between pumping energy and energy required to chill water or move air, and most of them involve trading first cost and operating cost, so interdependent sets of optimization variables are at work. For further discussion of the economics, see Appendix A.

Once the optimum flow is ascertained, the system should be designed to achieve it with the minimum possible pressure drop (since the required power is proportional to the product of flow and pressure). Small pressure drops are achieved by using large-diameter system components, smooth surfaces, and gradual bends and transitions in the elbows, tees, and so on. Again, there are tradeoffs between first cost and operating cost.

When the flow and pressure requirements are reduced to the lowest practical values (thus minimizing the theoretical power requirement), the overall efficiency of the pump or fan package (including the pump, the transmission between pump and motor, and the motor) should be scrutinized. The object here is to meet the required operating point with the minimum amount of electrical power. In industry jargon, the overall efficiencies of pumps and fans are known respectively as wire-to-water efficiency and wire-to-air efficiency.

As soon as the cost-effective minimum electrical power for meeting the design operating point is determined, a flow control scheme should be worked out for efficiently meeting system requirements at flows below design flows (since most systems operate below design load most of the time). For example, in the case of chilled water for cooling, the flow used for design is the one necessary to meet the peak cooling load on the hottest day of the year. The system could easily operate at lower flow rates for the rest of the time.

One of the best opportunities for cost-effective overall system optimization exists in new systems designed from the beginning to work with variable flow. However, many existing constant-flow systems can be converted to variable flow, with large potential savings. For example, constant-volume building ventilation systems can be converted to variable-air-volume systems. While most medium-size to large buildings are now being built with VAV systems, most buildings in the existing stock are likely to have constant-volume systems.

While optimization opportunities are specific to each case, some general opportunities for energy savings should be pursued for fan and pump systems, both new (including renovations) and retrofit. When installing fans and pumps in new applications, the following principles apply:

- Reduce restrictions in ductwork and fittings or in pipes and fittings (use larger sizes, gradual bends, and so on).

- Reduce flow in variable-volume heating or cooling systems, respectively, by increasing or decreasing the temperatures of the supply air, water, or both. This step will often involve a tradeoff in energy use between the fan and the boiler or chiller.

- Regulate pressure in variable-volume systems with a reset control, based on the actual needs of the worst-case zone of the system. For example, in a VAV system that provides cooling, the warmest zone would dictate the system supply pressure.

- Use an ASD control to vary the fan speed in VAV applications and the pump speed in variable-volume pumping applications (see the following discussion on ASD versus other control schemes).

When installing fans and pumps in retrofit applications, the following measures are recommended:

- Use an ASD control to vary the fan or pump speed either to convert constant-volume to variable-volume systems or to replace inlet vanes, discharge dampers, or throttling valves.

- Reduce pressure in variable-volume systems using worst-zone reset.

170

Variable-flow systems can be controlled directly or indirectly, i.e., the flow control components in the system may respond to feedback from a flow sensor or from another sensor reacting to other parameters, including pressure, temperature, and velocity. A few basic techniques, and variations of them, can vary flow—throttling devices (including discharge dampers on fans and throttling valves on pumps), multiple fans or pumps, and speed controls.

Throttling devices, which are essentially adjustable restrictions, operate by changing the system curve. Figure 5-6 illustrates the effect of throttle control on flow by steepening the system curve. Throttle control causes system and pump curves to intersect at a lower flow (the operating point is shifted to the left along the pump curve). The power required for throttled flow is generally somewhat less than for full flow since the flow reduction is a greater percentage than the

---

Figure 5-6

Throttling Operation in a Variable-Flow, Variable-Pressure Pumping System at 80% Flow

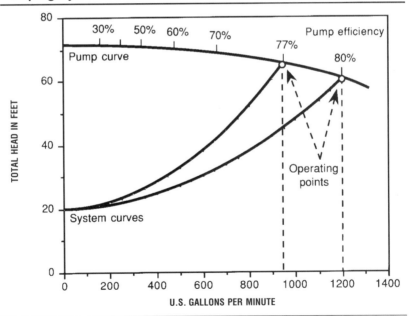

Note: Throttling makes the system curve more restrictive (i.e., steeper), which causes the intersection of the system and pump curves (i.e., the operating point) to occur at a lower flow. Throttling generally decreases the power requirement slightly, relative to full-flow operation.

Source: Adapted from Baldwin 1989

---

pressure increase percentage. The extent of this power reduction depends on the shape of the pump or fan curve. While throttling devices are relatively inexpensive and can give fairly precise flow control, using them is the least energy-efficient flow control technique. This is due to the fact that throttles dissipate flow energy provided by the fan or pump.

Another type of throttling device for fans, the inlet vane (also called the variable-inlet vane, or VIV), works by changing the fan curve. While these devices are more efficient than outlet dampers, they still control flow by dissipating energy across the control device.

Multiple fans or pumps can be used, in series or parallel (or both) to adjust the flow rate. This scheme works by changing the effective fan or pump curve. That is, the fluid-moving machine seen by the system is the combination of two or more fans or pumps. Thus, the operating point is changed, again resulting in flow control. This control scheme is not as precise or efficient as might be desired because it works in steps.

Speed control works by changing the fan or pump curve. For fans,

---

Figure 5-7

Typical Variable-Speed Pump Curve

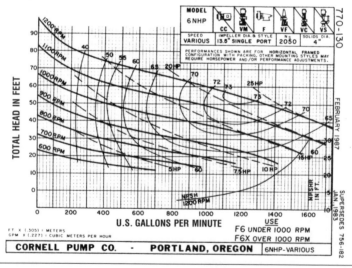

Note: The curve is similar in appearance to the constant-speed curve (e.g., Figure 5-4), but the characteristic pump curves are for different speeds rather than different impeller diameters. The axes are the same (pressure as a function of flow), as are the dashed shaft power requirements (in horsepower), the efficiency curves (labeled here with numbers from 40 to 73 as percentages), and the net positive suction head curve.

Source: Cornell Pump Company 1987

---

the usual curve or table demonstrates its effect. For pumps, special variable-speed (rather than variable impeller diameter) curves can be obtained for certain impeller sizes (see Figure 5-7 for an example). For other impeller sizes, or when variable-speed curves cannot be obtained, variable-speed curves can be calculated by several methods. These methods include interpolation between several available curves for the same pump at different fixed speeds (for example, 3,500, 1,750, and 1,150 rpm for 60 Hz motors and 2,900, 1,450, and so forth for 50 Hz), use of the affinity laws (though the assumption of constant efficiency may result in significant error), and approximating the speed curves using the impeller diameter curves. The details of constructing such custom curves are beyond the scope of this book: see Garay 1990 for further information. There are also several computer programs that can assist in this analysis (see the Annotated Bibliography for examples).

Speed control can be used with single- or multiple-fan or pump combinations, or for complete replacement for throttling control. Figure 5-8

Figure 5-8

Throttling Losses in a Variable-Flow, Variable-Pressure Pumping System at 80% Flow, Compared with Reduced-Speed Operation

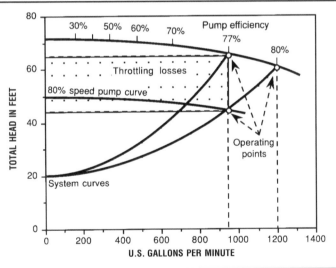

Note: This is the same system as depicted in Figure 5-6 but with the additional pump curve at reduced speed showing the alternative operating point at the same flow. The theoretical power loss due to throttling is proportional to the area with dots (since theoretical power is proportional to the product of pressure and flow).

Source: Adapted from Baldwin 1989

Figure 5-9

Typical Energy Consumption of a Centrifugal Fan System with Discharge Damper, Variable-Inlet Vane, Variable-Speed Eddy Current, and Variable-Frequency Drive Control

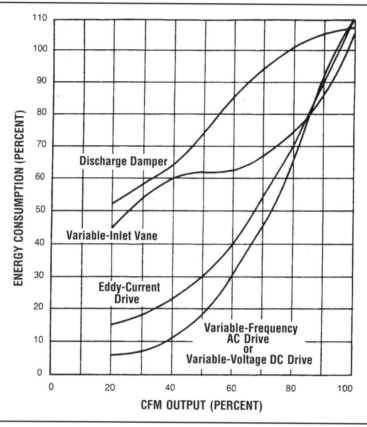

shows the comparison between a throttling control and a speed control application in terms of power requirements.

Speed control is generally the most energy-efficient flow control technique since it supplies only the amount of flow energy required (see Figure 5-9). In addition, it is the most suitable for retrofit applications where the fan or pump is already in place and varying flow with multiple-staged pumps is impractical. The equipment necessary to provide speed control, ranging from mechanical friction disks and adjustable pulleys to electronic ASDs, is covered in Chapter 4.

However, it is important to be sure that speed control is really warranted, because speed control devices add to the cost of the system.

They also introduce inefficiencies that must be offset by the energy savings realized from the speed control.

System Optimization Case Studies

Several ASD case studies were presented in Chapter 4, documenting the potential of this technology to save energy and money in fan and pump applications. Often, however, energy savings can be achieved in pump and fan systems using other techniques. Two such examples are given below.

Farm Irrigation

An irrigation system at a large farm consisted of a series of pumps that supplied water to irrigated fields. The pumps were arranged so that water for fields close to the water source was pumped directly from that source while water for distant fields passed through a series of booster pumps. The conservation strategy consisted of the following system changes:

- The sprinkler heads on the irrigation system were converted from high-pressure nozzles to low-pressure nozzles, halving the pressure needed to supply water to the fields.

- The nozzles at the end of each pipe run were equipped with small booster pumps. These pumps increased the pressure for the 5% of the water that went to the nozzles on the end of the system. As a result, the pressure in the main system could be lower.

- The decreased pressure requirements allowed the owner to trim pump impellers, thereby reducing the brake horsepower needed for pumping a given volume. (Note that significantly reducing the impeller diameter may decrease the pump efficiency to the point where it could be cost-effectively replaced with a pump better matched to the application.)

- Trimming the impellers caused many of the motors to be oversized for the application. These motors were marked and replaced, where physically possible, with smaller, more energy-efficient models.

- Traditionally, the fields were irrigated on a fixed schedule designed to provide enough water under worst-case hot-weather conditions. Instead, soil moisture sensors installed as part of the conservation package allowed the farmer to water fields only when they needed irrigation. As a result, less water was needed to maintain the quality and quantity of the crop.

- In the past, enough pumps had been run to supply more than an adequate amount of water to the system. After the system upgrading, pumps were scheduled to meet minimum flow requirements without excess water.

The results of the retrofit included annual energy savings of 34% of the base energy consumption for the system. The simple payback for the installation was 2.3 years, and there was less wear on the pumps due to the lower operating pressure and reduced use.

Pony Pump and Motor Addition

A school used a hot-water system to meet the space heating needs of the building. The system flow was designed to meet the worst-case needs of the building (the heating needs under the coldest expected condition of 5°F ambient temperature). However, the temperature during most of the year was far above the design criteria. The system had six hot water circulation pumps with a connected load of 95 hp.

The conservation retrofit entailed installing a second set of pumps (or pony pumps) in parallel with the existing set. These provided approximately 60% of the full-rated flow and 40% of the full-rated pressure of the system. Each set of pumps was equipped with a set of controls and valves that operated the main pumps if the ambient temperature fell below 35°F. If the ambient temperature reached 35°F or above, the main pumps suspended operation and pony motor pumps began to operate. All pumps were turned off if the ambient temperature was above 62°F. The total connected load for the pony motors and pumps was 20 hp.

The following data were retrieved from the building energy management system after the retrofit:

- With the original system, the hot-water circulation pumps ran for 5,800 hrs/yr. After the retrofit, the main pumps ran for 750 hrs/yr and the pony motors and pumps for 4,050 hrs/yr. Both pumps were off for the remaining period.

- Energy savings totaled 225,000 kWh, amounting to $14,700/yr, and offering a 1.6-year simple payback on the investment.

- The system provided added reliability since both sets of pumps were unlikely to break at the same time.

Compressed-Air Systems

Compressed-air systems are pervasive throughout industry and are also used by many commercial facilities. Compressed air is frequently referred to as the "third utility." Analogous to other utilities, compressed-air lines run throughout a facility, supporting air-driven hand tools, clamps, sprayers, and pneumatic motors, among other uses (Friedman et al. 1996). While it is the most expensive utility, averaging more than three times the cost of electricity, many plant-level staff remain unaware of the cost and treat it as though it were free (Aegerter 1999). As a result, the most important efficiency measure, perhaps, is awareness.

Like fan and pump systems, compressed-air systems are made up of an assemblage of components including the motor and drive, the air compressor itself, controls, air treatment equipment, piping, and often storage. Achieving peak compressed-air system performance requires addressing the performance of individual components and analyzing the supply and demand sides of the system, and assessing the interaction between the components and the system. This "systems approach" moves the focus away from components to total system performance. System opportunities have been shown to be the area of greatest efficiency opportunity. Typical compressed-air system wire-to-air efficiencies are about 10%. Experts have found that after they implement the measures identified in a thorough review of the system, either one or more compressors can be shut down or a compressor can be downsized, with energy savings frequently exceeding 40% (DOE 1998).

The DOE *Motor Challenge* program and the *Compressed Air Challenge* (CAC), both discussed in Chapter 9, developed *Improving Compressed Air System Performance: A Source Book for Industry* (DOE 1998). This reference guide provides a performance opportunity road map, factsheets to assist in system optimization, and lists of available resources. The book identifies the following types of interrelated actions:

- Establishing current conditions and operating parameters
- Determining present and future process production needs
- Gathering and analyzing operating data and developing load duty cycles
- Assessing alternative system designs and improvements
- Determining technically and economically sound options, taking into consideration all of the subsystems
- Implementing those options
- Assessing operations and energy consumption, and analyzing economics (i.e., validating performance)
- Continuing to monitor and optimize the system
- Continuing to operate and maintain the system for peak performance

As mentioned above, energy efficiency opportunities exist at both the component and system levels. These can be grouped into the following general categories:

- Leaks

- Inappropriate uses of compressed air

- System pressure level

- Air treatment

- Controls

- Distribution system

- End-use equipment

- Compressor package

- Automatic drains

- Air receiver/storage

- Heat recovery

Drawing on the DOE/CAC *Sourcebook*, we will briefly discuss some of these opportunities for energy efficiency. This section should not be considered comprehensive. For additional information, the reader is encouraged to consult the *Sourcebook* and the references appearing in it. Selected references also appear in the Annotated Bibliography at the end of this book.

Compressed-Air Leaks

Leaks can be a significant source of wasted energy, often accounting for 20–30% of compressor output. They can also contribute to other production problems. A drop in system pressure can adversely affect equipment performance and efficiency, and the increased compressor runtime (needed to satisfy the additional system demand created by the leak) will lead to increased equipment maintenance and unscheduled downtime.

Leak detection and repair is a critical element of a compressed-air system maintenance program. A good way to assess the condition of a system is with a "leak-down test." This test is performed during a plant shutdown. All equipment that uses compressed air is shut off and the system is pressurized. Then all compressors are turned off and the speed of the header pressure drop must be monitored. The pressure falling quickly would indicate significant leaks that demand immediate attention.

The most common sources of leaks are

- Couplings, hoses, tubes, and fittings

- Pressure regulators

- Open condensate traps and shut-off valves

- Pipe joints, disconnects, and thread sealants

Most large leaks are not in the occupied plant area, but in places where plant staffers go less frequently and leaks are less likely to be heard. Typical places to look include

- Leaking connections on compressor interstage piping

- Compressor, interstage, and condenser drains that are left open

- Compressor aftercooler knockout pot drains that are left open

- Overpurging of heatless air dryers (more is not better)

- Air dryer bleed-valves that are left open

- Air receiver bottom drain valves that are left open

Many of these leaks result from bleed-valves installed in place of failed water traps. The solution is to reinstall the traps (Aegerter 1999).

Typically, the worst leaks are in remote areas of the plant, such as abandoned equipment and roofs. An excellent time to check for leaks is during shutdowns, when the plant is quiet and they can be readily heard.

The best way to detect leaks is to use an ultrasonic acoustic detector, which can recognize the high-frequency sound associated with leaks. A simpler method involves applying soapy water with a brush to suspected leaks, and looking for bubbles.

Leaks occur most often at joints and fittings. Stopping them can be as simple as tightening fittings, but may require replacing a piece of equipment such as a hose, valve, or trap. In order to help avoid future leaks, only high-quality parts should be used to replace equipment, and they should be installed correctly with the appropriate thread sealant.

Unfortunately, even when leaks are identified and repaired, the job is not over. New leaks will develop over time. The best strategy for avoiding further problems is to set up a prevention program that monitors the system for new leaks and fixes them as they develop.

Inappropriate Uses of Compressed Air

Because compressed air is clean and usually readily available, many people choose it for applications without comparing it to more economical energy sources. Many operations would be better accomplished with other energy sources, including the following:

- Using small fans instead of compressed air to cool electrical cabinets

- Using a blower rather than a venturi to create a vacuum

- Using a blower rather than compressed air for low-pressure applications

- Using mechanical techniques (e.g., a brush) rather than compressed air for parts cleaning or debris removal

- Using mechanical techniques (e.g., a pusher arm) rather than compressed air for moving parts

While pneumatic tools have advantages, such as precise torque control, lower tool maintenance, and safety in flammable environments, many can be replaced with electric tools at a lower life-cycle cost. In particular, recent performance improvements and cost reductions in cordless electric tools make them attractive replacement options.

All applications should have proper regulators installed to minimize the demand on an air system. Compressed-air piping, like electrical wiring, is often left in place after the application is abandoned. However, unlike the wiring that is for the most part harmless, this unnecessary piping represents opportunity for leaks. Compressed-air flow to this equipment should be cut off as far back in the distribution system as possible.

System Pressure

A system's pressure level should be set at the lowest pressure that meets all requirements of the facility. Lowering the compressed-air header pressure by 10 psi reduces the air leak losses by approximately 5% and improves centrifugal compressor capacity by 2–5%. Maintenance staff usually requires 80 pounds per square inch gage (psig) to operate pneumatic hand tools (Aegerter 1999). Reducing system pressure will also decrease stress on system components, lessening the likelihood of future leaks.

The process of lowering system pressure must be approached with some caution, however, since this can cause the pressure at points in the system to fall below minimum requirements. However, appropriate use of storage and controls can resolve this issue. In addition, if a major piece of compressed-air-using equipment, such as a press, requires a higher pressure than the rest of the system, that piece of equipment should be evaluated for modifications that can reduce the required pressure. Replacing pneumatic presses with larger-bore products or reducing gear ratios can lower the pressure. The cost of most of the modifications is usually insignificant, compared to the large savings opportunities that can be realized from a reduction in system pressure.

Other Savings Opportunities

Removing water and oil from compressed air is necessary for most applications before the air can be used. Fouled compressed-air treatment equipment causes increased energy consumption while delivering poor-quality air that can harm equipment. For this reason, it is important to clean filters, and maintain and operate the drier, filters, aftercoolers, and separators per manufacturers' specifications.

Compressor Controls

One of the goals of system management is to shut off compressors in multi-compressor systems. To find out if one or more of the multiple plant air compressors can be shut down, it is first important to understand the theory of how different types of compressors are controlled. The following test can be used:

- Start by beginning to unload the smallest compressor while monitoring header pressure.

- If the compressor can be fully unloaded without the header pressure dropping, the compressor can be safely shut off.

Once a compressor has been shut down, it is important to turn off the purge air to that compressor's dryer to eliminate a now unnecessary compressed-air load (Aegerter 1999).

Controls match the air supply with system demand, regulating the pressure between two levels called the control range. They are one of the most important factors in determining the overall energy efficiency of a compressed-air system. Most compressed-air systems consist of several compressors delivering air to a common header. The objective is to shut off or delay starting a compressor until needed. To this end, the controls try to operate all units at full load except the one used for trimming (adjusting compressed-air supply based on the fluctuations in compressed-air demand).

In the past, control technologies were slow and imprecise. This resulted in wide control ranges and higher compressor set-points than needed to maintain the system pressure. Modern microprocessor-based technologies allow for much tighter control ranges as well as lower

Table 5-2

Characteristics and Applicability of Compressor Controls

Control Type	Types of Compressors	Applications	Characteristics
Start/Stop	Reciprocating and rotary screw	Low-duty cycle applications	Simple, consisting of a pressure switch that turns the compressor on and off. Should not be used with cycling loads that may result in multiple starts.
Load/Unload (Constant Speed)	All	May be inappropriate for rotary screws because they consume 15–35% of full-load power when fully unloaded	Motor runs continuously, but the compressor is unloaded when the pressure set-point is reached. Unload strategies are compressor manufacturer dependent.
Modulating (Throttling)	Centrifugal and rotary screw	Most appropriate for centrifugal because less efficient when used on positive displacement compressors (e.g., rotary screws); control range is, however, limited	Output is controlled by restricting the compressor inlet.
Multi-Step (Part Load)	Reciprocating and rotary screw compressors specially designed to operate in two or more partially loaded modes	Allows precise pressure control without requiring the compressor to start/stop or load/unload	Specially designed reciprocating compressors with three- or five-step control (0%, 50%, and 100% or 0%, 25%, 50%, 75%, and 100%). Some rotary screw compressors can vary compression ratios with special valves, frequently combined with inlet control to improve part-load efficiency and control accuracy.
ASD	All	Not widely applied, but a potential emerging control technology	

Source: DOE 1998

system pressure control points. Every 2 psi of pressure difference produces about a 1% change in energy consumption. In addition, a more constant pressure level can enhance production quality control.

The appropriate type of control technology needed is determined by the type of compressors being used and the demand profile. Controls for single compressors can be relatively simple, while multi-compressor systems are much more complex and sophisticated. The controls for an individual compressor in a multi-compressor system range from simple to very complex, depending on the type of compressor and the way it is integrated with the system. Table 5-2 describes the characteristics and applicability of different compressor controls.

System controls coordinate the operation of multiple individual compressors in order to meet the system requirements. Before the introduction of modern, automatic controls, systems were controlled using an approach known as cascading set-points. The set-points for each individual compressor would either add or subtract the compressor capacity to follow the system load. This approach led to wide swings in system pressure, as shown in Figure 5-10.

Figure 5-10

Impacts of Controls on System Pressure

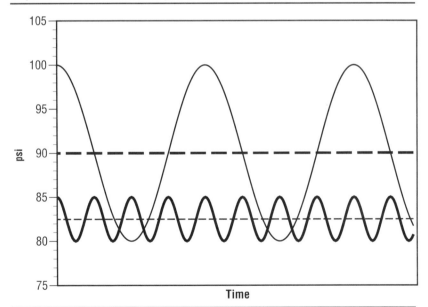

Source: DOE 1998

Modern automatic controls match system demand with compressors operated at or near their maximum-efficiency points. Two general kinds of system controllers exist: single-master (sequencing) controls and multi-master (network) controls. A single-master control meets system demand by sequencing or staging individual compressors. The control places individual compressors on- or off-line in response to demand while maintaining tight control around the system target pressure. This strategy, combined with appropriate storage, will frequently allow system pressure to be reduced.

Multi-master controls are the latest technology in compressed-air system control. They provide both individual compressor control and system regulation by means of a network of individual controllers. The controllers share information, allowing the system to respond more quickly and accurately to demand changes. One controller acts as the lead, regulating the whole operation. This strategy allows each compressor to function at a level that produces the most efficient overall operation. The result is a highly controlled system pressure that can be reduced to close to the minimum level required. Although more costly, these controls represent the most energy-efficient system available.

Storage plays a critical role in a compressed-air system, allowing it to maintain pressure while meeting surges in demand. In other words, storage serves to decouple the system demand from compressor operation. In addition, it can be located at critical pressure applications to ensure precise pressure regulation.

The pressure in the header can be maintained in a much narrower range than can the pressure of the compressor discharge to the primary receiver (the compressed-air storage tank located between the compressors and the distribution system). This shields the compressor from severe load swings. Reducing and controlling the system pressure downstream from the primary receiver can result in energy savings of more than 10%, although the compressor's discharge pressure remains unchanged.

Air Compressors

The two general categories of air compressors are called positive displacement and dynamic. In positive displacement compressors, a volume of air trapped within a mechanically reduced compression chamber causes a rise in pressure (see Figure 5-11). In dynamic compressors, the impeller imparts velocity to the airflow (see Figure 5-12), which is converted into pressure.

Within each category, there are several different types and designs within types (see Table 5-3). Each design also has a different full-load efficiency and part-load efficiency curve. For example, reciprocating compressors are about 10% more efficient than comparable screw compressors at full load. At part load, the screw's efficiency declines rapidly. When operating considerations do not dictate the

---

Figure 5-11

Schematics of Two Common Types of Positive Displacement Air Compressors: Single-Acting Reciprocating and Helical-Screw

---

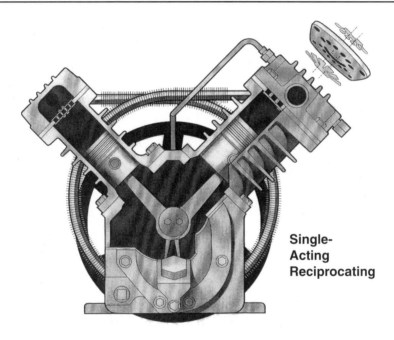

Single-Acting Reciprocating

Helical-Screw

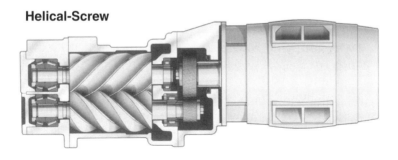

Source: Ingersoll-Rand 2000a

Figure 5-12

Schematic of a Centrifugal Air Compressor, the Most Common Type of Dynamic Compressor

---

Source: Ingersoll-Rand 2000b

---

choice, selection of a different type of compressor or a mix of compressors can reduce energy consumption (Elliott 1995).

Facilities frequently use several different types of compressors in order to take advantage of each design's unique operating characteristics. Many large manufacturing facilities will use centrifugal compressors for base load of general plant air and use positive displacement compressors to handle load swings.

Within each type of compressor, the potential also exists for purchasing a unit that is 5–20% more efficient. For example, a premium 100 hp reciprocating compressor is approximately 10% more efficient at full load than a standard unit. These more efficient units command a price premium of 10–30%. As with motors, part-load efficiencies are equally important. Screw compressors have particularly bad part-load performance. An internally compensated design can be purchased for a 10–15% premium that will significantly decrease part-load power consumption (Elliott 1995).

Air compressors are inherently inefficient devices. Wire-to-air efficiencies range from 21% down to 13%, depending on the compressor design and operation. The balance of the electricity goes to generate heat, of which 50–90% can be recovered. This heat can be used to provide space conditioning, process heat, or warm water, or preheat boiler makeup water.

Table 5-3

Types of Air Compressors

Positive Displacement	Characteristics
Reciprocating	A piston is driven through a crank and connecting rod in a cylinder. Available in sizes of 1–600 hp.
Single-Acting	Compression occurs in one direction of piston movement only. Typically smaller, air-cooled machines.
Double-Acting	Compression occurs in both directions. Typically larger, water-cooled machines. These are the most costly but most efficient compressors.
Rotary	Rotating formed shafts trap air in a cavity that decreases in volume as the shaft rotates, compressing the air. Generally larger machines.
Helical-Screw	The most common industrial compressor. Low initial cost, compact, and easy to maintain, but relatively low efficiency, particularly at part load. Available in both air- and water-cooled configurations from 3 to 600 hp.
Liquid-Ring	Less common
Scroll	Less common
Sliding-Vane	Less common
Dynamic	**Characteristics**
Centrifugal	Dominant dynamic compressor design, widely used in large industrial plant air applications. Has an impeller similar to a centrifugal fan's or pump's. Units available in the 100–20,000 hp range. Tends to be efficient at full load.
Axial	Has an impeller similar to a turbine. Generally high efficiency but restricted to very high flow capacities.

Sources: DOE 1998; Elliott 1995

Air-cooled, packaged rotary screw compressors are well suited for hot air applications such as space heating or heating process air. Since the aftercooler and lubricant cooler are generally located in enclosed cabinets with fans, only ducting and an additional fan to handle the duct loading are needed. When heating is not required, the air can be exhausted outside the building. Approximately 50,000 Btu/hr with 30–40°F of temperature rise are available for each 100 cfm of capacity.

Packaged, water-cooled reciprocating and rotary screw compressors are similarly excellent candidates for heat recovery. The hot water offers greater flexibility since the heat can be used for both water heating and space conditioning applications.

While carrying out a comprehensive compressed-air performance optimization may appear intimidating to plant staff, consultants are available to assist with the initial audit and training. Also, it is not necessary to do everything in order to realize huge savings, as the Equistar experience shows.

The Equistar Experience

Equistar Chemicals is a $6 billion producer of ethylene, propylene and polyethylene. It is headquartered in Houston, Texas, with 16 manufacturing facilities in the Midwest and along the U.S. Gulf Coast. These facilities contain large compressed air systems.

Bob Aegerter, an engineer with Equistar, has developed a practical strategy for managing air compressors at the company's plants that has had proven results. His principal strategy is to reduce compressed-air demand so that he can shut off compressors.

The first step toward achieving that goal is to eliminate unnecessary air demand by finding all leaks. First the staff looks for leaks in the compressor room containing the compressors and dryers. After the compressor and dryer leaks have been identified and repaired, the next step is to survey the rest of the plant for leaks. If a survey does not yield sufficient results, then plant staff can turn to ultrasonic leak detection. The third step is to isolate equipment and/or parts of the plant that are not in service, turn off all equipment in the area, and depressurize that portion of the system, which further reduces air demand.

Once the leaks are fixed, the staff sets the system pressure to the lowest level that meets operations and maintenance requirements. The final step in the process is for plant operators to find out if they can shut down one or more of the multiple plant air compressors. Equistar has been able to shut down at least one compressor in each of the plants where it has carried out this approach (Aegerter 1999). With air compressors being the largest single electric load, the plants have been able to reduce their compressed-air cost by more than a third by implementing this plan.

Other Motor Applications

Fans, pumps, and compressed-air systems are the largest users of motor energy and are therefore the main targets for energy-saving motor controls and system optimization measures. However, many other motor-driven loads exist in commercial and industrial applications. While this book cannot discuss in detail all end-use systems, several important additional systems are discussed to give the reader a sense of the conservation possibilities.

Most of these applications require custom engineering due to the special requirements of individual systems. The costs and savings are typically site-specific.

Centrifugal Compressors and Chillers

Centrifugal compressors and chillers can often benefit from "cube law" savings via speed control in much the same way that pumps and fans can. Wasteful throttling devices and frequent on-off cycling of the equipment can largely be avoided with precise speed control, leading to both energy savings and extended equipment life. Accurate control of centrifugal chillers is especially beneficial in buildings where space conditioning systems run regularly at partial load. The capability of motors with ASDs to operate at high speeds can eliminate the need for speed-increasing gearboxes, with corresponding savings in initial investment and in energy and maintenance costs.

Substituting AC Motors and ASDs for DC Drives

AC motors used in tandem with ASDs are supplanting DC drives for many applications in industry. As noted in Chapter 2, AC motors are favored by many industries in new applications because these motors require less maintenance and the maintenance that is required costs less: it is less expensive to rewind or replace an AC motor than to rewind or replace a DC motor. In addition to ASDs, some of the new AC motor designs discussed in Chapter 2 that allow precise speed control, such as switched reluctance, are competitive in some applications. However, a large number of applications still use DC motors.

Energy savings from converting to AC motors and ASDs accrue for several reasons:

- The greatest energy savings occur when the old DC system produces current using a motor generator set (see Chapter 4). In these applications, the overall system efficiency can be improved from approximately 55% to approximately 85% when an ASD and AC motor are installed.

- Substantial energy savings can also occur when rectifiers and DC motors are changed to ASDs because of the losses in the rectifier as well as the higher efficiency of the AC motor.

- AC motors with ASDs are currently being installed on systems where a constant mechanical tension is required, as in winders. With older DC systems, the tension was often adjusted by imposing friction on the system, thereby decreasing the overall system efficiency. Eliminating the need for such energy-wasting devices improves the system efficiency.

- DC drives were the standard for traction devices such as subway cars and ship propulsion. Such loads can be served instead by AC motors equipped with ASDs. This allows for regenerative braking, which puts power back into the system during braking. As a result, the overall energy use for the system decreases.

Examples of areas where AC drives can replace DC drives with resultant energy savings include mills and kilns (steel, paper, cement, and mining industries), traction drives (transportation), winders (paper machines and steel-rolling-mills), and machine tools and robotics.

Conveyors

The power used by a conveyor consists of the power needed to move the material on the conveyor and the power for moving the conveyor belt.

The power required to move the material is set by the rate of flow of the material. For example, a conveyor operating at its design capacity, measured in, for example, pounds per minute, uses 100 hp to move the material. When only half as much material is needed, it will require 50 hp to move the material, regardless of whether the belt is running at half load (in units, say, of pounds per foot of belt) and full speed or full load and half speed.

However, if the belt speed is slowed with an ASD, the power required to move the belt will decrease. For most short conveyors, the power needed to move the belt when it is fully loaded is less than 10% of the power needed to move the material. However, in long conveyors, such as those found at power plants and mines, the power to move the belt can be substantial.

As an example of energy savings at reduced speed, if moving the material on a fully loaded conveyor requires 150 hp, a belt needing 20 hp uses 12% of the power at full load. Running the belt at half load and full speed will use 95 hp (75 hp for the material and 20 hp for the belt). Running the belt at half speed and full load will use 85 hp (75 hp for the load and 10 hp for the belt), for a savings of 11%. Likewise, the savings at 25% of full load will be 26%.

Based on these calculations, the use of ASDs on conveyors shows some potential for energy savings. However, the savings are not as dramatic as the savings that are available when ASDs are used on pumps and fans.

Summary

Pumps, fans, and compressed-air systems provide a fertile area for energy conservation because of both the characteristics of fluid flow and the relative importance of these machines as a percentage of the total motor population. These applications are good candidates for system optimization techniques that reduce either the pressure or flow requirements of a system. For pumps and fans, installing speed control technology to match the pump or fan output to the system requirements offers the potential for large energy savings.

Other motor systems presenting conservation opportunities include cooling and refrigeration compressors, systems currently using DC drives for speed control, and conveyors.

The next two chapters estimate the aggregate conservation potential based on the technologies described in Chapters 2 through 5.

Chapter 6

A Profile of the Motor
Population and Its Use

More than half of all electricity produced in the United States (as well as most other nations) and about two-thirds of all U.S. industrial electricity flows through motors. (As discussed in Chapter 2, motors are in general very efficient, actually consuming only a small portion of their energy input; the rest is used by the driven equipment.) By way of comparison, U.S. primary energy input to motors exceeds fuel use in all highway vehicles. One would think that a class of devices responsible for such a sizable portion of world energy flows would be thoroughly profiled. Remarkably, however, less is known about the stock, performance, and usage of motors than about any other major category of energy-using equipment.

Data Sources and Limitations

Until recently, most of the literature published on the U.S. motor stock contained little or no reliable field data, including how many motors, of what types and sizes, are used for which purposes, for how many hours per year. Fortunately, in recent years a number of major studies have been conducted that provide substantial information on the U.S. motor population and its use. Particularly notable is a 1998 field study commissioned by the U.S. Department of Energy that collected detailed data on motors used in a representative sample of 254 manufacturing plants (XENERGY 1998). This study contains a wealth of primary data that provides a good foundation for analytic work on motor system energy use. Other major new data sources include a set of studies by Easton Consultants characterizing the U.S. motor, fan, pump, air compressor, and variable-speed drive markets (Easton Consultants

1996, 2000), a study by Lawrence Berkeley National Laboratory on small electric motors (LBNL 1996), a study by Arthur D. Little, Inc. on motors in the residential and commercial sectors (ADL 1999), and a study by XENERGY on the market for compressed-air efficiency services (XENERGY 2000a). These studies are primarily based on interviews with equipment manufacturers and retailers, although several of these studies include some field data. Finally, several analyses have been prepared that take the available data and use them to characterize motor system energy use in the residential, commercial, industrial, and utility sectors.

While data sources have significantly improved since the first edition of this book was prepared in 1991, there are still several data gaps. In particular, while the 1998 XENERGY study provides a wealth of data on the use of integral-horsepower motors in manufacturing plants, these account for only a minority of motor energy use. The data do not, however, extend to how the motor-driven equipment is used, which (as discussed in the previous chapter) is where the largest savings potential can be found. A similarly detailed source is not available on other industrial motor uses. Likewise, while the LBNL study on small motors is a good start, its estimates are highly approximate and based as much on educated guesses as on solid data. Furthermore, even the XENERGY motors study reports only the results of a limited number of analyses; further analysis of this database could produce substantial useful information, such as better data on the distribution of the motor stock by frame type, motor speed, and duty and load factors.

Given the enormity of motor energy use and the significant amount of missing data, governments, utilities, and other groups should place a high research priority on additional broadly based, well-designed field surveys of the motor stock. Policies and programs could be more easily directed to the largest and most lucrative savings opportunities if those opportunities could be better identified and quantified. Of course, there are limits to the value and applicability of general field data. Unlike electricity savings in commercial lighting, where there are millions of virtually identical fixtures operating under similar conditions, drivepower savings are far more application-specific. With these caveats in mind, let us now examine the data.

Motor Sales, Population, Distribution, and Use by Size and Type

In 1997, more than 200 million electric motors were sold in the United States (U.S. Census Bureau 1998). This is approximately a 30%

increase in the quantity of motors sold relative to 1989, with most of the sales increase in fractional-horsepower motors (motors under 1 hp in size) (see Table 6-1).

Based on these sales, and a variety of other assumptions discussed later in this chapter, we estimate that more than two billion electric

Table 6-1

U.S. Motor Sales: 1989 and 1997

Motor Type	Sales	
	1989	1997
Fractional-horsepower		
AC noncommutated		
Single-phase		
Shaded pole	68,265,514	90,446,231
Permanent split-capacitor	21,764,620	26,458,590
Other	27,685,392	14,165,846
Polyphase	828,852	1,747,024
AC commutated	(not available)	49,322,471
DC or universal	28,928,811	11,545,904
TOTAL	**147,473,189**	**193,686,066**
Integral-horsepower		
AC noncommutated		
Single-phase	1,798,343	1,919,713
Polyphase, induction		
1–5	986,679	1,232,412
5.1–20 hp	493,016	516,172
21–50 hp	145,826	174,532
51–100 hp	58,769	63,571
101–200 hp	37,833	35,561
201–500 hp	8,642	18,392
Over 500 hp	2,605	6,065
Synchronous	243	6,078
DC (includes motors and generators)	2,652,849	3,567,631
TOTAL	**6,184,805**	**7,540,127**
GRAND TOTAL (fractional & integral)	**153,657,994**	**201,226,193**

Note: Fractional-horsepower motor data exclude heretic and other rotating equipment, as well as motors used in automobile accessories, aircraft, spacecraft, toys, and clock-type timing. Universal motors can operate on either AC or DC power.

Sources: U.S. Census Bureau 1989, 1998b

Figure 6-1

Population and Energy Use of Electric Motors in the United States by Size Class: 1997

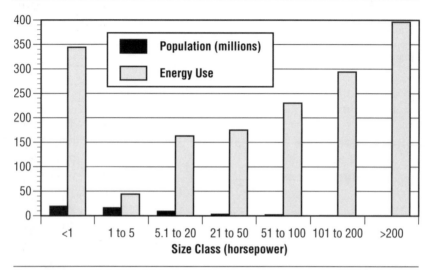

Source: Based on data in Table 6-2. Includes all fractional-horsepower motors and integral-horse-power polyphase induction motors (does not include other types of integral-horsepower motors since data on population by horsepower is not available)

motors were operating in the United States in 1997. One might well ask how an energy efficiency program can practically be applied to two billion motors. Luckily, the task is not quite as daunting as it may seem; relatively few large motors offer the lion's share of the motor savings potential. Less than 5% of all motors (those of 1 hp or larger) account for at least 80% of all motor energy input. The relationship of motor sales, population, and energy use by size class is depicted in Figure 6-1 and Table 6-2.

Less than 5% of drivepower energy used by integral-horse-power motors in the United States in 1997 went to DC and universal motors, while nearly all of the rest was used by AC induction motors. Thus, a program focusing on integral-horsepower induction motors has the potential to capture most of the available savings. Although the greatest savings potential in a systems sense lies with the largest motors, the opportunities with fractional-horsepower motors should not be ignored. Most energy use by fractional-horse-power motors occurs in major home appliances—notably refrigerators, freezers, furnaces, and air conditioners. Fractional-horsepower

Table 6-2
U.S. Motor Estimated Population and Energy Use: 1997

Motor Type	Average Life	Motor Stock (thousands)	Average hp	Average Load	Average Annual Op. Hours	Average Efficiency	Annual Elec. Use (TWh)
Fractional-horsepower	13	1,917,151	0.25	50%	1,250	65.0%	344
Integral-horsepower							
AC noncommutated							
Single-phase	13	23,378	1.5	50%	2,745	70.0%	51
Polyphase, induction							
1–5 hp	17	16,774	2.07	50%	2,745	80.2%	44
5.1–20 hp	19	9,367	11.9	50%	3,391	86.8%	163
21–50 hp	22	3,208	32.5	50%	4,067	90.3%	175
51–100 hp	28	1,646	65	50%	5,329	92.5%	230
101–200 hp	28	1,059	135	50%	5,200	94.3%	294
201–500 hp	29	251	300	50%	6,132	95.0%	181
Over 500 hp	29	76	1,200	50%	7,311	96.0%	215
Synchronous	29	76	200	50%	5,500	94.7%	82
DC (includes motors and generators)	13	34,487	2	50%	1,500	80.0%	72
TOTAL		**90,322**					**1,507**
GRAND TOTAL							
(fractional & integral)		**2,007,473**					**1,851**
As percentage of U.S. electric energy end-use							59%

Source: ACEEE calculations using 1989 sales from Table 6-1 to calculate motor stock; other data estimated by reviewing data from many different studies.

units receive an estimated 19% of all motor input energy and are considerably less efficient than larger models. Moreover, as discussed in Chapter 2, the difference in efficiency between standard- and high-efficiency motors is largest in the small size ranges. Therefore, the opportunity for improvement is great (LBNL 1996). As Table 7-1 shows, replacing standard-efficiency motors with EEMs would save more energy in fractional-horsepower units than in any other size class.

Additional information on the motor stock and motor energy use in the residential and commercial sectors is summarized in Figures 6-2 and 6-3. In the residential sector, the vast majority of motors are fractional-horsepower, but integral-horsepower motors still account for nearly half

Figure 6-2

Population and Energy Use of Electric Motors in the Residential Sector in the United States by Size Class: 1995

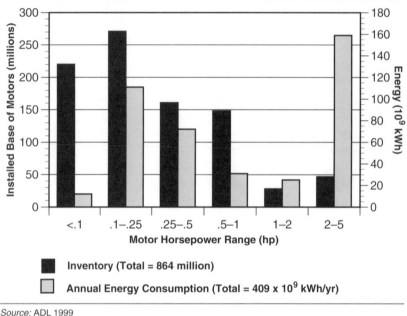

Inventory (Total = 864 million)

Annual Energy Consumption (Total = 409 x 10^9 kWh/yr)

Source: ADL 1999

of energy use. In the commercial sector, fractional-horsepower motors are also the most numerous, but integral-horsepower motors account for roughly two-thirds of energy use, with motors in the 10–25 hp range alone accounting for more than one-quarter of energy use.

Distribution by Design, Speed Frame, and Enclosure

As discussed in Chapter 2, high-efficiency, integral-horsepower motors are readily available in T-frame, NEMA Design B motors with open drip-proof and totally enclosed fan-cooled enclosures, for the most common speeds—1,200, 1,800, and 3,600 rpm. Price and performance vary among these categories; for this reason, as discussed in Chapter 9, several utilities offer different rebates for different speeds and enclosure types.

The availability of high-efficiency motors outside these standard categories varies. Some manufacturers make high-efficiency, explosion-

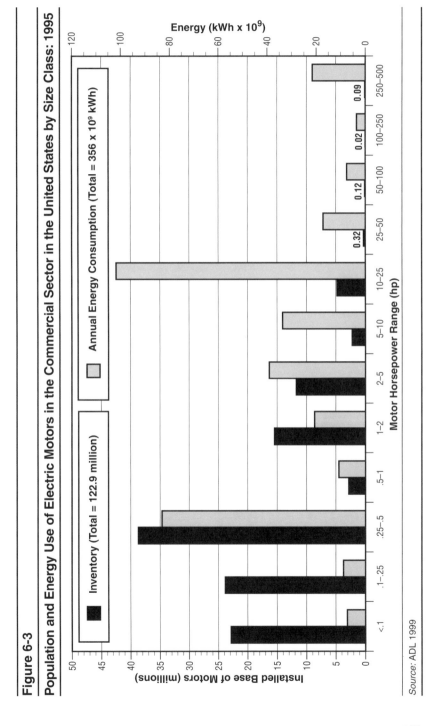

Figure 6-3

Population and Energy Use of Electric Motors in the Commercial Sector in the United States by Size Class: 1995

Source: ADL 1999

proof, C-face, and 900 rpm T-frame motors. High-efficiency replacements for some non-T-frame motors can be obtained with special modifications to the mounting hardware or other special-order modifications. In other instances, high-efficiency models are simply not available.

Analyzing the aggregate potential for efficient motors and designing programs to promote them require an estimate of the composition of the existing motor stock by design, speed, frame, and enclosure type. In terms of national data, the only available source is the XENERGY (1998) field study that estimates energy use for design B, other induction, DC, and other motor types. These results are summarized in Table 6-3 and show that Design B motors account for more than 60% of total motor system energy use. As for data on speed, frame, and enclosure type, we were unable to locate any national breakdown by these criteria, but several regional motor surveys offer some insight. These include an unpublished field survey of 2,641 motors in commercial and industrial facilities in Rhode Island (NEPSCo 1989); a proprietary survey of 405 motors in six Pacific Northwest factories (Seton, Johnson, and Odell 1983); an inventory of 106 facilities containing 28,000 motors, representative of the industrial sector in Wisconsin Electric Company's service territory (XENERGY 1989); and measured data on 97 HVAC motors in buildings at Stanford University (Wilke and Ikuenobe 1987). Two of the surveys have data on enclosure types. These data, summarized in

Table 6-3

Distribution of Motor Energy Use by Motor Type and Design

Motor Horsepower	Annual Consumption in Gigawatt-Hour per Year				
	Design B	Other Induction	DC	Other	TOTAL
1–5	21,244	150	1,321	5,092	27,807
6–20	47,561	795	3,406	8,360	60,122
21–50	55,565	1,064	4,658	11,824	73,111
51–100	56,293	2,519	6,646	7,466	72,924
101–200	64,853	2,189	9,621	6,436	83,099
201–500	49,691	3,927	14,035	23,166	90,819
501–1,000	38,896	450	12,862	25,030	77,238
1,001+	28,113	3,097	3,328	55,769	90,307
TOTAL	**362,216**	**14,191**	**55,877**	**143,143**	**575,427**

Source: XENERGY 1998

Table 6-4
Distribution of Enclosure Types

Survey	Open Drip-Proof #	Open Drip-Proof % of Total	Totally Enclosed #	Totally Enclosed % of Total	Explosion-Proof #	Explosion-Proof % of Total
Seton, Johnson, & Odell 1983	226	55.9	122	30.2	39	9.7
NEPSCo 1989	814	72.9	236	21.1	66	5.9

Table 6-5
Distribution of Motor Speeds

	Nominal rpm											
	3,600		1,800		1,200		900		720		Other	
Survey	#	%	#	%	#	%	#	%	#	%	#	%
Seton, Johnson, & Odell 1983	33	8	243	60	64	16	14	3	4	1	47	12
NEPSCo 1989	129	7	1,475	79	216	12	37	2	0	0	2	0.1
Wilke & Ikuenobe 1987	1	1	91	96	2	2	0	0	0	0	1	1

Table 6-6
Distribution of Frame Types

Survey	T-Frame #	T-Frame %	U-Frame #	U-Frame %	Other #	Other %
Seton, Johnson, & Odell 1983	223	55	160	40	22	5
NEPSCo 1989	1,359	87	159	10	44	3
Wilke & Ikuenobe 1987	52	56	41	44	0	0

Table 6-4, show that 56–73% of installed motors are models with open drip-proof housings, 21–30% with totally enclosed housings, and under 10% with explosion-proof housings.

Three of the surveys sort motors according to speed. As shown in Table 6-5, these data indicate that 60–96% of the motors operate at 1,800 rpm, 2–16% operate at 1,200 rpm, 1–8% operate at 3,600 rpm, and 1–12% run at various other speeds. Table 6-6 presents the distribution of frame types. T-frames clearly predominate, with a range of 55–87% in these three data sets. Nearly all remaining motors are U-frames.

Tables 6-3 to 6-6 suggest that most operating motors are types for which high-efficiency versions are readily available.

Duty Factor

One important consideration in determining the savings potential for a given motor is its duty factor, or how many hours per year the motor operates. Motors with high duty factors have the greatest savings potential. Fortunately, the XENERGY field study discussed above provides the first good national estimates of duty factors in manufacturing facilities. Data were collected for the manufacturing sector as a whole, as well as separately for the chemicals, paper, metals, petroleum, and food sectors. These data are summarized in Table 6-7.

However, the XENERGY study does not include commercial buildings or motors used in nonmanufacturing industrial applications such as

Table 6-7

Average Annual Motor Operating Hours in U.S. Manufacturing Plants

Motor Horsepower	28 Chem.	26 Paper	33 Metals	29 Petrol.	20 Food	Other	All SICs
1–5	4,082	3,997	4,377	1,582	3,829	2,283	2,745
6–20	4,910	4,634	4,140	1,944	3,949	3,043	3,391
21–50	4,873	5,481	4,854	3,025	4,927	3,530	4,067
51–100	5,853	6,741	6,698	3,763	5,524	4,732	5,329
101–200	5,868	6,669	7,362	4,170	5,055	4,174	5,200
201–500	6,474	6,975	7,114	5,611	3,711	5,396	6,132
501–1,000	7,495	7,255	7,750	5,934	5,260	8,157	7,186
1,000+	7,693	8,294	7,198	6,859	6,240	2,601	7,436
All Sizes	6,333	6,748	6,465	4,332	4,584	3,678	5,083

Source: XENERGY 1998

Table 6-8

Motor Duty Factors by Size Class

Horsepower	Average Hours of Operation/Year		
	Wisconsin	Rhode Island	Stanford University
0–5	2,979	6,195	—
5.1–20	4,132	6,161	5,156
20.1–50	4,132	504	6,190
>50	5,539	4,670	4,500

Sources: NEPSCo 1989; Wilke and Ikuenobe 1987; XENERGY 1989

mining and oil and gas extraction. A limited amount of data including these sectors is available from some of the smaller regional studies mentioned above, including one of commercial and industrial motors in Rhode Island (NEPSCo 1989), a Wisconsin survey on industrial facilities (XENERGY 1989), and data on 97 HVAC motors at Stanford University (Wilke and Ikuenobe 1987). These data are summarized in Table 6-8.

The Wisconsin data are broadly similar to the national data reported above. The Rhode Island and Stanford University data show significantly higher operating hours than the national data for motors less than 50 hp, indicating a need for further data collection on motors in nonmanufacturing applications.

Even more important than average values by size class, however, is an accurate profile of the distribution of duty factors within a size class—what percentage of motors of a given size are used 1,000 hrs/yr, 2,000 hrs/yr, and so on. This information is useful to utility program planners, among others, for determining eligibility criteria in motor rebate programs. New England Electric, for example, offers generous rebates for high-efficiency replacements for 100 hp or smaller motors operating at least 1,250 hrs/yr and for motors larger than 125 hp operating 2,500 hrs/yr. This rebate program is discussed further in Chapter 9.

Fortunately, XENERGY has recently completed such an analysis based on the data collected from its national field study. These data are summarized in Table 6-9 and show that the largest proportion of motors have low duty factors, but this is primarily because small

Table 6-9

Percentage of Motors Used in Manufacturing, Sorted by Motor Size and Duty Factor

Motor hp	≤1,000	1,001–2,000	2,001–3,000	3,001–4,000	4,001–5,000	5,001–6,000	6,001–7,000	7,001–8,000	Over 8,000
1–5 hp	20.4	11.2	5.1	4.0	5.0	1.9	4.1	1.7	5.2
6–20 hp	6.8	4.5	3.5	2.1	2.3	1.1	1.8	0.9	3.5
21–50 hp	1.6	1.5	1.3	0.8	0.9	0.5	0.9	0.6	1.1
51–100 hp	0.2	0.3	0.5	0.2	0.4	0.2	0.3	0.4	0.6
101–200 hp	0.1	0.1	0.3	0.1	0.2	0.1	0.2	0.1	0.5
201–500 hp	0.0	0.0	0.1	0.0	0.1	0.0	0.1	0.0	0.3
500+ hp	0.0	0.0	0.0	0.0	0.0	0.0	0.1	0.0	0.2
TOTAL	**29.0**	**17.7**	**10.8**	**7.2**	**8.9**	**3.9**	**7.5**	**3.7**	**11.3**

Source: Rosenberg 1999

Table 6-10

Percentage of Manufacturing Motor Energy Use, Sorted by Motor Size and Duty Factor

Motor hp	Duty Factor								
	≤1,000	1,001–2,000	2,001–3,000	3,001–4,000	4,001–5,000	5,001–6,000	6,001–7,000	7,001–8,000	Over 8,000
1–5 hp	0.2	0.4	0.3	0.3	0.5	0.3	0.6	0.3	1.0
6–20 hp	0.5	0.9	1.2	1.0	1.4	0.8	1.6	1.0	4.0
21–50 hp	0.3	0.8	1.2	1.0	1.5	1.1	2.1	1.6	3.5
51–100 hp	0.1	0.3	0.8	0.5	1.2	0.7	1.6	2.1	4.0
101–200 hp	0.1	0.3	1.2	0.5	1.6	0.9	2.4	1.2	6.1
201–500 hp	0.0	0.2	0.8	0.4	1.4	0.3	2.2	0.8	7.7
500+ hp	0.0	0.1	0.3	0.1	1.7	0.7	4.6	2.8	20.6
TOTAL	**1.2**	**3.2**	**5.8**	**3.8**	**9.3**	**4.8**	**15.1**	**9.8**	**46.9**

Source: ACEEE analysis based on data in Table 6-9 and average motor horsepower assumptions by size class from Table 6-2

motors tend to have low duty factors and small motors account for the largest number of motors. Viewed another way, we can estimate the proportion of motor energy use as a function of motor horsepower and duty factor. These data are estimated in Table 6-10. Since large motors account for the largest portion of motor energy use, and large motors tend to have high duty factors, the vast majority of motor energy use is by motors with high duty factors. Conversely, motors with duty factors below 2,000 hrs/yr account for less than 5% of motor energy use.

Load Factor and Motor Sizing

In addition to duty factor, the size of a motor relative to its typical load helps determine its operating efficiency and the potential for savings. Load factor is important because, as a motor operates further and further below 50% of its rated load, its efficiency drops. This decrease occurs more rapidly for standard-efficiency than high-efficiency units, as was illustrated in Figure 3-8. As discussed in Chapter 3, downsizing should be considered for any motors running below 40% loading.

The recent XENERGY national field study provides a good assessment of motor sizing practices in the manufacturing sector of the United States. In this study, instantaneous measurements were taken on a representative sample of hundreds of motors in the field

Table 6-11

Motor Loading by Horsepower

Part Load (% of full load)	Application				
	Air Compressor	Fan	Other	Pump	All
<40%	15%	30%	55%	39%	44%
40–120%	84%	69%	43%	56%	53%
>120%	1%	1%	2%	4%	2%

Source: XENERGY 1998

Table 6-12

Motor Loading by Application

Part Load (% of full load)	Horsepower Category					
	1–5 hp	6–20 hp	21–50 hp	51–100 hp	101–200 hp	200+ hp
<40%	42%	48%	39%	45%	24%	40%
40–120%	54%	51%	60%	54%	75%	58%
>120%	4%	1%	1%	0%	1%	2%

Source: XENERGY 1998

and the results were weighted to represent the motor population at large. Overall, 44% of motors were found to be operating at less than 40% of full load, indicating a very high proportion of oversized motors. With only a few exceptions, oversizing was not found to vary much as a function of motor size or application. The XENERGY data are summarized in Table 6-11 (as a function of motor horsepower) and in Table 6-12 (as a function of motor application).

These data should be interpreted carefully, however, since they are one-time instantaneous measurements taken on systems where load may vary substantially on an hourly or seasonal basis. In addition, the readings were subject to some measurement error.

For comparison, a survey of 22,300 motors in small and medium-size industries in Brazil provides the breakdown of motor loading shown in Table 6-13. This survey suggests somewhat better sizing practices than those reflected in the U.S. data just cited, as the survey indicates that only 22% of the motors operate below 50% load (Geller 1990).

As wasteful as motor oversizing can be in terms of energy and dollars, it is sometimes justified in terms of higher reliability, reduced downtime, and the flexibility to accommodate expanded process needs. Individual cases should be evaluated with an eye to the various tradeoffs involved in the sizing decision, including the power factor problems with lightly loaded motors, as discussed in Chapter 3. In general, it is probably not cost-effective from an energy savings point of view to downsize motors running above 40% of their rated load.

Motor Life

The life of a motor depends on how it is used and maintained and also depends somewhat on its construction. There is enormous variation in motor life, with some motors used in corrosive environments lasting less than a year, and other well-maintained motors lasting more than 40 years. While there is enormous variation, for purposes of analysis, average values are useful. A 1995 survey of motor repair shops (Schueler, Leistner, and Douglass 1994) collected information on motor life at time of failure. These data are summarized in Table 6-14.

Table 6-13

Motor Load Factors in Brazilian Industry

% Load	% of Motors
0–10	4.4
10–20	0.5
20–30	2.2
30–40	6.0
40–50	10.1
50–60	11.5
60–70	13.7
70–80	4.9
80–90	12.6
90–100	24.5

Source: Geller 1990

Table 6-14

Average Electric Motor Life

Horsepower Range	Average Life (years)	Life Range (years)
1–5	17.1	13–19
5.1–20	19.4	16–20
21–50	21.8	18–26
51–125	28.5	24–33
>125	29.3	25–38

Source: Schueler, Leistner, and Douglass 1994

Saturation of High-Efficiency Motors

In recent years, extensive data have been published on sales and saturation of high-efficiency motors in the United States. Since 1993, the U.S. Census Bureau has collected data on shipments of motors that meet the EPAct efficiency standards. Generally these data show a gradual rise in the sale of efficient motors, which accounted for 16% of 1–200 hp motor sales on a unit basis and 32% on a value basis in 1997 (see Figure 6-4). As was discussed in Chapter 2, in November

Figure 6-4

Efficient Motors as Percentage of Shipments: 1994 to 1997

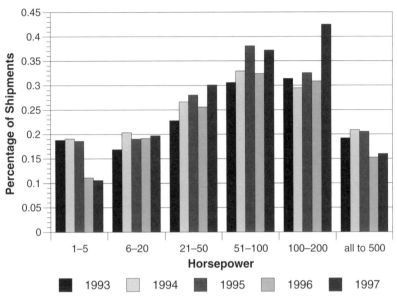

Sources: U.S. Census Bureau 1994, 1995, 1996a, 1997, 1998b

1997, the EPAct standards became mandatory for many types of motors. Therefore, the market share of these motors is likely to climb steeply.

The saturation of efficient motors in the motor stock will tend to lag behind the saturation among current sales because only a small portion of motors are replaced each year. Thus it takes several decades for the stock to turn over. This is borne out by the results of the XENERGY motor systems inventory, which found that efficient motors (generally defined in terms of EPAct efficiency levels) made up 9.1% of the integral motor stock in U.S. manufacturing plants in 1997. The saturation rate of efficient motors was higher among 51–100 hp (17% saturation) and 101–200 hp motors (25% saturation) than among other motor sizes. Efficient motors were particularly common in the chemical and paper sectors. These results are summarized in Table 6-15.

As was discussed in Chapter 2, and will be further analyzed in Chapter 9, manufacturers and utilities are increasing their promotion of so-called premium-efficiency motors that exceed EPAct motor standards by 1–3%. Unfortunately, no national data exist on the sales of

207

Table 6-15
Saturation of Efficient Motors in U.S. Manufacturing Plants: 1997

Motor Horsepower	28 Chem. (%)	26 Paper (%)	33 Metals (%)	29 Petrol. (%)	20 Food (%)	Other (%)	All SICs (%)	All SICs
1–5	7.8	12.0	2.1	4.7	6.6	7.5	7.2	523,735
6–20	15.1	17.3	2.0	8.3	12.4	10.3	10.4	340,735
21–50	21.6	21.9	4.3	11.8	13.2	7.8	11.3	127,111
51–100	27.9	27.2	8.4	2.1	28.3	15.3	17.1	62,234
101–200	32.7	17.0	0.1	7.0	7.4	37.6	25.5	56,247
201–500	19.8	4.2	0.0	19.6	5.2	48.4	17.7	15,346
501–1,000	1.3	0.0	0.0	0.0	0.0	9.5	1.3	352
1,000+	4.5	0.0	9.6	0.6	0.0	0.0	3.9	425
All Sizes	14.4	15.3	2.5	7.5	8.8	8.9	9.1	1,125,887

Note: EPAct applies only to motors up to 200 hp. For motors above 200 hp, XENERGY developed its own definition of efficient.

Source: XENERGY 1998

Table 6-16
Approximate Market Share of CEE Premium Motors

	New England/New Jersey 1998	New York 1998	Northwest 1997
1–5	21%	11%	
6–20	40%	22%	
21–50	59%	32%	
51–100	5%	37%	
101–200	63%	63%	
Total	29%	16%	12%

Note: "Premium" defined as meeting CEE efficiency levels (see Table 2-9).

Sources: Easton Consultants & XENERGY 1999a; PEA 1998

these motors, although some regional data are available from the Northeast (Easton Consultants and XENERGY 1999a) and Northwest (PEA 1998) United States. These data are summarized in Table 6-16. They show very substantial market shares for PEMs in New England and New Jersey for 1998 (29% of 1–200 hp motor sales), and somewhat lower market shares in New York and the Northwest (12–16% of 1–200 hp motor sales). However, the Northwest figures are from 1997, before the EPAct motor standards went into effect, and thus are not directly comparable to the 1998 figures. The authors of the Northeast study estimate that the national market share for PEMs is roughly similar to that found in New York; market shares are higher in New England and New Jersey due to utility promotions, above-average interest in energy savings, and high electricity prices. Past experience indicates that market shares for efficient motors also tend to be above the national average in the Northwest due to utility promotions and a high saturation of motor-intensive industries. Thus, if the saturation of PEMs is above the national average in the Northeast and Northwest, it stands to reason that the market share in other regions of the country is likely to be lower than the figures shown for New York. However, these estimates are very approximate; additional data collection is needed outside of the Northeast before more definite conclusions can be drawn.

Use of Adjustable-Speed Drives

In addition to energy-efficient motors, an energy-saving measure that has been widely promoted in recent years is the adjustable-speed drive. The XENERGY (1998) field study provides a useful snapshot of the use of these drives in manufacturing. Overall, XENERGY found that 9% of motor systems have ASDs, with their use highest among smaller motors (this finding is surprising but may reflect the wider availability of integrated motor/ASD packages in small horsepower sizes). The XENERGY data are summarized in Table 6-17.

ASD saturation has been increasing steadily in recent years, due to average annual sales growth of 24% during much of the 1990s (see Figure 6-5).

A 2000 study by Easton Consultants further estimates the potential and current markets for

Table 6-17

Saturation of Motor Systems with AC Adjustable-Speed Drives

By Horsepower	Motor Systems with ASDs	
	Number	% of Total
1–5	767,807	11
6–20	254,862	8
21–50	46,126	4
51–100	13,536	4
101–200	11,661	5
201–500	1,873	2
501–1,000	820	3
1,000+	644	6
TOTAL	**1,097,328**	**9**
By Application		
Pump	77,510	3
Fan	101,204	7
Compressed air	11,044	2
Other	907,570	11
TOTAL	**1,097,328**	**9**

Source: XENERGY 1998

Figure 6-5

U.S. ASD Sales: 1993–1998

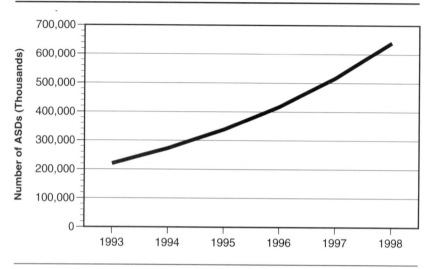

Source: Easton Consultants 2000

Table 6-18

U.S. Motors over 50 hp under ASD Control by Industry Segment

Industry Type/Segment	Total Number of Motors >50 hp	Percentage with Variable Load	Typical Variable-Load Applications	Percentage of Variable-Load Applications with ASDs
Industry Light/medium industry	176,000	27%	Machine control; HVAC; product handling	49%
Process industry	457,000	22	Pumps & fans; rolls & sheet handling; conveyors	33
Other heavy industry	441,000	35	Pumps & fans; reduction equipment; compressors	23
Water & wastewater	88,000	24	Pumps; aeration equipment	48
Commercial HVAC	436,000	20	Pumps and fans	6
Agriculture	85,000	10		25
TOTAL	**1,682,000**	**24**		**35**

Source: Easton Consultants 2000

ASDs by industry segment (e.g., light/medium industry, process industry, etc.). Overall, they find that about one-quarter of loads are variable loads, and of these, just over a third already have ASDs. These results are summarized in Table 6-18.

Motor Energy Input by Sector, Industry, and End-Use

While analyzing motor usage by type and size class is informative, most efficiency programs are designed and implemented by sector, industry, or specific end-use. It is thus useful to break out motor consumption by such categories. Probably the best analysis of motor system energy use is one by E Source that draws on data from the Electric Power Research Institute, Edison Electric Institute, and a host of other sources. This analysis is summarized in Figures 6-6 and 6-7 and concludes that motors account for 52.5% of U.S. electricity use, including approximately 70% of electricity used in the industrial sector, 38% in the residential sector, and 37% in the commercial sector. Of U.S. motor system energy use, the industrial sector accounts for approximately 44%, the residential sector for approximately 23%, the commercial sector for approximately 20%, and the utility sector for approximately 13%. The largest residential uses are appliances (particularly refrigeration) and space cooling. Space cooling, ventilation, and refrigeration dominate commercial-sector motor use. Pumps, blowers, fans, and compressors account for more than half of industrial and utility motor use.

Figure 6-8 shows where, within the large and diverse industrial sector, motor use is concentrated. This figure is based on data from the 1994 Manufacturing Energy Consumption Survey (EIA 1997). We also include the electric utility industry in this figure, based on data from Figure 6-6. Electricity use by motor systems is particularly large in the chemical, primary metal, electric utility, and paper industries and moderately large in the food, petroleum/coal, and mining industries. The large use of motor systems shows why utilities offer a prime test bed for efficiency improvements, as the EPRI case studies in Chapters 4 and 9 describe.

The XENERGY field study also provides useful information on the distribution of motor system energy use by application, including breakdowns for major manufacturing sectors and by motor size. These are summarized in Table 6-19 and Figures 6-9 and 6-10. The distributions show that material handling is a particularly important application in metal fabrication plants, pumps are particularly important in paper plants, and refrigeration and material processing

Figure 6-6

Drivepower's Share of U.S. Electricity Use, by End-Use

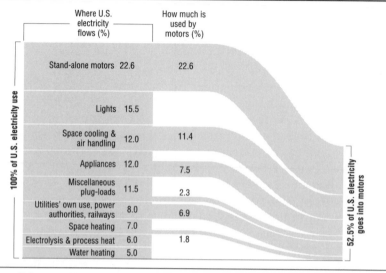

Source: E Source 1999

Figure 6-7

Drivepower's Share of U.S. Electricity Use, by Sector and Task

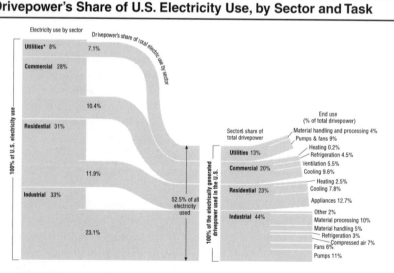

* Includes utilities' own use in generating stations, public authorities (water supply, waste water treatment, irrigation), and railways.

Source: E Source 1999

Figure 6-8

1994 Electricity Use by Industry, and Motors' Share of That Use

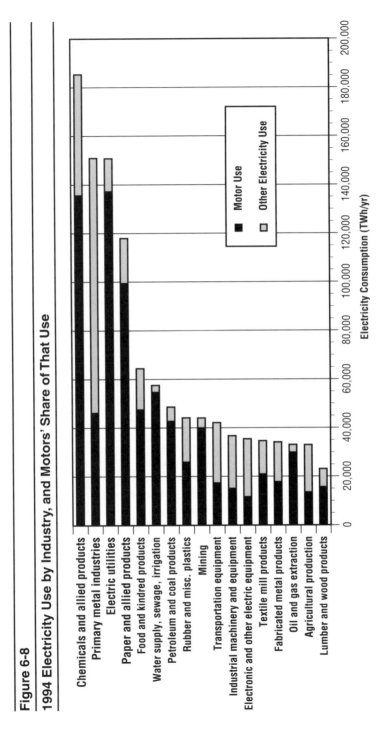

Sources: EIA 1997 for motor energy use; U.S. Census Bureau 1996b for electricity use; utility motor energy use based on E Source 1999; utility electric energy use from EIA 1995; and agricultural, mining, oil/gas extraction, and water supply data from XENERGY 1998

Table 6-19

Distribution of Motor System Energy Use by Application

Application	28 Chem. (%)	26 Paper (%)	33 Metals (%)	29 Petrol. (%)	20 Food (%)	Other (%)	All SICs (%)
Pump	26.0	31.4	8.7	59.0	16.4	19.0	24.8
Fan	11.9	19.8	15.3	9.5	7.5	13.5	13.7
Compressed Air	27.7	4.6	14.3	15.3	7.7	15.0	15.8
Refrigeration	7.7	5.0	0.1	0.7	29.4	7.1	6.7
SUBTOTAL **(Fluid Systems)**	**73.3**	**60.7**	**38.4**	**84.4**	**61.1**	**54.6**	**61.0**
Material Handling	1.4	7.4	47.1	2.6	6.1	10.3	12.2
Material Process	23.6	21.3	12.6	11.1	26.1	31.0	22.5
Other	1.8	10.6	1.9	1.9	6.7	4.1	4.3
SUBTOTAL **(Other Systems)**	**26.7**	**39.3**	**61.6**	**15.6**	**38.9**	**45.4**	**39.0**
ALL APPLICATIONS	100.0%	100.0%	100.0%	100.0%	100.0%	100.0%	100.0%

Source: XENERGY 1998

systems are especially important in food plants. Analogous information for the residential and commercial sectors was estimated by Arthur D. Little, Inc. (ADL 1999), and is summarized in Figures 6-11 and 6-12. Air conditioners and heat pumps account for nearly half of motor energy use in the residential sector. HVAC systems account for more than 70% of commercial sector motor energy use; refrigeration systems account for an additional 20%.

Data compiled by Easton Consultants (1996) on the fan, pump, and air compressor industries allow us to further characterize these applications. Centrifugal fans dominate the fan and blower market in the manufacturing sector, accounting for more than 90% of fan and blower energy consumption. An estimated 45,000–50,000 centrifugal fans are sold to the manufacturing sector each year. Axial fans (which are used primarily in commercial HVAC applications or for large-flow, clean air applications in the mining, utility, and transportation sectors) make up the remainder of the market.

Several types of pumps are used in manufacturing processes for moving different fluids under different pressures. For example, rotary pumps often handle higher-pressure fluid applications and reciprocating diaphragm pumps are used where preventing contamination is critical. Overall, centrifugal pumps dominate the process pump market in terms of annual unit and capacity sales,

Figure 6-9

Distribution of Motor Energy Use by Application: All Manufacturing and Selected Industries

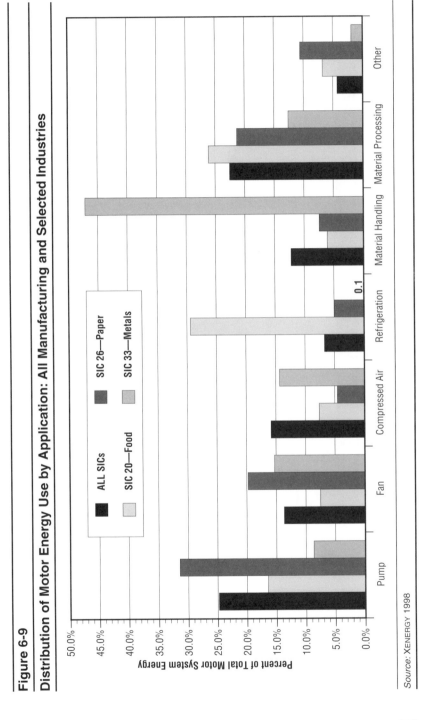

Source: XENERGY 1998

215

Figure 6-10

Distribution of Motor Energy Use by Application and Horsepower Class

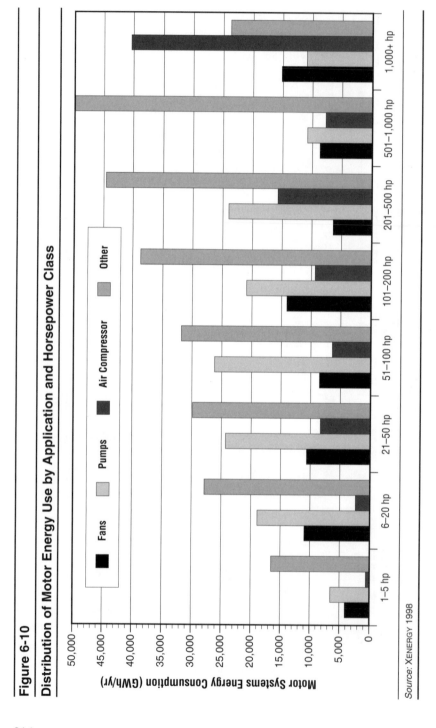

Source: XENERGY 1998

Figure 6-11

Distribution of Motor Energy Use in the Residential Sector by Application

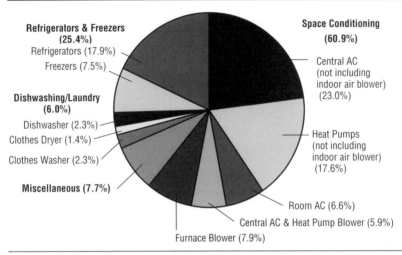

Refrigerators & Freezers (25.4%)
Refrigerators (17.9%)
Freezers (7.5%)

Dishwashing/Laundry (6.0%)
Dishwasher (2.3%)
Clothes Dryer (1.4%)
Clothes Washer (2.3%)

Miscellaneous (7.7%)

Space Conditioning (60.9%)

Central AC (not including indoor air blower) (23.0%)

Heat Pumps (not including indoor air blower) (17.6%)

Room AC (6.6%)

Central AC & Heat Pump Blower (5.9%)

Furnace Blower (7.9%)

Source: ADL 1999

Figure 6-12

Distribution of Motor Energy Use in the Commercial Sector by Application

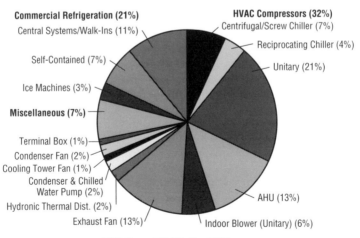

Commercial Refrigeration (21%)
Central Systems/Walk-Ins (11%)
Self-Contained (7%)
Ice Machines (3%)

Miscellaneous (7%)
Terminal Box (1%)
Condenser Fan (2%)
Cooling Tower Fan (1%)
Condenser & Chilled Water Pump (2%)
Hydronic Thermal Dist. (2%)
Exhaust Fan (13%)

HVAC Compressors (32%)
Centrifugal/Screw Chiller (7%)
Reciprocating Chiller (4%)
Unitary (21%)

AHU (13%)

Indoor Blower (Unitary) (6%)

HVAC Thermal Distribution (40%)

Source: ADL 1999

Figure 6-13

Distribution of Large Compressors by Horsepower Range

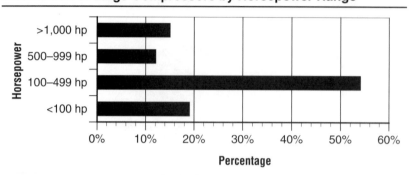

Source: Xenergy 2000a

Table 6-20

Presence of Compressed-Air (CA) Systems by Industry Group

SIC	Industry Group	No CA System	Small CA System	Full CA System
20	Food and Kindred Products	13	34	53
22	Textile Mill Products	0	24	76
23	Apparel and Other Textile Products	9	0	91
24	Lumber and Wood Products	8	10	82
25	Furniture and Fixtures	0	0	100
26	Paper and Allied Products	19	21	61
27	Printing and Publishing	95	0	5
28	Chemicals and Allied Products	8	7	84
29	Petroleum and Coal Product	0	16	84
30	Rubber and Miscellaneous Plastics Products	0	12	88
32	Stone, Clay, and Glass Products	0	0	100
33	Primary Metal Industries	5	13	82
34	Fabricated Metal Products	8	61	31
35	Industrial Machinery and Equipment	1	0	99
36	Electronic and Other Electric Equipment	9	0	91
37	Transportation Equipment	0	8	92
38	Instruments and Related Products	11	0	89
20–39	Overall Manufacturing	18	12	70

Note: CA systems account for less than 5% of motor system energy.

Source: Xenergy 2000a

installed units and capacity, and total electricity consumption (Easton Consultants 1996).

Slightly more than one million air compressors are sold to the U.S. market each year, of which 98% are 5 hp or smaller. These small compressors are sold largely to the commercial and residential markets.

Figure 6-14

Revenue Share for Single-Phase Fractional-Horsepower Motors by Motor Type: 1992

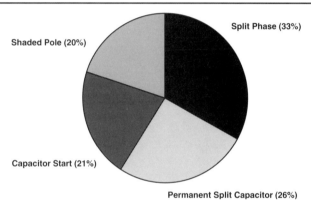

Source: LBNL 1996

Figure 6-15

Revenue Share for Single-Phase Fractional-Horsepower Motors by Application: 1992

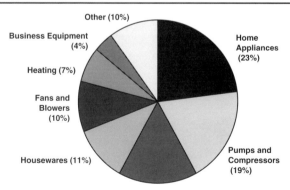

Source: LBNL 1996

219

Due to the very low hours of use, however, they account for only about 12% of annual electricity consumption by new units. Air compressors at or above 25 hp, while accounting for less than 1% of annual unit sales, make up an estimated 80% of annual electricity consumption by new units. Of these larger compressors, approximately 17,000 are sold annually, of which about 72% are single-stage, flooded rotary screw types (Easton Consultants 1996). A 1999 field study of 222 of these larger compressors used in manufacturing plants found that nearly one-half were in the 100–499 hp range (see Figure 6-13). This study also examined the proportion of manufacturing enterprises that have compressed-air systems and estimated that 70% of manufacturing plants have plant-wide compressed-air systems, 12% have small compressed air systems, and 18% have no compressed-air systems. Data on this distribution by industry type are summarized in Table 6-20.

Additional Data on Small Motors

Most of the preceding discussion focuses on integral-horsepower motors. However, nearly 20% of motor energy use is accounted for by fractional-horsepower motors. Research by Lawrence Berkeley National Laboratory (LBNL 1996) has begun to characterize the small-motor market. For example, as shown in Table 6-1, the vast majority of fractional-horsepower motors are single-phase. Of single-phase, fractional-horsepower motors, four types dominate—split phase, permanent split capacitor, capacitor start, and shaded pole. A description of the characteristics of these motor types can be found in Chapter 2. Roughly, these types have equivalent market shares, as shown in Figure 6-14. These motors are used for a wide variety of applications, of which home appliances, pumps/compressors, and refrigeration/air conditioning are the largest (see Figure 6-15).

These data are a useful start, but given the substantial amount of energy used by small motors and the many diverse types, applications, and sizes of these motors, much more data are needed to characterize small-motor sales and applications and to identify those market segments with the greatest energy use and opportunity for energy savings.

Chapter 7

Estimating the National Savings Potential in Motor Systems

Most analyses of the potential for improved efficiency in motor systems have focused on only two technologies: high-efficiency motors and adjustable-speed drives. These are indeed very important technologies, but (as the preceding chapters make clear), as part of an overall systems approach for reducing motor energy use, many other measures deserve attention as well. These include

- Optimal motor sizing

- New and improved types of motors

- Better motor repair practices

- Improved controls in addition to ASDs

- More efficient motor-driven equipment including fans, pumps, and compressors

- Motor system design improvements including better selection of equipment and system components matched to the application

- Reduced waste of compressed air and other fluids moved by motor systems

- Electrical tune-ups, including phase balancing, power-factor and voltage correction, and reduction of in-plant distribution losses

- Mechanical design improvements, including optimal selection and sizing of gears, chains, belts, and bearings

- Better maintenance and monitoring practices

Only by paying close attention to all of the above parameters and to the synergism among them can designers and users of motor systems truly optimize efficiency and reliability. We now examine the major efficiency measures in turn and evaluate their savings potential. We start with the motor itself and then proceed to motor speed control, opportunities for system improvements, and electrical and mechanical measures.

Induction Motor Improvements

As discussed in Chapter 6, induction motors draw an estimated 53–59% of all U.S. motor input, or approximately 1851 terawatt-hours (TWh)/year. How much of this energy could be saved if readily available EPAct-efficiency and premium-efficiency induction motors were substituted for standard units? Estimated savings are in the range of 100–140 TWh/yr, equivalent to the output of 18–25 1,000 MW power stations (assuming typical 70% capacity factor and 8% grid loss). This estimate includes savings from the use of three-phase EPAct and CEE premium-efficiency integral-horsepower motors as well as savings from more efficient single-phase motors and fractional-horsepower three-phase motors.

The savings estimate for integral-horsepower motors is based on the average efficiency of standard-efficiency, EPAct-efficiency, and CEE premium-efficiency motors as determined in a detailed 1998 inventory of motors used in manufacturing in the United States (XEN-ERGY 1998). Savings are further adjusted to account for the current use of EPAct and premium-efficiency motors in the current motor stock (as discussed in Chapter 6) and to limit the savings to cost-effective applications (estimated based on the proportion of motor energy use consumed by motors operating 2,000 hours or more annually per data in Table 6-9). Overall, we estimate a savings potential of 49 TWh/yr from increased use of EPAct motors and 30 TWh/yr from increased use of CEE premium-efficiency motors. Savings are higher for EPAct motors because of the greater efficiency spread between standard and EPAct motors than between EPAct and premium motors.

For single-phase and fractional-horsepower motors, the energy savings potential was approximated by Lawrence Berkeley National Laboratory at 5–24% (LBNL 1996). We apply this range of savings to the estimated energy use of fractional-horsepower motors from Table 6-2, resulting in a savings rate of 20–47 TWh/yr. The uncertainties associated with this estimate are substantial, but overall the potential savings from higher-efficiency fractional-horsepower motors are similar to the savings from integral-horsepower CEE premium motors. Savings are large due to the many fractional-horsepower and single-phase motors in use and the relatively high difference in efficiency between less efficient and more efficient motors. Capturing these potential energy savings will not be easy given the many diverse motor types and applications represented. However, most small motors are purchased by original equipment manufacturers. This provides an opportunity to influence thousands of motor purchases by convincing a few decision-makers. The potential size of the savings in this area suggests the value of gathering improved data on sales and operating characteristics of these motors; the LBNL study was a good start but is only the first step in efforts to better characterize this important sector of the motor market.

The Cost of Savings from Induction Motor Improvements

As discussed in Chapter 6, the typical motor lasts 13–29 years before it must be replaced, depending on motor size. When models under 10 hp fail, they are typically replaced; larger motors are often repaired. With the passage of EPAct, what used to be called high-efficiency motors are now standard practice for new motors; as old motors are gradually replaced, the stock efficiency will gradually increase. In addition, as Figures 2-26, 2-27, and 2-29 through 2-36 show, in new installations (including replacement of failed units), CEE premium-efficiency motors in typical operating regimes (4,000 hrs/yr) have simple paybacks of 6 years or less relative to EPAct motors; paybacks are 4 years or less for ODP motors of 5 hp and above and 3 years or less for TEFC motors of 2–25 hp. Relative to repairing an existing standard-efficiency motor, CEE premium-efficiency motors have simple paybacks of 2 years or less in sizes up to about 200 hp for ODP enclosures and 40 hp for TEFC designs. Assuming a typical motor life of 20 years and a real discount rate of 6%, a 2- to 6-year payback corresponds to a cost of saved energy of 1–3 cents/kWh. Savings from motors with low duty factors will cost more than this aggregate analysis suggests, and motors with very high duty factors will yield cheaper savings.

Similarly, the LBNL analysis on single-phase and fractional-horsepower motors estimates a typical simple payback of 2.5–10 years, varying with motor type, size, and application. Assuming an average motor life of 13 years, these simple payback periods correspond to a cost of saved energy of 1–6 cents/kWh.

These analyses ignore the sizable savings from correcting for past rewind damage and oversizing.

Correcting for Past Rewind Damage

As discussed in Chapter 2, standard rewind techniques that bake motor cores in high-temperature (>600°F) ovens can damage the motor core and cause other problems that reduce the motor's efficiency. In

Chapter 2 we estimate that this damage equals 1% of total motor input or about 13 TWh/yr. At a retail electric rate of 6 cents/kWh, these losses are valued at about $1 billion/yr and equal the output of about 2 gigawatts (GW) of installed capacity. Full replacement of the existing motor stock with new, higher-efficiency motors would thus not only save energy equal to the difference in nameplate efficiencies but also eliminate the losses associated with operating a "wounded" motor fleet, essentially for free. Such rewind damage could be avoided in the future by adopting improved motor repair practices, as discussed in Chapter 2.

Optimal Motor Sizing

When replacing the current motor stock, care should be taken to correct for past oversizing with the caveat that oversizing in some cases is justified by reliability and flexibility benefits (Baldwin 1989). As noted in Chapter 6, a recent field study estimates that 44% of integral-horsepower motors are running at or below 40% of rated load (XENERGY 1998). Assuming, as noted in Chapter 3, that this oversizing is causing, on average, a 5% efficiency penalty in the affected motors, the savings from correcting oversizing would be on the order of 38 TWh. For the various reasons discussed in Chapter 3, not all motors running lightly loaded should be downsized, so we reduce the 38 TWh by one-third to account for installations where downsizing is seemingly attractive but in fact not practical or desirable. The remaining savings potential is thus 26 TWh.

Since the oversized motors will be replaced eventually, and smaller, high-efficiency replacements will often cost less than standard-efficiency replacements of the original size, the marginal cost of the new, downsized motors may be negative in some cases.

Motor Speed Control

As explained in Chapters 4 and 5, the range of potential cost-effective applications for improved motor speed control is vast. In many processes, loads vary over time and the use of electronic adjustable-speed drives and other speed control techniques (e.g., multispeed motors, pony motors, or staged motors) offers large opportunities for energy savings. Opportunities for speed control are particularly great with centrifugal machinery and other applications where losses drop as speed is reduced. ASDs are becoming increasingly popular for flow control in new industrial and large commercial HVAC installations. The nationwide field study discussed extensively in Chapter 6 (XENERGY 1998) found that 9% of motor systems in the manufacturing sector (representing 4% of manufacturing motor energy use) are controlled by ASDs. This study also estimated that motor systems representing 29% of motor system energy are candidates for ASDs, of which 14% of total motor system energy is the "prime market" for ASDs (defined as fluctuating centrifugal loads above 20 hp and over 2,000 operating hrs/yr). These figures indicate that significant savings are being achieved from systems that are already installed, but also indicate that a clear majority of potential applications for ASDs are still available.

The XENERGY figures discussed above help identify opportunities for motor speed control in manufacturing but do not consider other industrial applications or commercial applications. To estimate potential applications for motor speed control (including ASDs as well as other speed control techniques discussed in Chapter 4), we go back to the data in Figure 6-9, which indicates that approximately 47% of motor energy is used for fans, pumps, conveyors, and HVAC equipment—the prime applications for motor speed control (we deal with compressed-air systems separately below). There is great uncertainty about the proportion of these applications that are good candidates for motor speed control, but based on discussions with several industry experts, we estimate that between one-third and two-thirds of these applications are appropriate. From these figures we subtract the 4% of motor energy use that is currently controlled by ASDs (discussed above), leading us to estimate that motor speed controls can be applied to 12–27% of motor energy use. This range is very similar to the 14–29% range that XENERGY estimates for manufacturing.

Energy savings vary widely from application to application, but, as shown in Chapter 4, savings of 15–50% are common. While savings of 40–50% are not unusual, the average across many applications is likely to be lower. To be conservative, we estimate that typical savings are likely to be in the range of 15–30%.

Applying these figures to our estimates of motor systems energy use leads to an estimated savings potential of 30–139 TWh annually. This is a broad range since there are a large number of uncertainties regarding average savings and appropriate applications. But even at the lower end of this range, savings are similar to those shown above for premium-efficiency motors. At the upper end of this range, savings from motor speed control are higher than those of any other measure examined in this chapter.

The cost-effectiveness of ASD applications varies widely. Numerous case studies cite simple paybacks of 2 years or less, corresponding to costs of saved energy of $0.01–0.025/kWh saved (Greenberg et al. 1988; PEAC 1987). These studies are probably skewed toward the most attractive installations, however, with many other installations having paybacks of 2 to 5 years. A 5-year payback corresponds (assuming

$0.07/kWh electricity and 10-year equipment life) to a cost of saved energy of almost $0.05/kWh. We adopt this value as the upper bound of the cost of savings from ASD installations.

Other Controls

In addition to ASDs, there are many other types of controls that can be used to reduce motor system energy use. For example, as discussed in Chapter 4, microprocessor-based controls that monitor multiple system variables can often reduce system energy use substantially. Likewise, power-factor controllers can offer up to 10% overall energy savings in certain applications driven by small (15 hp or less) motors that run most of the time at zero or near-zero loading. As discussed in Chapter 5, there are several specialized control systems specifically designed for compressed-air systems. Also, energy management systems and other controls can be used in commercial HVAC and refrigeration systems to reduce energy use by 10% or more in typical operations. However, many of these controls are application-specific, so rather than discuss opportunities for other controls across all uses, we proceed now to discuss specific savings opportunities in some of the major types of motor applications—fans and pumps, compressed air, space cooling, and refrigeration.

Opportunities for System Improvements
Fan and Pump Systems

As explained in Chapter 6, approximately 31% of motor system energy is used to power fans and pumps. In addition to motor speed control, there are multiple opportunities to reduce energy use in fan and pump systems, including installing more efficient fans and pumps and improving system design. Regarding equipment, energy can be saved through improvements to equipment components (e.g., cast versus stamped fan impellers) and the use of more efficient equipment types (e.g., airfoil fans tend to be more efficient than backward-inclined fans, and radial tip fans tend to be more efficient than standard radial fans). Modest prospects for improving equipment components are illustrated by the difference in efficiency between standard and high-efficiency models offered by several manufacturers, as well as dissimilarities in efficiencies among manufacturers. An example of such an efficiency spread is illustrated in Figure 7-1, which shows the spread in efficiencies among different models of a given fan type. Opportunities to trade up to more efficient equipment types are illustrated in Figure 7-2. However, they are limited by the fact that the

Figure 7-1

Efficiencies of Four Manufacturers' Most Efficient Backward-Inclined Fans

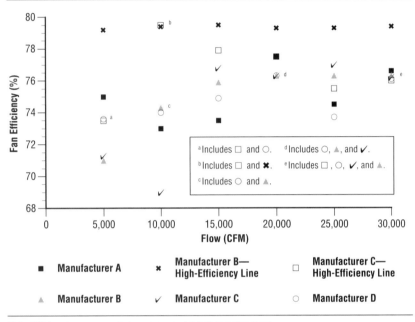

ªIncludes □ and ○.   ᵈIncludes ○, ▲, and ✓.
ᵇIncludes □ and ✖.   ᵉIncludes □, ○, ✓, and ▲.
ᶜIncludes ○ and ▲.

| ■ Manufacturer A | ✖ Manufacturer B—High-Efficiency Line | □ Manufacturer C—High-Efficiency Line |
| ▲ Manufacturer B | ✓ Manufacturer C | ○ Manufacturer D |

Note: Efficiencies measured at a pressure of 6 inches water gage.

Source: Easton Consultants 1996

Figure 7-2

Typical Fan Efficiency Levels

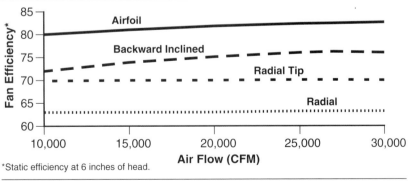

*Static efficiency at 6 inches of head.

Source: Easton Consultants 1996

more efficient equipment types are not appropriate for all applications; for example, airfoil and backward-inclined fans can only be used in relatively clean environments.

Easton Consultants (1996) estimates that, overall, more efficient fans and pumps have the ability to decrease energy use by 5–15% in typical applications. We reduce this range to 5–10% because their estimate includes savings from more efficient motors, which we have already examined separately. Approximations of the cost of these improvements are not readily available. Easton (1996) notes that backward-inclined fans cost only 5–10% more than standard fans. A 15 hp fan operating 4,000 hrs/yr implies a simple payback of roughly 1 year (at $0.06/kWh) and a cost of saved energy of less than 1 cent/kWh (assuming a 6% real discount rate and a 15-year fan life).

Easton (1996) also examines potential savings from improved fan and pump system design, and estimates that, overall, improved design can reduce energy use 10–20% in typical applications. This is on top of the savings that can be achieved from improved equipment and motor speed control. These estimates include application of the optimization steps discussed in Chapter 5. Costs for these improvements vary widely, but Hanson (1997) notes an average simple payback of 1.2 years for several projects. Assuming a 6% real discount rate and a 10-year time span before the system is reconfigured, the average cost of saved energy is 1 cent/kWh.

Compressed-Air Systems

As discussed in Chapter 5, compressed-air systems offer considerable potential for energy savings. These systems are complex, and the savings opportunities result from a wide range of measures that we have grouped into two categories. Reductions in air that is wasted due to inadequate maintenance, leaks, and inappropriate uses can save 20–30% of compressed-air energy. A further 15–25% savings is available from various system design improvements and more efficient equipment selection (Easton 1996). Many of these measures focus on reducing the discharge pressure of the air compressors, and include installation of lower-pressure-drop equipment and advanced compressor controls. Both of these classes of measures have a wide range of costs, but Suozzo and Nadel (1998) estimate an average cost of saved energy of 1.5 cents/kWh, although costs will likely vary from near zero to more than 5 cents/kWh, depending on the measure and the facility. Costs for reduced waste will tend to be at the lower end of this range, and costs for equipment and design improvements in the upper portion of this range.

Space Cooling Systems

As with fans, pumps, and air compressors, opportunities for lowering cooling system energy use include equipment efficiency improvements as well as enhanced system design. Improved efficiency of packaged and component air conditioning equipment is readily available across a wide range of system sizes. Options to improve overall system design are also available for large and small systems.

At the residential end of the spectrum, central air conditioning systems with seasonal energy efficiency ratios (SEER) of 13 or more are readily available and have been successfully promoted by many utilities (CEE 1997). Energy savings are more than 20% relative to a typical baseline system with a SEER of just over 10. Savings relative to the existing stock of equipment are often 30% or more since, prior to federal efficiency standards that took effect in 1992, SEERs averaged approximately 9 (ARI 1993) and many of these older systems are still in use. The Energy Center of Wisconsin (1997) found an average incremental cost of about $600 for going from SEER 10 to SEER 13. A national analysis for DOE (1999) estimates average annual savings of about 670 kWh for this improvement (including for heat pumps and cooling-only units) and an 18-year average unit life, resulting in a cost of saved energy of 8 cents/kWh. While this value may appear high, it should be compared to the price of peak electrical power (which is often 10–20 cents/kWh) and not an annual average price.

Home air conditioning energy use can also be reduced with improved attention to system installation and maintenance including proper refrigerant charge, proper airflow across the coil, and suitably sealing ducts to decrease leakage of conditioned air. Neme, Nadel, and Proctor (1999) reviewed multiple field studies on proper installation and maintenance and estimated that, on average, savings of about 24% are possible relative to typical installation and maintenance practices. Incremental costs for these improvements are on the order of $250–400 (assuming that repairs are done during a regularly scheduled service call and that approximately half of the homes need duct sealing) (Suozzo and Nadel 1998). Assuming a 10-year average measure life and average air conditioner and heat pump annual energy use of 3,251 kWh (EIA 1995), the cost of saved energy is approximately 4–6 cents/kWh. Most of these savings occur at times of high electrical demand, when electricity prices are high.

Overall, combining 20–30% savings from improved equipment efficiency with approximately 24% savings from proper installation and maintenance, and adjusting for overlap between these two measures, gives typical savings of 25–40%.

At the large equipment end of the spectrum, state-of-the-art new water-cooled centrifugal chillers have efficiencies of less than 0.50 kW/ton of cooling capacity at design conditions and 0.35 kW/ton under typical operating conditions (Nugent 1993). These efficiencies represent 20–44% savings relative to a typical new large chiller (efficiency of roughly 0.63 kW/ton of cooling capacity) and 33–53% savings relative to a typical existing large chiller (0.75 kW/ton). Houghton et al. (1992) estimate that these savings are available at 1–4 cents/kWh saved.

Refrigeration Systems

The energy efficiency of refrigeration systems can also be increased substantially. A 1996 analysis by Arthur D. Little, Inc. (ADL 1996), examined 16 different measures to improve the efficiency of supermarket refrigeration systems, noting the proportion of supermarkets that had already adopted each measure. Based on this analysis, they estimated average available energy savings of 31%. For a typical store, the improvements will cost $117,000 but save 683,000 kWh annually. The cost of saved energy for these improvements is 2 cents/kWh assuming a 10-year equipment life. The ADL study also found similar available savings and costs for improvements to packaged refrigeration equipment such as ice makers, vending machines, beverage merchandisers, and reach-in refrigerators and freezers.

Improvements to Synchronous and DC Motor Systems

As illustrated in Chapter 6, we estimate that the combination of synchronous and DC motors accounts for only about 8% of total motor input, or 154 TWh/year, specifically about 72 TWh for DC units and 82 TWh for synchronous motors.

Synchronous motors are large, efficient (96–98%) units made to order. We assume no significant savings potential because they are already so efficient. We conservatively assume no penalty from oversizing since they are carefully specified. However, synchronous motors are subject to the same potential rewind damage as induction motors. We estimate this damage to be about 1% of the energy input to induction motors and we assume the same for synchronous equipment. Thus, perhaps 0.8 TWh/yr of excess losses could be corrected virtually for free by adopting better repair practices as the motor stock is replaced.

The savings potential of DC motors is considerably larger. Some DC motors are driven by motor-generator (M-G) sets, wherein a

90%-efficient AC motor turns an 85%-efficient generator to produce DC power for an 85%-efficient DC motor. These values are imprecise but representative, and they yield an overall efficiency of only 65%. Replacement of such systems wherever practical with AC motors controlled by ASDs, or other variable-speed technologies such as switched-reluctance motors, can potentially improve efficiencies by tens of percentage points. Also, the economics of replacing instead of repairing is even more attractive for DC systems because rewinding costs more for DC motors than for AC units.

M-G sets are not only inefficient, but also create large idling losses in numerous applications. For instance, they are commonly found in traction drives for such uses as elevators, cranes, and hoists. When the drive is idle, the DC motor shuts down, but in many cases the M-G set continues to run. In other instances, DC motor efficiency can be markedly improved by installing more sophisticated solid-state controls or using more efficient electronically commutated DC motors.

DC motor losses from oversizing are probably not as vast a problem because they are more carefully sized, due to their higher cost. Rewind damage in DC motors is probably comparable to that in AC models, where we estimate an overall 1% penalty.

Combining the essentially zero-cost 1% savings from correcting for past rewind damage in both DC and synchronous motors with the sizable savings available from replacing or better controlling DC motors, we assume that 5% of synchronous and DC drivesystem energy use can be saved for less than $0.03/kWh. Again, the value is imprecise, but the term is so small that it matters little to the overall analysis.

Applying this 5% improvement to the estimated 154 TWh/yr used by synchronous and DC motors yields a savings potential of 8 TWh/year.

Electrical Tune-Ups

As mentioned in Chapter 3, phase unbalance, voltage variations, low power factor, and poor supply waveforms can reduce motor efficiency and damage equipment. No precise calculations of the aggregate efficiency loss caused by such problems exist, but one study has estimated the range of potential savings from their mitigation at 15–179 TWh (corresponding to a 1–15% savings), with a cost well under $0.01/kWh (Lovins et al. 1989). The wide range of uncertainty in this appraisal underscores the need for far more field data. To be conservative, we adopt a 1–5% savings range for our analysis, recognizing that the actual value may be higher.

Better Mechanical Drivetrain Equipment and Lubrication

Significant but much overlooked opportunities for inexpensive savings are improvements in drivetrain design, equipment lubrication, and maintenance. For example, as discussed in Chapter 4, synchronous and cogged V-belts can often be substituted for conventional V-belts, with potential efficiency gains of 0.5–8 percentage points.

Proper selection of speed reducers (or increasers) is another critical area since different types, with the same capacity and design life and suited to the same task, can vary in efficiency by up to 25 percentage points. Moreover, the most efficient option is not necessarily the most expensive. For example, a Dodge APG size 4 helical geartrain listing for $579 with 94% efficiency has comparable capacity, design life, and speed ratio to a Dodge Master WM4O worm gear drive costing about four times more and yielding 68% efficiency (Lovins et al. 1989).

Premium lubricants have yielded 3–20% energy savings in various devices, from wire-drawing machines (Ibanez 1978) to automobiles (Gutman and Stotter 1984; Milton and Carter 1982) to gear reducers, compressors, and motors (Kent 1989). A U.S. Navy–sponsored case study of one specialty lubricant in a 350 hp compressor reduced electricity use by 2.6% at a negative cost of saved energy. The lubricant lasted four times longer, more than compensating for its higher price (Kent 1989).

Improved maintenance is also critical to decreasing costly downtime and keeping efficiencies at optimal levels. An EPRI study stated, "The efficiencies of mechanical equipment in general can be increased typically 10 to 15 percent by proper maintenance" (Ibanez 1978). A motor diagnostics program involving a vibration tester and a surge tester cut motor failures in one plant by about half (Kochensparger 1987).

In some cases, less maintenance of the wrong kind is called for. Southwire Company's energy manager Jim Clarkson has observed a tendency of plant managers to have workers paint motors and other equipment to spruce up a facility when important visitors are expected. Since extra paint makes motors run hotter, thereby reducing their lifetime and efficiency, Clarkson only half in jest issued a policy requiring workers to first strip the old paint before repainting any electrical equipment (Clarkson 1990).

Potential savings from improved drivetrain technologies and maintenance practices are difficult to estimate. Lovins and his colleagues (1989), citing numerous case studies, suggest that optimal practices in these areas could save 3–10% of all drivepower input, with the cost of saved energy reduced to nearly zero by improvements in reliability and equipment life. Most of the cited studies of the belt, chain, and gearbox systems show paybacks from energy savings alone of under 2 years. The EPRI study by Ibanez estimates 10–15% savings from better maintenance alone.

It is very difficult, however, to extrapolate the full savings potential from case studies. First, the relative shares of various transmission types are poorly documented, as are typical lubrication and maintenance practices. Second, case studies, particularly from vendors, tend to be skewed toward the most attractive applications. Nevertheless, anecdotal evidence suggests that considerable room for improvement exists in this area. More objective and well-documented case studies are needed.

Given the lack of data, we attempt no independent analysis here. The Lovins et al. and Ibanez estimates are plausible but difficult to verify. To be conservative, we adopt a lower range of 3–7% as an almost certain opportunity for inexpensive savings.

Indirect Savings

Reduced Distribution Losses

An estimated 6% of energy input is lost in the distribution wires between the customer's meter and motor terminals (see discussion of wire sizing in Chapter 3). As net input at the meter decreases, distribution losses will fall in relation to the ratio of the square of the current.

HVAC Bonus

More efficient drivesystems produce less waste heat and thus reduce the need for cooling in buildings and factories. They also increase the heating load. Since space cooling uses three times the energy of space heating in the commercial sector and twice as much in the industrial sector, reducing internal waste heat released in those sectors results in a net HVAC bonus by allowing the cooling system to save more than the heating system must make up. Only in the residential sector, where (nationally) heating loads are nearly twice cooling loads, will reduced heat from motor systems add to space conditioning energy.

Some researchers have documented HVAC bonuses in the 40% range for commercial spaces in hot climates (California Energy Commission 1984; Linn 1987; Treadle 1987). A typical HVAC bonus in New England is 27% (Jackson 1987), the value we use as representative. The HVAC effects in other energy-use sectors are smaller and not as well documented. One estimate puts the average effect at about 4% in industry and –5% in residences (Lovins et al. 1989).

Applying these values to the sectoral shares of motor input energy shown in Figure 6-7 (23% residential, 20% commercial, and 57% industrial plus other) yields an input-weighted HVAC bonus for waste drivepower heat of around 6.5%. To avoid double counting, this free savings in net space conditioning energy must be discounted by the proportion of savings achieved by prior measures before being applied to the cumulative savings.

Adding Up the Savings

We now compile in Table 7-1 all of the savings estimates discussed above. Note that not all the savings are additive; savings percentages are applied only to the input remaining to motors after savings from prior measures have been subtracted.

We thus estimate that 28–42% of all U.S. motor input energy (15–25% of all U.S. electricity) can be saved by full application of the measures described above, most of which cost less than 3 cents/kWh saved. Most of this savings potential remains untapped. Furthermore, even this estimate is most likely conservative, because major redesign of industrial processes are not considered. To achieve the greatest total energy savings, energy use must be looked at on a system basis. With a system approach, each process needs to be studied for its energy use and its relation to the operation of the plant as a whole (examples of this approach are discussed in Chapters 4 and 9). While this would be a resource-intensive process, it would also yield benefits (such as waste minimization, improved product quality and materials utilization, and improved productivity) in addition to energy-cost reductions.

Many measures contribute to the savings potential estimated here. No single measure dominates. The largest savings are available from cooling equipment and system efficiency, motor speed control, and fan and pump system design. The contribution of these and other measures to the overall efficiency potential are illustrated in Figure 7-3.

Six other recent studies have presented a roughly similar savings potential from electric motor systems, as shown in Figure 7-4. The most aggressive estimate, of 28–60%, was made by E Source (E Source 1993). This estimate is for the motor population as a whole, and includes many measures to improve system efficiency. The lowest calculation is 11–18%, which was made by XENERGY (1998) in a study for DOE. However, the XENERGY study is very conservative in that it only looked at measures that have a 3-year simple payback on a retrofit basis. The latter criterion is particularly limiting, because a substantial portion of the efficiency potential is cost-effective only when existing equipment is being replaced, a consideration the XENERGY methodology ignores.

Table 7-1
Summary of Savings Potential: 1997

	Typical Savings (%)	Proportion of Motor Load for Which Measure Appropriate (%)	Savings (TWh/yr)	Remaining Input (TWh/yr)	Cost of Savings (cents/kWh)	Notes
Induction motors				1,851		
Replacement with EPAct motors	4.3%	87%	49		0 for new	Applies to integral motors operating at least 2,000 hrs/yr that are not already efficient
Replacement with premium motors	2.4%	93%	30		1–3	Applies to integral motors operating at least 2,000 hrs/yr that are not already efficient
Higher-efficiency small motors	5–12%	100%	20–47	1,712–1,740	1–6	LBNL (1996) considered average savings of 5, 10, and 12%
Elimination of past rewind damage	1%	84%	13		0	Applies to motors >20 hp
Correction of previous oversizing	5%	29%	26	1,687–1,724	0	For motors operating at <40% of full load
Motor speed control	15–30%	12–27%	30–139	1,548–1,685	1–5	Applicable in 1/3–2/3 of fan, pump, HVAC equipment, and conveyor applications; 4% already implemented
Fan & pump equipment efficiency	5–10%	31%	26–48		typ. 1	Applies to fan and pump loads
Fan & pump system design	10–20%	31%	50–86		1 on avg.	Applies to fan and pump loads
Reduce compressed-air waste	20–30%	7%	24–32		0–1.5	Average across all compressed-air loads—includes leaks and inappropriate apps.
Air compressor equip. & system effic.	15–25%	7%	14–19		1.5–4	Average across all compressed air loads
Cooling equipment & system efficiency	25–40%	17%	67–117		1–8	Applies to cooling loads
Refrigeration equip. & system efficiency	30%	8%	35–38	1,207–1,469	2	Applies to refrigeration loads
DC and synchronous motors				154		
More effic. equip., correct rewind damage	5%	100%	8	146	<3	
Remaining input to all motors				1,353–1,616		
Electrical tune-ups	1–5%	100%	16–68	1,285–1,599	?	
Drivetrain, lubrication, and maintenance	3–7%	100%	48–90	1,195–1,551	<0	
Savings						
Indirect savings						
Reduced distribution loss	~8%[a]		33–65	1,131–1,518	0	
Reduced HVAC effect	6.5%[b]		26–30	1,101–1,492	0	
Total Savings			**513–780**			
Savings as percentage of original input			28–42%			

Notes:

a. $6\% \times$ initial input = initial distribution loss. Initial distribution loss $\times \left[1 - \dfrac{(\text{remaining input after efficiency measures})^2}{(\text{initial input})^2}\right]$ = reduction in losses.

b. $6.5\% \times \dfrac{\text{remaining input}}{\text{initial input}} \times$ cumulative savings = reduction in HVAC effect.

Figure 7-3

Distribution of Midpoint Electric Motor System Conservation Potential

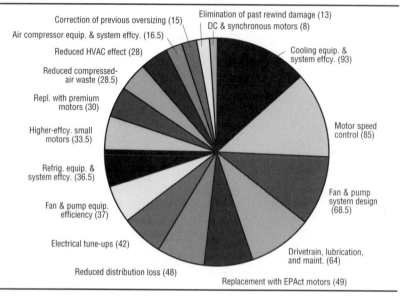

Correction of previous oversizing (15)
Air compressor equip. & system effcy. (16.5)
Reduced HVAC effect (28)
Reduced compressed-air waste (28.5)
Repl. with premium motors (30)
Higher-effcy. small motors (33.5)
Refrig. equip. & system effcy. (36.5)
Fan & pump equip. efficiency (37)
Electrical tune-ups (42)
Reduced distribution loss (48)

Elimination of past rewind damage (13)
DC & synchronous motors (8)
Cooling equip. & system effcy. (93)
Motor speed control (85)
Fan & pump system design (68.5)
Drivetrain, lubrication, and maint. (64)
Replacement with EPAct motors (49)

Figure 7-4

Estimates of Motor System Electricity Conservation Potential from Recent Studies

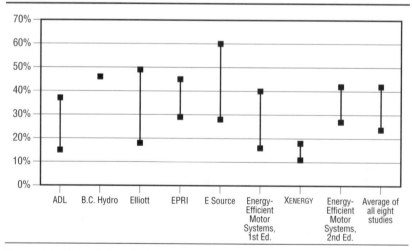

Sources: Elliott 1995; XENERGY 1998

Most of the other studies estimate a savings potential very similar to the estimate made in this chapter (ADL 1999; Elliott 1995; Faruqui et al. 1990; Jaccard, Fogwill, and Nyboer 1993). Interestingly, our estimate of 28–42% available savings is somewhat higher than the 16–40% savings potential estimated in the 1991 edition of this book. Our current calculation is higher because, unlike the 1991 estimate, the current analysis includes savings in fan, pump, compressed-air, and cooling systems. The average savings range for all eight studies is 23–41%.

Some savings require replacement of existing equipment, a process that will take years. Barriers from lack of information to lack of incentives must be overcome. Fortunately, some innovative programs aiming to capture drivepower savings have already begun. The lessons learned from these efforts along with ideas for further implementation and research strategies are discussed in the chapters that follow.

The Motor Market

Designing programs and policies that successfully promote motor system efficiency requires more than familiarity with the relevant technologies—it calls for an understanding of the key players in the motor market, their motivations, and the challenges they face. Successful programs must appeal to the key decision-makers and fit into their current ways of doing business.

Given the many players involved, various approaches are needed when developing a program design, each one suited to particular market segments. For example, the decision-makers, decision processes, and economic criteria involved in purchasing motors for new applications differ from those involved in purchasing motors for replacement applications. Programs or policies can be better targeted if they reflect an understanding of the differences between these two markets.

How to tailor programs and policies to such differences is the subject of Chapter 9. Here, we lay the foundation for that discussion by identifying the players, their interactions, and the factors that influence their decisions. Among the key players are

- End-users
- Motor manufacturers
- Motor distributors and repair and rewind shops
- Original equipment manufacturers such as fan, pump, and compressor producers
- Consulting engineers and design-build contractors
- Electronic motor control equipment manufacturers, distributors, and representatives

Figure 8-1

Interactions among Major Players in the Electric Motor Market

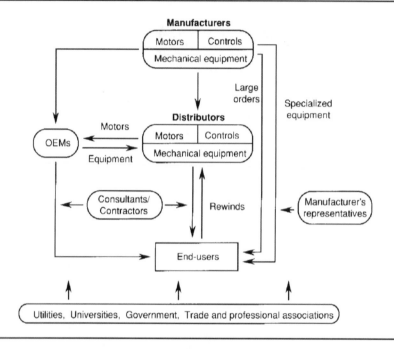

- Mechanical equipment manufacturers, distributors, and representatives
- Electric utilities, universities, government agencies, and trade associations

The interactions among these players are illustrated schematically in Figure 8-1.

The material discussed in this chapter is based on a number of published reports (the most useful ones are listed in the Annotated Bibliography), as well as on a series of discussions with knowledgeable people in the motor field, including manufacturers, distributors, end-users, and representatives of utilities and trade associations.

End-Users and Customers

Customer needs, attitudes, decision criteria, and decision-making processes vary widely. This section summarizes many of the important issues facing drivepower equipment users.

Decision-Makers

Maintenance staff often make purchase decisions for replacement equipment. The exception to this tendency is in small plants, where there is often not a maintenance manager and instead decisions are frequently made by the general manager or other senior staff. These patterns are illustrated in Table 8-1, which summarizes the results of a recent nationwide survey in which motor purchase decision-makers were identified as a function of facility size. In companies with multiple locations, motor purchase decisions are mostly made at the plant level (more than 90% according to one survey [XENERGY 1998]), and not at headquarters.

By contrast, purchase decisions for new equipment and replacements for ASDs usually involve engineering staff or consultants. The level of engineering analysis varies widely but generally increases with the project's size. Sometimes engineers oversee bidding and installation. In other cases, engineers write specifications and then the purchasing department takes over. Large companies are more likely to do a thorough analysis than are small companies.

Aversion to Downtime

In some manufacturing facilities, if a production line is shut down for several hours, the value of the lost production capacity can be greater than the value of many years of energy savings. End-users are thus very

Table 8-1

Position of the Motor Purchase Decision-Maker as a Function of Facility Size

	Size Categories					
	Large	Medium/Large	Medium	Small/Medium	Small	Total
Plant Manager	0%	17%	0%	12%	14%	13%
Maintenance Manager	41%	43%	72%	5%	3%	9%
Purchasing Manager	0%	0%	0%	20%	0%	2%
Plant Engineer	16%	8%	12%	2%	4%	5%
Chief Electrician	23%	4%	4%	1%	0%	1%
President or General Manager	0%	0%	4%	35%	47%	40%
Other	20%	24%	8%	25%	31%	29%
(Blank)	0%	4%	0%	0%	0%	1%
TOTAL	100%	100%	100%	100%	100%	100%

Source: XENERGY 1998

concerned about downtime and reliability, particularly when replacing equipment. For this reason, there is a need to document case studies on equipment reliability and to provide this information to end-users.

Concerns about downtime also cause customers to demand quick delivery of equipment and avoid purchasing high-efficiency equipment if it will take longer to obtain. Downtime costs also make end-users reluctant to replace operating equipment, even if it is inefficient. Thus, the only time many end-users will consider high-efficiency equipment is during the brief period between the failure of old equipment and the hasty purchase of new components.

To minimize downtime, many companies stock spare motors and mechanical drivetrain parts such as gearboxes. To reduce the number of items that must be stocked, some companies standardize on a limited number of equipment sizes, such as 50 and 100 hp motors. For applications that require an intermediate size of, say, 75 hp, the next largest stocked motor is used. This practice results in systematic oversizing. In recent years, some large companies have made arrangements with local distributors for them to stock and quickly ship replacement equipment under medium- or long-term contracts. The advantages of this arrangement for the customers are that their internal inventory needs are reduced, they get quick service, and they generally get good prices as part of the long-term contractual relationship. Since distributors serve multiple customers, they can stock more items and oversizing problems due to stocking a limited number of items are reduced (Brithinee 1999; Darby 1997).

Purchase Practices

Existing equipment is usually replaced or repaired without engineering analysis, and is often replaced with the same size, brand, and model number. For example, a recent survey found that 55% of manufacturing customers "always" select the same size motor as the one being replaced, and an additional 31% of customers do this "most of the time" (XENERGY 1998). Only in the case of large motors (over approximately 250 hp) with high operating costs does an engineering or economic analysis usually precede decisions concerning replacement equipment.

End-users can obtain information on motors and related equipment from manufacturers' catalogs, trade publications, manufacturers' representatives, distributors, contractors, and professional organizations. Customers commonly believe that motors under approximately 200 hp and other drivepower components such as gearboxes, bearings, belts, chains, and lubricants are commodity items, meaning that models produced by different manufacturers are interchangeable. Given this outlook, purchase decisions are made based primarily on reliability, price, and availability, not on efficiency (ADL 1980). A recent survey in the northeastern United States found that the decision-maker is generally aware of energy saving in the abstract but not cognizant of the specifics. As a result, energy cost saving is a factor in decisions, but not a primary concern (Easton Consultants and XENERGY 1999a).

Some large companies (and a few smaller ones) have formal motor purchase policies that address motor efficiency; however, most do not. For example, a 1998 nationwide survey of manufacturers found that 20% of large customers have a motor efficiency policy, but among all customers only 3% have such a policy (XENERGY 1998).

Repair or Replace?

Most end-users replace small motors and repair large ones in the event of burnout because repairing is generally more expensive than replacing a small motor and less expensive than replacing a large one. Different companies use different guidelines for deciding whether to repair or replace. In most cases, motors of 10 hp or less are replaced (Seton, Johnson, and Odell 1987b), while motors of 40 hp or more are repaired (Marbek 1987). Decisions on 15–30 hp motors vary extensively among firms. These tendencies are illustrated by data from a recent survey of manufacturing customers and summarized in Table 8-2. A few firms will routinely replace motors as large as 100 hp instead of repairing them (Lovins et al. 1989). Repair-or-replace decisions are generally made at the plant level, although a few large corporations have established guidelines for their plants.

Table 8-2

Percentage of Motors Repaired by Horsepower Category

Horsepower	Percentage Repaired
1–5	20%
6–20	61%
21–50	81%
51–100	90%
101–200	91%

Source: XENERGY 1998

There are a few exceptions to the "replace small motors, repair large motors" rule:

- *Specialty motors and old, odd-dimension motors.* Since new motors of these types are hard to find, they are generally repaired when they burn out.

- *Situations where the old motor is damaged and cannot be repaired.* In these cases the motor is replaced.

- *Situations where considerable time can be saved by either repairing or replacing an old motor.* When order time for a new motor is long, the motor is usually repaired. When repair shops are busy and cannot provide quick service and replacement motors are readily available, a new motor may be purchased.

End-users select repair shops primarily on the basis of price and speed of service. Most motor repair shops do not provide the customer with any evaluation of the motor to be repaired or recommendations on replacement options unless the motor is severely damaged. To encourage competition and responsiveness, most end-users use more than one repair shop. Unless consistent reliability problems are encountered, the quality of the shops' repairs is not considered (Schueler, Leistner, and Douglass 1994).

Southwire Company's Rewind Policy

Southwire Company, a major wire and cable firm with annual sales in excess of $1 billion, has a general policy of replacing all standard-efficiency motors of 125 hp or less instead of rewinding them. Southwire has calculated that replacement is cost-effective in nearly all applications. For larger motors, the firm compares the costs and savings of rewound versus new motors and purchases a new motor if the net present value of savings over 5 years exceeds the financing costs. The general policy is to replace motors when the repair cost exceeds 40% of the purchase price of a new motor.

Southwire has experienced reliability problems with rewound motors and is also concerned about efficiency losses from the rewind process. Therefore, the firm prefers new motors unless it determines that the incremental cost of a new motor is excessive. The policy of buying large numbers of new, high-efficiency motors has enabled the company to negotiate a price for high-efficiency motors that is only 5% above the cost of standard-efficiency motors. To achieve this price, Southwire generally buys all motors from one supplier and expects that supplier to have high-efficiency motors in stock at all times (Clarkson 1990).

Adjustable-Speed Drives

For ASDs and major design decisions, the most prevalent criteria that the designer must take into account are equipment reliability, features, and performance (e.g., impact on production control and quality); equipment and installation costs; and savings in operation and maintenance (e.g., reduced maintenance, increased equipment life, or reduced demand charges due to the soft-start capabilities of many ASDs). End-users often treat ASDs differently from other types of equipment because they are a relatively complex and unfamiliar technology. A number of early ASD systems had problems that have made some end-users distrustful of even the new, improved versions. Due to their high price and high savings, ASDs usually receive engineering attention from either in-house or consulting engineers. However, it can be difficult and expensive to precisely determine how much energy and money a particular ASD installation will save. Many simplifying assumptions are often made, adding to the range of uncertainty for the end-user. One recent market study found that most ASD purchases are made by in-house staff that are responsible for identifying cost saving or process improvement opportunities. Other ASDs may be purchased at the recommendation of engineering firms or consultants, typically as part of a larger project such as a plant construction project. And finally, some ASDs are purchased as part of an OEM package, such as a chiller (Easton Consultants 2000).

ASDs are not generally treated as a commodity product. There are important differences among models that warrant careful comparisons. Furthermore, many plants use only a particular brand of motor switchgear. In these cases, engineers will generally try to specify ASDs made by that manufacturer. Another important consideration is matching the motor with the drive, which often means purchasing a matched motor-drive set from the same manufacturer. Where the brand is not critical, particular features and capabilities can be specified and more than one product allowed to compete. This is particularly true in government facilities, where competitive bidding is generally required.

When an ASD unit fails, it can usually be repaired by replacing a circuit board rather than the whole unit. However, the rapid evolution of ASDs means that new features are regularly entering the market and customers desiring these new features will sometimes replace the whole equipment. In the rare cases when replacement parts for early units are no longer manufactured, customers must buy new equipment when an existing component fails.

Maintenance Practices

Motor maintenance practices are generally limited to what is needed to keep equipment running rather than optimizing performance and saving energy. Most industrial plants and large commercial firms have full-time maintenance staff who regularly lubricate (and often overlubricate) motors, listen for bearing noise (a sign of wear or misalignment), and check and tighten belts as needed. Few firms do any more sophisticated monitoring or maintenance work on motor systems. For example, during a major nationwide field study on a representative sample of motors and motor-users:

> [T]he field engineers noted repeatedly the limited resources available for motor system monitoring and maintenance. The priority for facilities management and maintenance staff was to ensure continuity and consistency of mechanical options. It was very difficult for facilities' management staff to break away from their jobs long enough to answer a few questions or to provide escorts for the field engineers. There was clearly little slack in their schedule for the additional tasks required for active motor systems management—at least without considerable guidance concerning the most worthwhile allocation of resources. (XENERGY 1998)

According to some industrial observers, the time available for maintenance is becoming even more limited in some firms due to industrial company downsizing over the past decade (Hamer 1999).

Many large industrial and commercial firms have some type of tracking system for motor system maintenance involving log books, file cards on each motor (sometimes kept in a central file and sometimes attached to individual motors), or computer-based records.

Small firms without maintenance staff primarily rely on outside equipment contractors to maintain HVAC systems, conveyor belts, and other drivepower components. Some large firms have begun to hire outside maintenance contractors as well, since corporate downsizing programs have left some companies short of maintenance staff. With outside service providers, maintenance frequency varies widely. A few firms schedule regular service calls. Others wait for problems before calling someone in. In many of these cases, the call is placed too late, as equipment has been damaged beyond repair and must be replaced.

Overall, motor maintenance practices in the United States are less than optimal. In Japan, on the other hand, the responsibility for maintaining individual motors is often assigned to specific mechanics, who receive extensive training. When a motor fails on the production line, the problem can be attributed to its mechanic. This provides a strong impetus to conduct preventive maintenance (Johnston 1990). Approaches for encouraging improved maintenance are discussed in Chapters 9 and 10.

Other Factors Influencing Decision-Making

Several other factors, in addition to those related specifically to motor systems, influence most efficiency-related investment. Some of the more important ones are discussed below.

Limited Information

As noted above, most maintenance managers and other decision-makers are very busy, leaving little time to research new opportunities, including opportunities to save energy. This lack of time generally causes knowledge of energy-saving options to be limited. For example, in a survey of motor purchase decision-makers, only 22% were aware of premium-efficiency motors and only 25% knew of tools for helping to select new or replacement motors. Only among large companies were the majority of decision-makers aware of premium-efficiency motors or decision-assisting tools (XENERGY 1998). Adding to this confusion is publicity surrounding the EPAct motor standards, leading many users to mistakenly conclude that all motors are efficient and that they no longer need to pay attention to efficiency. Also, the lack of a standardized definition for the term *premium efficiency* and the resulting inconsistent definitions developed by different manufacturers make it difficult for some users to understand what is meant by the term.

To our knowledge, similar survey data are not available for other energy-saving measures, such as optimization of fan, pump, and compressed-air systems. Given the fact that these other opportunities are usually more complicated than purchasing improved-efficiency motors, the lack of information is likely to be even more of a problem for these other opportunities.

Limited Access to Capital

The average end-user is more restrictive with capital than with operating funds (ADL 1980; Comnes and Barnes 1987). Generally, capital expenses are closely scrutinized and require approval at multiple levels in a company. To minimize capital outlay, companies tend to choose the least expensive equipment that will do the job satisfactorily.

Operating funds, on the other hand, are relatively easy to obtain, since they are required for production. Operating budgets are typically based on expenses in previous years and are only seriously examined when out of line with expectations. Moreover, unlike capital costs, operating costs are paid with pretax dollars.

Payback Gap

It is a curious fact that most firms look for a simple payback period of 2–3 years or less on energy projects and other operations and maintenance investments (Marbek 1987), even though longer paybacks are often considered when investing in new product lines. This difference, known as the payback gap, makes it difficult to implement all but rapid-payback energy-saving measures, although measures with longer paybacks will sometimes be considered as part of a major facility upgrade designed to improve the long-term competitiveness of the firm (Comnes and Barnes 1987). The payback gap is most pronounced when viewed from the societal perspective—individual firms pass up energy-saving investments with paybacks of 3–4 years, while utilities invest in distribution lines with economic returns equivalent to 10- to 20-year paybacks.

Low Priority Assigned to Energy Matters

For the average industrial firm, energy costs represent only a small percentage of total costs; labor and material costs are usually far greater. For example, in 1998 the U.S. Census's Annual Survey of Manufacturers estimated that, on average, electricity accounts for a little over 1% of manufacturing costs. Since motors make up about 70% of manufacturing electricity use (see Chapter 6), they make up about 1% of total costs for the average industrial firm. Since energy costs represent a small proportion of an average end-user's total operating costs, motor and other energy-related operating costs are rarely examined in reviews of operating expenses. Breaking this log-jam will require creative approaches, some of which are discussed in Chapters 9 and 10.

Transaction Costs

Contributing to the low priority that energy matters are given is the fact that many energy-saving measures, including motor measures, have substantial transaction costs. Comparing equipment or optimizing a system takes time, which is a commodity in short supply in many firms. For example, Ostertag (1999) notes that "search costs" for high-efficiency motors will often be greater than for standard motors because it is the standard motor that is the known quantity for many purchasers. For larger projects, outside engineers can be brought in to help with project design and implementation, but for small projects, if existing staff are short on time, decisions are commonly made based on expediency rather than economic merit.

Misplaced Program Emphasis

Since they generally have full-time maintenance staff or energy managers, large firms are more likely to be interested in energy efficiency. Even in firms with energy managers, however, motor systems historically have not received much attention because of (often incorrect) perceptions that motor system improvements have high capital expense, low rates of return, and low percentage savings. Energy managers tend to focus on low capital cost measures with high savings (ADL 1980). While this approach is reasonable during the start-up stages of an energy management effort, many firms have not moved beyond high-savings, low-cost measures. Moreover, many drivepower-saving measures are relatively inexpensive.

Lack of Internal Incentives

For many companies, energy bills are paid by the company as a whole and not allocated to individual departments. This practice gives maintenance and engineering staff little incentive to pursue energy-saving investments because the savings in energy bills show up in a corporate-level account where the savings provide little or no benefit to maintenance and engineering decision-makers. As is discussed in Chapter 10, mechanisms to improve internal incentives have been put into place in some facilities.

Differences between Industrial Sectors

Our discussion to this point treats all motor users the same and does not distinguish between customers except on the basis of size. However, large differences between customers do exist regarding energy efficiency, including differences in their available opportunities, knowledge, receptivity, and prior activities. Some of these differences, though far from all of them, relate to industrial sectors—several sectors tend to have available capital (e.g., microelectronics); others do not (e.g., lumber and wood products). Some use motors heavily and some do not (see Figure 6-8). A recent study prepared for the Northwest Energy Efficiency Alliance looked at these differences and made the following conclusions regarding these criteria (Easton Consultants and XENERGY 1999b):

- The process industries (pulp and paper, chemicals, petroleum, and primary metals) rate high in the relative ease of achieving motor and motor systems energy savings. Their large, easily identifiable facilities are generally financially healthy and receptive to energy-saving initiatives. The fact that these facilities have already taken steps to implement energy-saving improvements is a principal barrier because additional savings may be more difficult and expensive to obtain. Further, many facilities will require site-specific engineering and process changes to realize such savings.

• Other manufacturing categories rate lower in these criteria than the process industries. They are less concentrated, less healthy financially (except microelectronics), and less receptive to energy conservation. They do have many opportunities for savings, however.

• Mining is viewed by experts as having many opportunities, considering that the handful of underground mines are major users of energy. There are barriers, however, as mine managers are reported to be less receptive to energy conservation, and many of the opportunities require site-specific engineering.

• Water and wastewater facilities have opportunities in pump systems, motor upgrades, and new aeration technologies. Government

Table 8-3

Availability of Savings by Industry Sector

Industry	Concentration of Target	Financial Health of Industry	Relative Receptivity of Energy Efficiency	Relative Ease in Achieving Savings	Summary
Process Industry					
Pulp & Paper	1	2	2	4	2
Chemicals	2	2	2	4	2
Petroleum	1	2	1	4	2
Primary Metal	2	2	3	3	3
General Manufacturers					
Food	3	3	2	3	3
Lumber & Wood	3	5	4	3	4
Microelectronics	1	1	3	3	2
Aircraft Parts	2	2	4	4	3
Mining	1	3	3	4	3
Water & Wastewater	2	3	3	4	3
Irrigation	5	3	3	2	3

Note: The ratings should be interpreted as follows:
1 indicates that the factor is a positive contribution to achieving energy savings
3 indicates factor is neutral toward energy saving
5 indicates factor is an important barrier to achieving energy savings

Source: Easton Consultants and XENERGY 1999b

ownership and regulation present barriers to significant investment and often limit changes to times when major construction or renovation is planned.

• Irrigation is judged to be a very difficult sector in which to accomplish energy savings, although the potential opportunities are significant. The sector is hard to reach as it is made up of many (mostly small) farms, and receptivity to conservation is low, due in part to low power costs (Easton Consulting and XENERGY 1999b). Some programs that have successfully reached this sector are described in Chapter 9.

While these findings apply to the Pacific Northwest, most are likely to apply to some extent in other regions of the United States. Table 8-3 summarizes these findings.

We have covered some of the main factors influencing the decisions of motor system users. Now we turn to another of the key players, the motor manufacturer.

Motor Manufacturers

In 1977, eight major manufacturers made up over 75% of the North American motor market (ADL 1980). Since then, a few new manufacturers, including Wegg, Toshiba, ABB, and Siemens, have entered the market, while a few old ones have left (MagneTek bought both Century and Louis Allis, and Westinghouse sold its motor business to Reliance and Siemens). Even so, the total number of major manufacturers has changed little (see Appendix D for a list of the major manufacturers and their product lines). Many experts predict that additional consolidation in the industry is likely, as well as a further increase in the presence of international firms in the U.S. market.

A number of manufacturers, such as General Electric, make motors ranging from fractional-horsepower up to custom motors of thousands of horsepower. Others are more specialized. U.S. Motors produces models ranging from 0.75 to 350 hp, while Siemens tends to emphasize motors above 250 hp. Types of motors sold by the different major manufacturers are listed in Appendix D. In general, motors below 250–350 hp are regularly produced and stocked by manufacturers, while larger motors are custom built, sometimes to standard specifications or to custom specifications, including desired efficiency levels.

Motor Efficiency

Within the 1–200 hp range, most major manufacturers produce two lines of motors—an EPAct-efficiency line and a premium-efficiency line. Of the premium-efficiency motors produced, some meet the Consortium for Energy Efficiency's efficiency thresholds (discussed in Chapter 2), while others do not. In addition, most manufacturers produce efficient motors in sizes about 200 hp, with some premium lines extending as high as 500 hp.

Premium-efficiency three-phase motors are generally available in T-frames for 1,200, 1,800, and 3,600 rpm nominal speeds for both TEFC and ODP enclosure types. Premium-efficiency motors are sometimes produced in EXP (explosion-proof) enclosure types, C-face and D-flange mountings (commonly used for pump and other specialty applications), vertical-shaft designs, and 900 rpm models; however, because consumer demand for these products is low, only a few manufacturers produce them. It is technically straightforward to produce premium-efficiency versions of most types of motors. In fact, for a price premium, increases in efficiency can be ordered for nearly any motor. Thus, availability of premium-efficiency motors primarily depends on manufacturers' perceptions of likely customer demand.

Efficiencies for both EPAct- and high-efficiency lines vary among manufacturers. However, differences in efficiency among manufacturers are much smaller for EPAct-efficiency motors than for premium-efficiency models since many EPAct-efficiency motors tend to cluster just above the EPAct-mandated efficiency levels (see Figure 2-25). With these motors, manufacturers compete primarily on quality and price. In the premium-efficiency lines, manufacturers compete on efficiency ratings in addition to these other two factors. This variation among manufacturers suggests that buyers should comparison-shop.

Manufacturer Strategies

Individual manufacturers employ various strategies to differentiate themselves from other manufacturers. Some emphasize efficiency and quality but charge higher prices; others stress low prices, while the rest use intermediate strategies. Current strategies for each of the major manufacturers are summarized in Table 8-4.

Distribution Channels

Manufacturers sell motors in two major ways—through local and regional distributors, and directly to large national companies and original equipment manufacturers. OEMs are firms, such as air conditioning manufacturers, that incorporate motors into the equipment they make.

Most distributors are independent; i.e., they have no formal link with the manufacturer(s) whose products they sell. A few manufactur-

Table 8-4

Manufacturer Price, Quality, Efficiency, and Distribution Strategies

Manufacturer	Price/Quality Strategy	Position on Energy Efficiency	Distribution Strategy
Baldor	Strong premium emphasis, above average (5–10%)	Leader in developing a CEE-qualified line	Many distributors supported by well-stocked local warehouse; no direct sales
USEM	Solid, broad line; competitively priced	Most of premium line meets CEE standards	Large number of distributors; many carry a second line
MagneTek	Low price line	Only partly CEE-qualified	Lower price used to fill in distributors' primary brand
Reliance	High-quality premium emphasis; higher price	Full line of CEE efficiencies; some hp sizes only offered as CEE-qualified	Two-tier distribution with strong support to top distributors
General Electric	Strong premium emphasis, average pricing	Most of line meets CEE; offers some of the highest efficiencies and highest guaranteed minimums	A few very large electrical supply distributors; GE Supply the most important and Grainger increasingly important
Leeson	Low price line	Only partly CEE-qualified	Lower price used to fill in distributors' primary brand
Lincoln	Low price line	Only partly CEE-qualified	Second price line for some distributors
Toshiba	Higher price, high-premium emphasis; short line	Mostly CEE-qualified; emphasis on quality	A small number of very loyal exclusive distributors who are well supported
Siemens	Specialty line, generally larger motors	Only partly CEE-qualified	A few specialty distributors
Dayton	Standard line, moderate price	Only partly CEE-qualified	Private brand for WW Grainger manufactured mainly by USEM, declining as GE takes over

Source: Easton Consultants and XENERGY 1999a

ers, including Reliance, have their own networks of regional distributors. Sales from distributors to customers are discussed later in this chapter.

Large national companies and OEMs have sufficient buying power to demand the lowest possible price from manufacturers. To achieve these low prices, manufacturers sell directly to large customers, avoiding any distributor markup. Depending on the size of the order and the customer, these prices can be 50–70% off of suggested list prices (Seton, Johnson, and Odell 1987a).

Motor Distributors and Repair-Rewind Shops

Recent market research studies (e.g., Easton Consultants and XENERGY 1999a) divide distributors of small integral-horsepower motors (those who sell motors up to 200–400 hp) into four categories:

- Independent, value-added distributors

- Independent, "order-taker" distributors

- Electrical product chains

- Motion product chains

The independent, value-added distributors often offer a broad line of motors and motor-related accessories (e.g., belts, pulleys, and bearings). They usually provide repair consultation and service as well, with motor sales and repairs accounting for approximately equal portions of revenues. These distributors act as sales consultants, informing customers of their options (including efficiency upgrades) and offering technical assistance. Relative to other types of distributors, these distributors tend to obtain a higher proportion of their total income from motor sales (an average of 37% according to a study in the northeastern United States) and sell a higher proportion of premium-efficiency motors. According to the study, this type represents just under 20% of distributors and accounts for just under 20% of integral motor sales (Easton Consultants and XENERGY 1999a).

Independent, "order-taker" distributors are the most numerous type of distributor (nearly half of all distributors in the Northeast) and account for an even higher proportion of motor sales (65% in the Northeast). These distributors emphasize prompt order filling and price orientation. Like the value-added distributor, they offer a line of motors and related accessories (although often not as broad as the value-added distributor's) and provide rewind services. They tend to emphasize repairs and rewinding more than do the other distributor types; in the Northeast repairs/rewinding account for approximately 50–60% of their revenues, while motor sales represent about 25% of revenues.

Electrical product chains are companies like GE Supply and Grainger that sell a broad range of electrical products and components, including motors. They tend to sell at competitive prices, but generally do not offer motor repairs, except sometimes on a subcontract basis as a convenience for customers. In the Northeast, motors are only about 4% of their total sales. Unlike the independent distributors, who sell primarily to industrial accounts, electrical product chains also have significant sales to contractors and small OEMs (in the Northeast about 20% of electrical chain sales are to contractors and an additional 20% are to OEMs). In the Northeast, electrical product chains account for about 12% of motor sales (Easton Consultants and XENERGY 1999a).

Motion product chains are companies like Kaman and Eastern Bearing that sell a broad range of drivetrain products (e.g., transmissions and belts), including motors. Like electrical chains, they have a significant OEM customer base and do little motor repair work. They offer technical advice, and the best chains provide many value-added services, such as plant motor inventory and rationalization services. In the Northeast, motor sales represent about 9% of motion product chains' business, and the chains account for about 5% of integral motor sales (Easton Consultants and XENERGY 1999a, 1999b).

Distributors generally stock common motor types and sizes, such as standard-efficiency ODP and TEFC motors from 1 to 100 or 200 hp. Less common sizes and such specialized lines as explosion-proof and hollow-shaft motors are usually handled as special orders. Premium-efficiency models are stocked by some distributors, quickly drawn from upstream supplies by a few dealers and left to special order by others. Data on how many distributors fall into each category are not available; however, research in the Northwest and Northeast indicates that in highly developed regions such as the Interstate 5 corridor, the vast majority of dealers fall into the first two categories, while in less developed areas the third category is significant (Easton Consultants and XENERGY 1999a, 1999b).

Manufacturers generally publish suggested list prices on motors, but only small orders to very small customers are actually sold at list price. Distributors sell motors at a discount that varies with the order size, how valued the customer is, and how many other distributors are competing for a particular order. Discounts are typically 30–50% (Easton Consultants and XENERGY 1999a; Seton, Johnson, and Odell 1987a), although high-volume dealers can sometimes provide discounts as high as 60% in bid situations (Easton Consultants and XENERGY 1999a; Stout 1990).

Dealers may compete intensely on orders that go out to bid. In these situations, unless premium-efficiency motors are specified in the bid documents, pressure to come in with the low bid often means that the successful bid will be for EPAct-efficiency motors. Utility rebate programs (discussed in the next chapter) can help overcome this problem.

For the repair of existing motors, distributors compete primarily on speed and price. When a motor breaks down, a whole process line may be shut down, so customers generally want repairs done "the day before yesterday." As discussed in Chapter 2, unless customers test each repaired motor, they cannot easily assess the quality of a repair job, so distributors do not usually compete on quality. A few repair shops have tried to differentiate themselves on quality in recent years by obtaining third-party certification (e.g., ISO 9000, EASA-Q, or Advanced Energy; these certifications are discussed in Chapter 2). However, only a few large customers look for these enhanced services, and as a result dealer participation in quality certification initiatives has been minimal.

Due to the variety of products they carry, many distributors have neither the knowledge nor the time to provide detailed information on whether a high-efficiency motor is appropriate for a particular application. The Marbek study (1987) found that 25% of Canadian distributors surveyed had misconceptions about the technical reliability and applicability of high-efficiency motors.

Sales of high-efficiency products are also hampered by the way orders are handled. First, many orders are processed over the telephone, which provides little opportunity for a distributor to explain high-efficiency products. Recently, some distributors have begun to take orders over the Internet, which provides even less opportunity for interaction. Second, orders are often placed by maintenance or purchasing staff who are concerned primarily about availability and price and have neither the technical background nor the interest to consider high-efficiency products. Third, most order-desk personnel at the distributors have only as much technical knowledge as is published in manufacturers' brochures and catalogs. Clearly, improved dealer education and support are needed so that dealers can better work with customers to improve motor system efficiency (specific ideas are discussed in the next chapter).

Original Equipment Manufacturers

OEMs incorporate motors into many types of equipment, including pumps, compressors, fans and blowers, air handling and HVAC units, industrial machine tools, and conveyor systems.

Figure 8-2

Motor Sales by Sales Channel

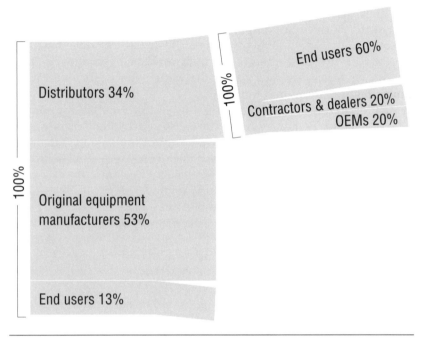

Source: E Source 1999

A study of the U.S. motor market completed in 1995 estimates that OEMs account for about two-thirds of sales, including those directly from manufacturers to OEMs as well as sales from distributors to OEMs (see Figure 8-2) (Easton Consultants 1996). A previous study found that OEM sales account for an even larger share of small motors, with the OEM share of the market gradually decreasing as motor size increases (ADL 1980). Viewed by sector, OEMs buy nearly all the motors used in residential equipment, most of the motors used in commercial buildings, and a sizable portion of industrial motors. Viewed by motor type, OEMs account for approximately 85% of ODP unit sales because both OEM sales and ODP motors dominate in the residential and commercial sectors (Boteler 1999). OEMs thus play a key role in the market penetration of high-efficiency drivepower equipment.

The reality is, however, that OEMs operate in a highly competitive market that encourages them to keep costs down. Because few customers purchasing OEM equipment are aware of or concerned

about efficiency, OEMs generally use standard-efficiency components. An exception to this trend is appliances and cooling equipment that are covered by minimum-efficiency standards: in these
cases, use of improved-efficiency motors is a common strategy for
reaching the minimum-efficiency requirements. Even so, Easton
Consultants' 1995 national study estimates that in 1992 5–10% of
motors purchased by OEMs were energy-efficient models, far lower
than the 20–25% of all motor sales that were energy efficient and the
40–45% of distributor motor sales that were energy efficient (Easton
Consultants 1996).

OEMs generally have engineering staff who evaluate motors
and other components and provide a list of acceptable products to
the purchasing department. Besides basic equipment characteristics
such as size and speed, the most important technical factor in
motor evaluations is reliability—efficiency is usually not considered. For all but the largest motors, engineering staff are usually
not involved in purchasing decisions after a list of acceptable products is developed. The purchasing department selects motors from
the approved list based primarily on price and delivery terms
(ADL 1980; Marbek 1987).

In many cases, particularly with small equipment orders, OEMs
are unable or unwilling to use high-efficiency motors, even when requested by the customer. The most common reasons are small differences in motor technical characteristics and difficulties in obtaining
high-efficiency motors through supply channels geared to supplying
large volumes of standard-efficiency motors. At other times, efficient
motors are available as an option, but at a premium price.

These restraints do not mean that customers have no chance to influence OEM practice. A Canadian study found that approximately
half of the Canadian OEMs surveyed will use high-efficiency motors
in a small proportion of their products, either when specified by customers or with some types of large, specialized industrial equipment
(Marbek 1987).

While OEMs have been slow to use high-efficiency motors,
some have adopted improved gears and belts. As discussed in
Chapter 3, high-efficiency gears and belts are sometimes stronger
than conventional equipment. Added strength can extend service
life and reduce maintenance requirements—important marketing
advantages in some product lines. Stronger materials also allow
components to be downsized, thus reducing material costs at the
same time that efficiency increases.

One final aspect of the OEM market is worth mentioning. When
motors in OEM equipment fail, in about 60% of the cases there are

Compaq Computers Pushes OEMs to Improve Equipment Efficiency

The Compaq Computer Corp. is a large manufacturer of personal computers. It has over 20 facilities throughout the world and spends millions of dollars on energy each year. A large portion of their energy use is for HVAC applications in Compaq offices and manufacturing clean-rooms. Compaq often purchases the most efficient HVAC equipment on the market because in the company's highly competitive market, minimizing long-term operating costs is an essential part of efforts to maintain or increase market share and profits.

However, for major equipment, Compaq is not always content to purchase even the most efficient equipment on the market. Its analyses showed that chillers and cooling towers with even higher efficiency could be produced through changes such as improved motors and heat exchangers. Armed with this information, Compaq approached major chiller and cooling tower manufacturers about producing such units. Manufacturers were originally skeptical, but when Compaq went to Japan to purchase a cooling tower with a higher efficiency than any produced in the United States, domestic manufacturers began to produce the higher-efficiency units too. In fact, due to the competition, the price of this equipment to Compaq has declined.

Compaq's former energy manager of facility resource development, Ron Perkins, offers this advice for companies seeking to improve the efficiency of OEM equipment:

- Conduct some research (using technical bulletins) to estimate what efficiency improvements are possible.
- In approaching manufacturers, make the case that there will be a market for the new, high-efficiency equipment, either due to the purchaser's own market size or due to the fact that the purchaser is a leader in its industry or trade association.
- Publicize success stories and offer testimonials to responsive manufacturers. These efforts will make manufacturers more responsive in the future (Perkins 1989).

restrictions regarding which replacement motors can be purchased, making it difficult to upgrade motor efficiency. The common restrictions are listed in Table 8-5.

Table 8-5

OEM Restrictions on Replacement Motors

Restriction	Percent of Customers Reporting
Replacement motors available only through OEM	22%
Replacement motors available only through one manufacturer	14%
Replacement with motors from unauthorized vendors voids warranty	7%
Replacement motors not available in premium-efficiency models	18%
Other problems	10%
Not applicable to motors in facility	6%
No problems reported	33%

Note: Customers could name more than one restriction.

Source: XENERGY 1998

Fans, Pumps, and Compressors

As noted in Chapter 6, fans and blowers, pumps, and air compressors together account for the majority of industrial motor energy use and are thus perhaps the most important types of OEM equipment. A DOE report (Easton Consultants 1996) discusses these markets in some detail; the remainder of this section is based largely on this work.

All four types of equipment have several different classes (dozens in the case of fans/blowers and pumps), each designed for separate types of applications. These markets tend to be very competitive and not every manufacturer produces all types of equipment. At the manufacturer level, the fan/blower, pump, and compressor markets are largely distinct—few manufacturers serve more than one of these three markets.

Industrial Pumps

Industrial pumps are primarily sold through manufacturers' representatives and distributors. Manufacturers' representatives are independent sales agents whose primary function is to assist in equipment selection for the job, often using manufacturer-provided manuals, pump curves, and software. However, they do not install or service equipment or systems and therefore have little stake in system efficiency. Representatives and distributors provide the primary link between the manufacturer and the contractor or end-user, though representatives and distributors vary widely in sophistication and the service they provide. Some provide design, repair, and maintenance services, while others simply order and obtain equipment for the end-user or contractor.

In some cases, pumps are sold directly to the end-user through the manufacturer. One example is the process pump, which is engineered specifically for a particular end-use application. As for design and installation, consulting engineers design nearly all pump systems in new facilities and may also play a role in system renovations and retrofits, while mechanical contractors handle the installation. Some end-users design smaller system renovations themselves. Larger end-users, particularly in the chemical and petroleum industries, often employ internal process engineers to do system design work.

Fans and Blowers

For industrial fans and blowers, the market is fragmented, with no manufacturer accounting for more than a 12% share. Manufacturers sell fans and blowers to other OEMs for applications that include dust collection, HVAC, oven vents, boilers, and pollution control equipment. OEMs and end-users generally purchase motors separately from the fan or blower. Representatives provide varying levels of design assistance, but contractors generally install most fan and blower systems (including fans, ductwork, and ancillary equipment). Design and specification engineers work with the contractor, end-user, and manufacturers' representative to design the system for reliability, low noise, and efficiency, and may recommend specific equipment models or their equivalents. However, they are not responsible for the efficiency of installed equipment, relying instead on post-installation air-balancing by independent firms. These firms may be called on to test the system after installation and certify that it meets design criteria, but they generally do not focus on energy per se.

Air Compressors

For air compressors, the market is concentrated, with a few manufacturers making up more than 75% of sales. Sales by distributors dominate the market, accounting for 85–90% of all sales. However, distributors have little motivation to offer energy-efficient equipment because most of their revenue is derived from parts and service. This results in a tendency to offer low-cost, low-efficiency compressor models in the distributors' bids in an attempt to win business and then establish a lucrative service relationship. An exception to this rule is Ingersoll Rand, the largest compressor manufacturer, which has its own sales force that markets not only compressors but also a variety of value-added services. In a search for higher profits, some other compressed-air distributors have also begun to offer value-added services such as auditing and maintenance. A few vendors have even begun to offer metered compressed-air contracts in which the vendors maintain compressed-air systems on customers' premises and bill their customers based on the cubic feet of compressed air delivered.

By contrast, compressor "air-end" manufacturers and packagers are involved in the whole process—component design and manufacturing, package design and assembly, and (in certain cases) distribution. As a result, the manufacturers and packagers determine the effort applied to compressor engineering and design and the overall level of efficiency of the compressor package. Some manufacturers also rebuild compressors. Manufacturers of motors and other ancillary compressed-air system components as well as foreign air-end manufacturers also supply components to compressor OEMs.

Distributors, manufacturers' representatives, and consulting engineers (including compressed-air specialists) all have an impact on the compressed-air market. Three types of distributors serve the compressed-air market. They vary by level of design and selection assistance: compressed-air specialists offer extensive assistance; general industry distributors offer limited assistance; and warehouse distributors offer little or no technical support. Manufacturers' representatives serve the market for large compressors (approximately 1,000 hp and higher) and generally do not get involved with the plant air market (with the exception of Ingersoll Rand). Consulting engineers design and plan compressed-air systems for ease of maintenance, low noise, and reliability; efficiency is rarely a primary concern. Few engineering firms have compressor system specialists on staff, and expertise is rare. Engineers focus primarily on ensuring that the system can deliver sufficient airflow at the required pressure to all point-of-use locations. Finally, air compressor audit firms troubleshoot compressed-air systems and recommend solutions to equipment, system, and O&M problems.

Additional information about the compressed-air systems market can be found in a report prepared for the DOE and the *Compressed Air Challenge* (XENERGY 2000).

Consulting Engineers and Design-Build Contractors

Consulting engineers prepare designs and specifications and help oversee the bid and construction process, but they leave the construction work to in-house staff or outside contractors. Design-build contractors, on the other hand, handle both design and construction, thereby providing turnkey facilities, and are typically large firms.

Consultants are generally hired by end-users to assist with large projects such as new plants, new production lines, and major new equipment such as large ASDs. These projects are infrequent and large enough to make hiring temporary consultants more cost-effective than hiring permanent staff.

Successful consulting practices are based on satisfied customers. For this reason, many consultants use decision criteria similar to those considered by end-users: the consultants stress reliability and are notorious for oversizing motors so as to provide a wide safety margin (Van Son 1989). Consultants sometimes recommend high-efficiency equipment, but if clients resist the idea, the consultants will usually drop the suggestion. Even when high-efficiency equipment is specified, it is commonly one of the first items to be cut from a project if bids come in higher than expected, which is a common occurrence.

When equipment unfamiliar to installation contractors is specified (ASDs, for example), installation prices often include a "risk factor" to cover the cost of unanticipated problems. In essence, the customer is paying for the contractor to learn how to install and troubleshoot a new type of system. On the other hand, ASDs are becoming common enough that customers have multiple vendors to choose from and are pressuring consulting engineers as well as equipment manufacturers to compete on price (along with other factors).

Design-build contractors are increasingly being paid through fixed-cost contracts (Ontario Hydro 1988). A fixed-cost contract provides protection against cost overruns and thereby allows the end-user to correctly budget for a project. It also places great pressure on design-build contractors to keep initial costs down. Under these conditions, high-efficiency motors and other high-efficiency measures with increased costs are rarely specified.

OEM representatives (discussed in the next section) often provide design services similar to those provided by consultants. These representatives are not paid directly for this work: their fees are incorporated into the price of OEM equipment. The smaller the project, the more likely an OEM representative will be the primary designer.

Control Equipment Manufacturers, Distributors, and Representatives

Manufacturers of ASDs and other electronic controls include some large firms (both motor manufacturers and independent control manufacturers) as well as many small specialty firms. Many of the major firms serving the U.S. market are listed in Table 8-6. Of these companies, about one-third each are broadline electrical equipment

Table 8-6

ASD Sales in Broadline Electrical Equipment Companies, Electric Motor Manufacturers, and Automation Product Companies

	Name and Location of ASD Operation	Est. ASD Sales (million US$)	Est. Market Share	Product Emphasis	Industry Strength
Broadline Electrical Equipment Companies	ABB, New Berlin, WI	200–240	20%	Full range of products from fractional to a recently introduced intermediate voltage unit to over 5,000 hp.	Chemical, pulp, HVAC, water, and wastewater
	General Electric (GE Industrial Control Systems Group), Salem, VA	80–90	7.5%	A range of products in the lower voltage range—torque control, vector drives. Joint venture with Fuji provides deep small unit offering. Expanding a higher voltage line.	Refineries, HVAC, water, and wastewater
	Siemens, Albany, NY	50–60	5%	Products in the low voltage range primarily list drives up to 5,000 hp+.	Refineries, metals
Electric Motor Manufacturers	Baldor, Fort Smith, AZ	50–60	5%	Broad line of products for low voltage.	Across industry
	MagneTek (A.O. Smith), Milwaukee, WI	50–60	5%	Broad line of low-voltage product—private brands and Yaskawa Products.	Automotive
	Emerson Electric, St. Louis, MO	50–60	5%	Has acquired a broad line of electrical controls—servos, AC and DC controls from small to 1,500 hp.	Metals, process control, HVAC
	Toshiba, Houston, TX	50–60	5%	A higher cost line in the low-voltage range.	Refinery, metals
Automation Product Companies	Rockwell Automation (Allen Bradley–Reliance), Mequon, WI	230–250	22%	Extremely broad range of ASDs up to 400 hp.	Automotive, chemical, pulp and paper, HVAC, food and beverage, water
	Robicon, New Kensington, PA	50–60	5%	Range of products including growing line of intermediate-voltage units.	Water and wastewater, food and beverage
	Danfoss, Rockford, IL	30–35	3%	Range of products generally oriented to machine control; has a heavy-duty line to 500 hp.	HVAC, food and beverage
	Cutler Hammer (Eaton Corp.), Oldsmar, FL	20–25	2%	480 volt units from small to 1,100 hp; sold with motor starters, overload devices, etc.	HVAC

Source: Easton Consultants 2000

companies, control-based companies, and motor-based companies. As of this writing, all major motor manufacturers now sell ASDs also, with many of these companies having acquired a control manufacturer in recent years. Motor-based manufacturers have been steadily increasing their share of the market in recent years, partly by developing and marketing matched control/motor sets that are optimized for each other and thereby minimize incompatibility problems in the field. Most manufacturers specialize in particular sizes or types of electronic controls. Many of these company-specific emphases are summarized in Table 8-6.

Adjustable-speed drives and other electronic controls are generally sold through distributors or sales representatives, most of which handle more than one type of equipment. In addition to ASDs, for instance, they may also sell power-factor correction or uninterruptible-power-supply equipment. A few handle more than one product line for a particular type of equipment. Distributors stock products, while representatives generally place orders as needed with the manufacturers. Representatives and distributors will often perform design work for end-users in an effort to sell equipment the representatives and distributors represent. The overall structure of the ASD market is illustrated in Figure 8-3.

Electronic controls are generally purchased in small quantities, so steep discounts are the exception rather than the rule. Because

Figure 8-3

Schematic Illustration of the Structure of the ASD Industry

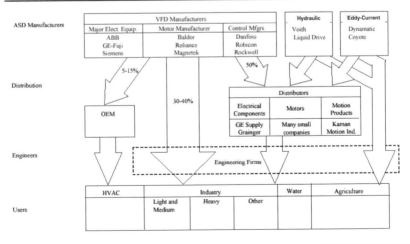

Source: Easton Consultants 2000

installation costs can be substantial, equipment and installation costs are often considered together in evaluating system economics.

Mechanical Equipment Representatives and Distributors

Manufacturers of gears, belts, chains, bearings, and lubricants also range widely in size. Large firms involved in mechanical equipment include motor manufacturers (such as Reliance, a major producer of gears), rubber companies (which manufacture belts), and oil companies (which manufacture lubricants). Many small specialty firms are also involved in the manufacture of mechanical equipment. A growing number of foreign firms are exporting mechanical equipment, especially belts, to the North American market.

Mechanical equipment manufacturers primarily sell to small buyers through mechanical equipment distributors, and sell to OEMs and large end-users directly. Sometimes a motor distributor will also stock mechanical equipment. Smaller firms often use regional representatives to promote their products, with stocking and distribution handled by the central office.

Lubricants are generally sold by large oil companies through local petroleum product distributors. These distributors do the majority of their business in the automotive field—commercial and industrial business represents only a small portion of their annual sales. Small lubricant companies that manufacture premium-grade (high-efficiency and long-life) lubricants tend to sell through a network of product representatives. In a previous section we noted how some motor distributors are now offering lubrication services; these services are offered by certain lubrication distributors as well.

Mechanical equipment is usually priced similarly to motors. Suggested list prices are published, but most purchasers receive a substantial discount that varies with the size of the order, the size of the customer, and other factors. With lubricants, published price schedules include volume discounts. These prices are generally followed except in bid or other special situations (Hudson 1989; Kent 1989).

Electric Utilities, Universities, Government Agencies, and Trade and Professional Associations

These groups have become increasingly active in the motor market. Many provide educational materials, seminars, and technical assistance on motor system issues. For example, as discussed in Chapter 9, DOE is undertaking the *Industrial Best Practices: Motors* (formerly known as *Motor Challenge*) program to educate end-users about opportunities to improve their motor systems and strategies for taking advantage of these opportunities. In addition, trade associations such as NEMA and groups of government agencies and utilities (e.g., the Consortium for Energy Efficiency and the Canadian Electrical Association) develop standards for testing, classifying, and labeling motor system products.

Utilities have become increasingly involved in improving power factor, power quality, and efficiency; they have also begun promoting use of electrotechnologies among customers. Utility efforts are primarily educational, although several utilities provide rebates for the purchase of premium-efficiency motors and other energy-saving measures.

Further information on the activities of these players in the motor market is provided in Chapter 9.

Summary

The motor market is quite complex in that it involves many different players and many different decision criteria. As we have discussed, the market may be segmented many ways, including by system type (e.g., fan), customer size, and industry type (e.g., petroleum refining and paper and pulp). However, a particularly important way to segment the market is by market event—that is, whether the purchase is for replacing existing equipment, retrofitting existing equipment, or new applications. The market for replacement equipment is very different from the retrofit and new applications markets. Key attributes of these markets are summarized in Table 8-7. Planners and managers of programs to promote motor system improvements need to understand these different markets in order to shape programs that will successfully serve them.

The replacement equipment market involves frequent purchases of small quantities of equipment, primarily from local or regional distributors. Decisions are made quickly by maintenance or purchasing staff, primarily on the basis of availability, cost, and reliability. Engineering analyses of such replacement purchases are rare; when old equipment fails, it is often replaced with identical components.

Programs designed for this market need to either reach the decision-maker at the time the purchase is made or promote standard policies that determine in advance which equipment to replace with high-efficiency models. The Southwire case study earlier in this chapter exemplifies how this can be done.

Purchases of equipment for new applications occur infrequently, but when they are made, they generally involve an engineering analysis.

Table 8-7

Key Attributes of the Markets for Replacement Equipment and New Applications

	Replacement Equipment Market	New Application Market and Retrofit Market
Frequency of end-user purchase decisions	Continual	Infrequent
Order size	Generally small	Frequently medium or large
Decision-makers	Maintenance and production staff	Engineering staff (sometimes at head office)
Time spent making decisions	Hours or days	Weeks or months
Key factors	Availability, reliability, cost	Cost and reliability for commodity products; and operating cost savings for noncommodity products and design decisions
Engineering analysis	None, except for large motors and OEM equipment or for setting general purchasing guidelines	Done by either in-house staff, outside consultants, or OEM representatives
Purchased from	Distributors or (for very large companies) manufacturers	Design-build contractors, OEMs, manufacturers, or (for small projects) distributors
External influences	Distributors, service shops	Engineering consultants, contractors, OEM representatives

These analyses often balance reliability, initial cost, delivery terms, and performance and operating cost considerations. Due to pressure to minimize initial costs, future savings are frequently heavily discounted. But, because new application decisions often involve large quantities of equipment, selection and design decisions are made over a period of months and there is time to reach the decision-maker.

The retrofit market is similar to the new application market in that retrofit projects generally occur infrequently; but when they do, an engineering analysis is frequently involved. On the other hand, retrofit projects are usually much smaller than new application projects due to an aversion to downtime, limited capital availability, and other considerations. Furthermore, unlike new application projects that are generally undertaken to address key organizational objectives such as increasing production capacity or other services, retrofit projects are much more difficult to sell to management unless the benefits of the project are really compelling.

Programs and policies to reach these different markets are discussed in the next chapter.

Programs to Promote Motor System Efficiency Improvements

The opportunities for increasing motor system efficiency are substantial, but so are the barriers to such improvements. Programs designed to overcome these barriers and foster more efficient drive-power systems have changed substantially since the first edition of this book was published 10 years ago. In this chapter we emphasize current program paradigms but also discuss some history and results of past efforts.

Program Paradigms

Energy-saving programs, including programs to improve the efficiency of motor systems, are commonly offered by utilities, government and nonprofit agencies, and private industry.

These program operators have many rationales for operating energy efficiency programs, but today three paradigms tend to dominate: market transformation; customer service; and resource acquisition.

Market transformation initiatives seek to remove market barriers that impede specific energy-saving practices and, over time, make these practices common and self-sustaining. As discussed below, in recent years many utility and government programs have emphasized the market transformation perspective and developed strategies for removing market barriers to efficient motors and other motor system products and services.

Customer service has been a goal of the private market and many utilities for a long time. Private companies provide customer service in order to promote sales of their products. In recent years, with the onset of competition in the utility industry due to restructuring, some utilities have begun to use energy efficiency services as part of their efforts to provide value-added services to customers. These services can be provided by traditional utilities or they can be provided by competing power retailers as part of their efforts to build or retain market share. Frequently these services are provided on a reduced-cost basis, although they can also be offered on a full cost-for-service basis.

Resource acquisition programs traditionally sought to reduce energy use whenever conservation is less expensive per kilowatt-hour than available electricity supplies. In recent years resource acquisition programs has been commonly used to indicate the direct and quick acquisition (relative to market transformation) of energy savings in order to reduce power plant emissions, help address power reliability problems, and defer the need for expensive (and sometimes controversial) upgrades to the distribution system.

These paradigms are not competing—in fact they can complement each other. For example, customer service is frequently a part of market transformation and resource acquisition–focused programs. With motor systems, it is even possible to combine market transformation and resource acquisition by conducting short-term resource acquisition (for example, by promoting leak detection and repair in compressed-air systems) in ways that have direct and long-term impacts on market barriers.

In this chapter we discuss program approaches that address each of these perspectives. However, given the emphasis placed on market transformation in recent years by many program operators, we emphasize this perspective while keeping the other perspectives in mind.

Market Transformation

Why Market Transformation?

In the past, governments, utilities, and other program operators have commonly used a broad array of regulatory and voluntary mechanisms to promote energy-saving investments and actions that are in the public's interest. These mechanisms have included education and technical assistance programs, utility rebates and other demand-side interventions, building codes, equipment labeling, and minimum-efficiency standards. However, in many cases, these past efforts have focused on short-term objectives and not on addressing underlying market barriers that hinder the long-term adoption of cost-effective energy-saving measures. And many of these activities have been conducted in isolation from similar activities by others.

In order to address these limitations with traditional program approaches, a growing number of practitioners and policymakers are adopting a "market transformation" framework that attempts to incorporate the best features of, and improve the coordination between, market-based and regulatory approaches. For example, as part of utility restructuring policies, quite a few states have embraced the market transformation concept and a growing number of states have established special funding for new market transformation programs (Nadel and Latham 1998).

What Is Market Transformation?

As noted above, market transformation means reducing, in a sustained manner, market barriers to the adoption of cost-effective energy efficiency products and services. If the most important and relevant market barriers have been addressed to the point where efficient goods and services are normal practice in appropriate applications, and these changes are sustained over time, then a market is transformed.

Due to the substantial effort required, generally a market transformation strategy for a particular measure is designed to promote comprehensive changes across many parts of a market, not just at the margins. Measures are chosen for which substantial increases in market share appear achievable. Choosing measures in this way can maximize savings while making efficient use of limited resources. The real benefits of market transformation are achieved when multiple activities are combined into coordinated initiatives.

Market transformation efforts are different from most traditional utility demand-side management (DSM) programs in several respects. The primary difference is that the fundamental goal of market transformation is to change markets, not save energy in the short term. By changing markets, market transformation initiatives are designed to save substantial amounts of energy in the long term. As a result, market transformation activities are devised in direct response to identified market barriers. In fact, understanding the particular market barriers to widespread adoption of a technology or service is essential for developing and implementing successful market transformation activities. In addition, market transformation initiatives generally are broader and longer term than typical DSM programs. A market transformation initiative may have several phases, many players, and a variety of activities. Coordination among the relevant players is thus necessary to ensure that a market transformation initiative or strategy is effective and the broad goals are accomplished. Since the primary goal of market transformation is to change markets, the evaluation of market transformation programs emphasizes progress made in addressing market barriers and not precise measurements of program energy savings. While many traditional DSM programs include some of these attributes, few include all of the attributes that typify market transformation programs. However, market transformation is not a label that uniquely identifies certain energy efficiency program designs to the exclusion of others. It is instead an objective that all energy efficiency programs have at least a theoretical potential to achieve, although some programs are clearly more effective at achieving this objective than others.

Frequently, a market diffusion, or "S", curve is used to illustrate the market transformation process (see Figure 9-1). The market diffusion curve shows an idealized version of the process by which a new technology or practice evolves from market introduction to mass-market or wide-scale adoption. The market history of many technologies (such as microwave ovens, VCRs, etc.) can be represented using this type of curve. Market transformation initiatives typically include activities designed to accelerate the market adoption of a particular energy-saving measure so that it becomes (and hopefully remains) common practice much sooner than it would otherwise. Accordingly, market transformation initiatives often include activities designed to (1) stimulate the development and market introduction of new energy-efficient models; (2) strategically build the market share of these new products until they attain a niche position in the market; and then (3) change consumer purchasing practices in order to further expand the market adoption of these measures so that they reach mass-market status and eventually become common practice.

Different activities, or "tools," are appropriate at different points along this market diffusion curve since barriers are often a function of product/market maturity (see Figure 9-1). For example, research and development and technology procurement efforts may be employed in the early stages of an initiative in order to stimulate the introduction of new high-efficiency measures. Rebates and targeted outreach to large purchasers (e.g., bulk purchases) may be used to strategically increase market penetration until the measure achieves "niche" status. Consumer education, loans/rebates, and other promotional activities such as ENERGY STAR labeling may be used to expand a measure's market share to its full mass-market potential. And codes and minimum-efficiency standards (or in some cases voluntary standards) can be used to complete the transformation process by removing clearly inefficient products and practices from the market.

274

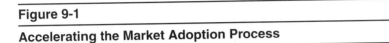

Figure 9-1

Accelerating the Market Adoption Process

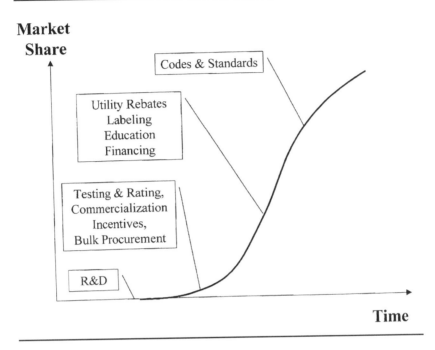

Market Transformation Example

To illustrate the market transformation concept, let us examine a market transformation initiative that took place in the 1980s and successfully transformed the market for energy-efficient motors (meaning motors near the EPAct-efficiency levels) in the Canadian province of British Columbia (B.C.) and ultimately in all of Canada.

The B.C. initiative consisted of four components: (1) educational efforts to provide customers and dealers with information on energy-efficient motors (specifically, their economics and availability); (2) customer incentives to pay part of the incremental cost of energy-efficient motors; (3) vendor incentives to encourage vendors to routinely stock and promote energy-efficient motors; and (4) support for efforts to enact provincial and national minimum-efficiency standards. As a result of the first three components, energy-efficient motors had a 70% share of the new motor market in 1993, up from approximately 5% in 1987. In 1992 and again in 1993, the utility reduced the incentives by just over 10%; market penetration still held since by then dealers routinely stocked (and many customers routinely requested) efficient motors. In 1993 provincial efficiency standards were adopted and B.C. Hydro was able to phase out their motor activities (Nadel 1996). In 1994, these same standards were adopted by the Canadian federal government.

Elements of a Market Transformation Initiative

Drawing from this experience, in general terms, a market transformation initiative or strategy for a specific market segment or end-use will often involve

- A careful analysis of the overall market, including identification of the particular barriers that are hindering the development, introduction, purchase, and use of the targeted measure. In the case of the B.C. initiative, special attention was devoted to improving local availability and reducing the initial cost of energy-efficient motors, concentrating initially on large customers and large motor distributors.

- A clear statement of the overall goal of the initiative or strategy as well as the specific objectives that will be accomplished along the way by the different initiative activities. In the B.C. initiative, the strategy was to raise the local market share in order to make provincial, and ultimately federal, minimum-efficiency standards uncontroversial.

- The development of a set of coordinated activities that will achieve the desired objectives and systematically address each of the identified barriers. In B.C., the initiative included education and incentives for vendors and customers, and advocacy for minimum-efficiency standards.

- Successful implementation of the individual activities, including periodic evaluations and adjustments designed to respond to actual experience. In B.C., as a result of evaluation results, program operators decided to adjust incentives downward several times.

- Development and execution of a plan for transitioning from extensive market intervention activities toward a largely self-sustaining market, i.e., an "exit strategy." In B.C., the exit strategy was minimum-efficiency standards; for other market transformation initiatives, other strategies may be employed. However, B.C.'s "exit" applies only to sales of efficient motors in new construction and replacement situations; promotion is still needed for sales in retrofit situations and for sales of premium-efficiency motors in new construction and replacement situations.

Limitations to the
Market Transformation Approach

While the market transformation approach offers many advantages, it also has some limitations and as a result it is not appropriate for all situations. First, the market transformation approach is most appropriate for measures whose benefits to the customer are substantial and whose barriers can be readily identified and addressed. Market transformation also tends to be more successful when a mandatory code or standard is ultimately enacted to complete the transformation process or the inefficient technology can be effectively eliminated from the marketplace by customer preferences. Conversely, for measures with many complex barriers (some of which may be very difficult to overcome), market transformation will be difficult. Also, some measures are appropriate in only a few applications and thus do not lend themselves to mandates. For example, a market transformation approach may be very appropriate for premium-efficiency motors and improved motor management practices since these measures face relatively well-defined barriers that, as discussed below, can ultimately be addressed by clearly defined corporate purchase and management policies. On the other hand, properly optimizing fan and pump systems is difficult to do, faces many complex barriers, and is too application-specific for standards to be of much help, with the result that partial rather than full transformation of the market is a more likely outcome.

Second, market transformation is a long-term process that typically provides only limited benefits in the early years—the major benefits tend to occur after a 5–10-year period. For example, in the B.C. program discussed above, it took 8 years from initial program planning to the effective date of new standards. Where substantial savings are needed in the short term, resource acquisition strategies will generally be more appropriate. Market transformation also typically requires extensive coordination among many diverse parties (in B.C., for example, motor distributors, motor manufacturers, end-users, the utility, and the provincial government) as well as periodic refining of the program approach as the market evolves over time. In cases where such coordination and program refinements are not possible, other program approaches will be needed.

Finally, market transformation initiatives typically require extensive efforts and considerable expenses. Market barriers are often substantial and do not disappear with minimal effort. Some people have mischaracterized market transformation as meaning primarily low-cost educational activities. In most cases, significantly more intervention will be needed, such as one-on-one technical assistance and financial incentives, particularly in the early years of a program. For example, B.C. Hydro started its motor program with substantial incentives for high-efficiency motors and then gradually reduced and finally eliminated the incentives.

Market Segments and Market Events

With the market transformation approach, as well as with other program approaches, it is important to understand and segment the market. Appropriate strategies will be different for different market segments; a program that applies a "one size fits all" approach will be doomed to failure.

There are many ways to segment the motor systems market, including by type of product (e.g., motors, fans, and compressed-air system optimization services), customer type (e.g., small vs. large, single-site vs. multi-site, and owner-occupied vs. tenant), sector (e.g., commercial, institutional, and industrial), and industry (e.g., mining, chemicals, and paper). In the sections below, we will discuss some of these different segmentation schemes where appropriate. However, one overarching scheme that needs to be considered in designing most programs is segmentation by market event.

Market events are times within the life of a motor system in which decisions are made to procure motor equipment and services. For program planning purposes, three market events are particularly important—new construction, equipment replacement, and retrofits. Each of these market events typically involves different decisions, decision-makers, and economic returns. Programs either can be designed to serve one of these markets or can serve multiple markets if they include different program design features to serve each market.

New construction involves installing equipment for the first time, such as installing a new process line or building a new building. New construction projects (as well as major renovation projects including design of new process lines) occur infrequently, but when they occur, major design decisions are made that involve a large investment of money and have a substantial impact on energy use. These decisions are frequently made by design engineers and other design professionals. In new construction projects, the cost of efficiency projects is the cost increment between standard and efficient equipment and design practices. In new construction projects, budgets are typically tight and thus efficiency investments often compete with other design ideas (e.g., a nicer lobby) for a share of the budget. If efficient equipment or designs are not chosen, the cost to retrofit these improvements later is generally much higher. Hence new construction and major renovation opportunities are often called a "lost opportunity" resource; if the efficient practice is not adopted at the time of construction, the opportunity may be lost forever.

The replacement market involves replacing existing equipment with new equipment when it wears out or fails. Many failures happen suddenly, so decisions must be made quickly to get equipment back on-line as soon as possible. Replacement situations are often a good time to promote high-efficiency equipment and proper equipment sizing, but due to the tight time schedule and the fact that only one part of the system is being replaced at a time, it is usually not possible to implement major system design changes. On occasion, staff may anticipate that equipment will fail soon and undertake a more orderly planned replacement project (planned replacement strategies are discussed in Chapter 2), but in most businesses, planned replacement is more the exception than the rule. In the replacement market, decisions are frequently made by maintenance or purchasing department staff; frequently an engineer is not involved. As with the new construction market, since equipment must be purchased, the cost of efficiency investments in replacement situations is the difference in cost between standard and efficient practices.

Retrofit projects involve the planned removal of equipment before it needs to be replaced and tend to disrupt normal operations. They also tend to be relatively expensive since there is no credit for the cost to purchase and install standard equipment, as is the case with new construction and replacement projects. For these reasons, retrofit projects are generally undertaken when they produce large benefits, such as large energy bill savings and/or productivity or product improvements. It is often these latter benefits that are needed to sell the project, because retrofit projects are typically organized by in-house engineers who plan the projects and then solicit approval and funding from management.

Building on the preceding discussion of program approaches and market events, we now turn to a discussion of program strategies for promoting more efficient motor systems. There are many possible ways to organize such a discussion. However, given our emphasis on the market transformation approach, we begin our discussion of program strategies by discussing strategies for different types of products because the marketplace today is primarily ordered by product (e.g., motor manufacturers compete with other motor manufacturers across multiple market segments), and it is the market that market transformation initiatives seek to change. However, many strategies cut across these different product categories and seek to promote integrated applications of multiple products. Thus, following the discussion of strategies for specific product segments, we proceed to a discussion of strategies that cut across many of these segments.

Program Approaches for Specific Products

In discussing program approaches for different products, we define the term *product* from a marketing perspective: it refers to anything that is for sale, including hardware, software, and services. In the following sections, we discuss programs to promote the following products: efficient motors; good motor management and motor repair practices; compressed-air system optimization; fan and pump system optimization (including irrigation systems); improved air conditioning systems; and refrigeration system efficiency improvements.

Motors

Over the past two decades there have been many efforts to promote purchases of high-efficiency and premium-efficiency motors. In this section, we briefly review some of these past efforts and then discuss current programs to promote premium-efficiency motors in more depth.

Past Efforts

Efforts to promote high-efficiency motors began with education, technical assistance, and labeling programs by manufacturers, government agencies, and utilities. These efforts included educational publications (e.g., NEMA 1999), motor labeling (discussed in Chapter 2), calculation tools to estimate energy and economic savings of more efficient motors (e.g., slide rules and computer programs), seminars, and one-on-one technical assistance with individual customers. These programs were seldom evaluated, but as of the late 1980s the limited data available indicate that high-efficiency motors accounted for only about 3% of the motor stock (Gilmore 1989; XENERGY 1989), suggesting that these programs were having only limited impact.

In the late 1980s many utilities began offering incentive programs to encourage the purchase of high-efficiency motors. Most of the programs offered rebates for purchase of high-efficiency motors, where "high-efficiency" was defined in terms of a utility-developed table that indicated qualifying efficiency levels as a function of motor type (ODP or TEFC), speed, and horsepower. Some utilities used the NEMA definition of high-efficiency motors (NEMA 1999), but most used their own definition, feeling that the NEMA definition was too low. The resulting confusion prompted NEMA to develop a new definition labeled "suggested standard for future design" (NEMA 1999) that many utilities then adopted. Typical rebate programs provided rebates of about $10/hp for qualifying motors. Rebates were commonly paid to the customer, although some utilities used dealer rebates in addition to or instead of customer rebates. Programs were marketed both directly to customers and through motor dealers. Most programs, however, had only limited marketing efforts, and limited participation rates as a result (Nadel et al. 1991).

A few programs, such as the B.C. Hydro program discussed above, included extensive marketing and education efforts, which tended to be much more successful. In addition to the B.C. Hydro program, another very successful program was a 1986 pilot program operated by Niagara Mohawk Power Corp. in which 33% of targeted customers participated. The program targeted large customers with long operating hours, provided extensive personal attention, offered a free computer assessment of costs and savings, and provided high ($25/hp) rebates—sufficient to pay more than half the cost of a new replacement motor in many applications (Niagara Mohawk 1987).

By the early 1990s, high-efficiency motors accounted for approximately 20% of new motor sales. While this was a significant accomplishment, after more than a decade of work the vast majority of sales were still standard efficiency. At the same time, minimum-efficiency standards on residential appliances were going into effect and having dramatic impacts on sales of high-efficiency appliances. California and the federal government had looked at setting motor standards in the 1980s but had not proceeded on this path due to lack of time (California) or a feeling that sales of high-efficiency motors would take off in the absence of standards (the federal government) (Nadel et al. 1991). With the success of appliance standards, states and the federal government again began to look at setting motor standards, starting with Massachusetts and then extending to a Massachusetts congressman who introduced federal legislation. This proposed legislation spurred negotiations between motor manufacturers and energy efficiency advocates, resulting in an agreement to jointly support modified federal legislation. This legislation, included in the Energy Policy Act of 1992, took effect in October 1997. It covers the most common types of motors (further details are provided in Appendix B). At the time of passage, it was estimated that the new standard would result in savings of approximately 13 billion kWh by 2013 and cumulative net benefits to motor users of nearly $5 billion. This analysis is summarized in Table 9-1.

The legislation establishing minimum efficiency standards in the United States calls for DOE to update standards every 5 years, with new standards set at the maximum levels that are technically feasible and economically justified. However, DOE is unlikely to begin a rule-

Table 9-1
Estimated Savings from Motor Efficiency Standards in the United States

Motor Horse-power	Annual Sales[1]	Sales for Which Efficient Motors Are Available[2]	Avg. Motor Size[3]	Avg. Efficiency[4]		Avg. Annual Op. Hours[5]	Avg. Ann. kWh Savings per Motor[6]	GWh Savings[7]		MW Savings[8]		Avg. Cost Premium for High-Efficiency Motor[9]	Net Dollar Savings (millions)[10]			15-Year Benefit/Cost Ratio[10]
				Standard	High			1st Year	15th Year	1st Year	15th Year		1st Year	15th Year	15-Year Cumulative	
1–5	1,154,483	738,869	2.07	79.8	85.0	2,352	165	122	1,828	56	838	$43	($18)	$84	$496	2.3
6–20	470,211	300,935	11.9	86.3	90.7	2,928	868	261	3,917	96	1,443	$100	($8)	$211	$1,521	5.2
21–50	144,658	92,581	32.5	89.9	93.1	3,568	1,970	182	2,736	55	827	$223	($6)	$148	$1,066	5.3
51–125	70,298	44,991	86.7	92.0	94.7	4,163	4,977	224	3,359	58	870	$431	($2)	$186	$1,380	6.9
126–200	14,661	9,383	212	93.8	95.6	4,163	7,899	74	1,112	19	288	$1,534	($7)	$55	$361	3.1
Total	1,854,311	1,186,759						863	12,952	284	4,266	$98	($41)	$684	$4,824	4.5

Notes:

[1] From U.S. Census Bureau 1988.

[2] Based on the following estimates: 87% of motors have a T-frame × 94% are TEFC or ODP × 98% are 2–6-pole × 80% are foot-mounted and of NEMA Type A or B. Figures from Gilmore (1989) on over 1,000 motors in Rhode Island.

[3] From Argonne National Laboratory (1980).

[4] Average nominal efficiency for motor nearest in size to average motor size. Based on average nominal efficiency for 1,800 rpm ODP and TEFC motors as listed in Tables A-1 and A-2.

[5] Average of values estimated by Arthur D. Little (ADL 1980) and XENERGY (1989). Values from Rhode Island study of commercial and industrial motors (see Note #2 above) are even higher.

[6] Motor hp × .746 kW/hp × .75 average load × operating hours × [1(80% × standard efficiency + 20% × high efficiency) − (1/high efficiency)]. 20% and 80% are estimates of present sales shares of high-efficiency and standard motors in the United States (see Chapter 6).

[7] First-year savings are the product of kWh savings/motor and annual sales to which standards apply. Fifteenth-year savings are 15 × 1st-year savings. For these simple calculations, growth in the motor stock over the next 15 years is ignored.

[8] GWh savings/annual operating hours × 1.096 T&D loss factor × 1.23 reserve margin factor × 80% of motors assumed to be operating at time of maximum peak demand. T&D and reserve margin factors estimated from New England Power Pool data. Peak coincidence factor is an ACEEE estimate.

[9] Average cost difference for motor nearest in size to average motor. Based on 1,800 rpm TEFC and ODP motors as summarized in Nadel et al. (1991).

[10] In 1990 dollars based on $.06/kWh. Includes allowance for increased purchase price on 80% of motor sales that are presently standard-efficiency models. For this simple calculation, real fuel price inflation and discounting are ignored.

making process until at least 2005 since a rulemaking prioritization process is now in effect and revisions to other standards (e.g., appliances) are likely to save more energy than revised motor standards.

At the same time the United States was adopting minimum-efficiency standards, similar efforts were taking place in Canada. The first Canadian standards took effect in 1993 in British Columbia and Ontario. These standards were not as stringent as the U.S. standards but took effect more than 4 years earlier. These standards became national standards in 1994 and were subsequently revised to require the same efficiencies as the U.S. standards, effective 1996 in British Columbia and Ontario, and October 1997 nationwide.

Current Efforts

Following the effective date of the EPAct motor efficiency standards, efforts to promote improved motor efficiency have focused on premium-efficiency motors, with most programs using the CEE definition of premium-efficiency motors that is summarized in Table 2-8. Some of these efforts also promote good motor management practices (discussed in the next section). Most of today's programs were designed from a market transformation perspective and seek to address barriers to the use of premium-efficiency motors—such as limited end-user familiarity, limited local stocking, and high prices—with the long-term goal of establishing a significant market penetration for premium-efficiency motors. In addition, many of today's programs attempt to use promotion of efficient motors as an entree to working with customers to help them begin to consider more comprehensive motor system improvements. Most of today's programs include educational and promotion efforts and many include financial incentives as well, with the majority paying incentives to the dealer.

An important foundation for many recent programs is the *Motor-Master+*® software package supported by DOE and discussed in Chapter 2. This package includes a very comprehensive database of the motors on the market, their efficiency, and suggested list price. The program also includes a savings calculator and several motor management tools such as inventory management, maintenance logging, and energy accounting features (WSU 1999). This information is an important tool for planning programs and offering technical assistance to customers and dealers.

One of the more successful of today's programs is the Northeast Premium Efficiency Motors Initiative, administered by the Northeast Energy Efficiency Partnerships (NEEP). The NEEP program has more than 20 participating utilities across New England and New Jersey and builds on previous programs that were operated by individual utilities for many years. Each participating utility serves on a working group that oversees program implementation. Program implementation is largely done by a private contractor hired by the working group and includes marketing, several "circuit riders" (staff whose job it is to regularly visit distributors in the region and encourage and assist them in participating in the program), and rebate processing. Rebates cover about half of the cost difference between EPAct and CEE premium motors. Individual utilities provide additional marketing and field support in their service areas. The NEEP program began in 1998; in 1999, 2,300 rebates were issued (CEE 2000c). Rebates in 2000 were running significantly ahead of 1999 levels (Gordon 2000).

A 1999 evaluation of the program (Easton Consultants and XENERGY 1999a) found that the market share of premium-efficiency motors is approximately 30% in New England, about 10 percentage points above the national average and those of several neighboring states. The study found that 85–90% of premium motors are purchased without application for a rebate due to such factors as the small size of the rebate and the fact that rebate payments are seldom applied to the decision-maker's budget or account. On the other hand, the evaluation found that the NEEP was having a critical impact on the market by bringing attention to the value of premium motors and also by inducing some motor distributors to pressure manufacturers to design and stock qualifying motors. In addition, the study found that the relatively high market share for premium motors in New England is also due in part to past motor programs that have provided customers with experience and comfort with improved-efficiency motors. The evaluation recommended several improvements to the program, including broadening the promotion message to include premium motor benefits besides energy savings and using the program to promote model motor purchase policies (these are discussed further below).

Another interesting program is the Express Efficiency Program operated by Pacific Gas & Electric. The PG&E program pays incentives to motor distributors, which they define as companies that buy directly from manufacturers. Earlier versions of the program provided incentives to vendors (companies that sell to end-users) and were not successful because there are many more vendors than distributors, making a vendor-based program hard to market. The program now pays incentives to the distributor for each motor sold; incentives range from $35 to $630 per motor, varying with motor size. The program also includes a Web-based energy savings calculator and other Web-based information, tips for distributors on selling premium motors, print advertisements in key trade publications, direct mail marketing to distributors and end-users, visits and telephone contacts with distributors, and several special promotions. In 1999, 2,400 motors were sold, and in 2000 participation rates were up significantly. A 1999 evaluation of the program found that participating distributors reported increases in premium-efficiency motor awareness, stocking, and sales. Overall, 88% of the distributors surveyed thought that the program had a somewhat significant or very significant effect on their sales. The program contractor attributes their relative success to their incentives and marketing, noting in particular extra marketing, incentives, and personal attention to get distributors to submit their first rebate applications. According to the contractor, "after receiving their first incentive check, distributors more clearly see the program merits for their customers and for themselves and are likely to continue to participate." Still, they note that rebates account for on the order of 5% of motor sales in the northern California market and that significant work remains to be done (Barbour 2000; Barbour, Kulakowski, and Harwick 2000).

Another program worth mentioning is the Energy $mart program in New York State, offered in most regions of the state by the New York State Energy Research & Development Authority (NYSERDA). This relatively low-budget program includes promotion, circuit riders, and distributor incentives to encourage distributors to stock and salespersons to sell premium motors. The distributor incentives are $40/motor, regardless of motor size. Participation has been low (242 motors in the first year), but, as with the New England program, the market share of premium motors is significantly higher than implied by the number of rebated motors—in 1998, premium motors had a 16% market share in New York, similar to the national average (Easton Consultants and XENERGY 1999a). The New York and New England programs differ in two significant respects: rebates and prior program experience. In New England, rebates are paid directly to the customer and (for all but the smallest motors) rebates are higher. Also, while New York utilities did offer motor programs in the past, these programs were generally not as extensive as those in New England, and these programs ended several years ago, providing a gap between the old and new programs. Therefore, the new program in New York had less of a foundation to build on than New England's did. There is no available evidence to indicate the relative importance of these two factors in explaining the market share differences between New York and New England.

Another interesting program is Premium-Efficiency Motors in the Northwest, offered by the Northwest Energy Efficiency Alliance (NEEA) in 1997–1998, which was the first regional program to concentrate on premium-efficiency motors, building on previous programs operated by the Bonneville Power Administration and local utilities. The goal of the program was to increase the quantity of premium-efficiency motors purchased by increasing dealer awareness and product availability as well as increasing customer awareness. The program included dealer incentives, in-person visits by "circuit riders" targeted at dealers and key industrial customers, and other dealer and customer promotions. In addition, several local utilities offered additional incentives paid directly to customers. After 7 months of field activity, the program resulted in 451 rebates, although, as in other programs, the market share of qualifying motors was significantly higher (approximately 12% in the Northwest at that time). At this time an evaluation of the program (PEA 1998) concluded that it was having little influence on motor sales, stocking, or promotion. This was in part because promotion efforts were just getting going, due to both limited incentives (incentives averaged 22% of the incremental cost of qualifying motors) and limited availability of qualifying motors from manufacturers (as discussed in Chapter 2, this latter problem has since been largely resolved). The evaluation also found that motor distribution practices were changing and that, in and near major cities, distributors could obtain premium motors quickly, even if they did not stock them. Based on these findings, the NEEA board decided to cancel the program before promotion efforts could begin in earnest, concluding that the program was primarily designed to affect stocking patterns, and since stocking was no longer a major issue, different program approaches were needed (these are discussed below).

Information on these and other current premium-efficiency motor programs is summarized in Table 9-2.

All in all, premium-efficiency motor programs have had some successes, but many programs are struggling. Major successes include improved availability of premium motors from manufacturers and distributors. These programs have also probably contributed to a significant market share for premium motors in some regions of the country. However, direct participation rates in these programs have been small.

Based on discussions with several program managers, it appears that major barriers to premium motors are confusion among end-users about the different types of motors on the market following EPAct (e.g., confusion about the difference between EPAct and "premium" efficiency), limited customer understanding of the reliability and economics of premium motors, the relatively modest savings (in percentage terms) between EPAct and "premium" motors, and the relatively

Table 9-2

Summary of Information on Premium-Efficiency Motor Programs

Program Operator	State(s)	Program Start	Incentives/Motor (TEFC)		Who Receives Incentive	Other Program Components	No. of Motors Rebated in 1999
			5 hp	50 hp			
Eugene Water & Elec. Board	OR	1997	$50	$375	Customer		60
Northwest Energy Efficiency Alliance	ID, MT, OR, WA	1997 (ended 1998)	$30	$100	Dealer	Printed materials, circuit riders	451 in 1st½ of 1997
Northeast Energy Efficiency Partnerships	CT, MA, NH, NJ, NY, RI,	1998	$55	$220	Customer	Printed materials, Web circuit riders. In 2000 an additional $25/motor dealer incentive was offered.	2,267
NY State Energy Research & Development Authority	NY	1999	$40 (plus bonus, which in 2000 effectively doubled incentive)		Dealer	Printed materials, circuit riders	242
PG&E	CA	1999	$50	$200	Distributor	Web site, print ads, printed materials, personal contacts	2,426
Riverside Public Utilities	CA	1999	$50	$200	Customer	Brochures	24
Sacramento Municipal Utility District	CA	1999	$50	$200	Contractor or customer	Dedicated field staff provide tech assistance	Very limited
San Diego G&E	CA	1995	$50	$200	Dealer		220
So. Calif. Edison	CA	2000	$127	$511	Dealer		NA

Sources: CEE 2000c; PEA 1998; personal communication with program managers

high costs for premium-efficiency motors (as motor manufacturers seek to recoup the costs of their EPAct- and premium-motor-related investments). These factors result in many end-users not paying attention to premium motors.

To address the end-user confusion issue, many motor program managers have also encouraged EPA to begin an ENERGY STAR labeling program for premium-efficiency motors. The ENERGY STAR brand is now being used to identify high-efficiency office equipment and residential appliances; extension of the program to include motors could build on this brand recognition and make it easier for purchasers to identify the most efficient motors on the market. EPA is now consider-

---

Opportunities and Programs on Fractional Motors

We have a good understanding of the efficiency of integral-horsepower, polyphase motors and how they are used in commercial and industrial applications. However, integral-horsepower motors are substantially outnumbered by fractional motors. These motors are produced in both single- and polyphase configurations and are ubiquitous in commercial applications from small pumps and fans to compressors and conveyors. In the industrial sector, estimates of the proportion of drive energy represented by fractional-horsepower motors range from 0.5% to 1.5% (EPRI 1992; Rosenberg 1996). On the other hand, almost three-fifths of the motor energy use in the residential and commercial sector is by motors of 1 hp or below (ADL 1999). Our knowledge of the efficiency of these motors and their applications is much more limited than for their larger siblings.

Studies have hinted that energy consumption in this market segment is huge and that the opportunities for efficiency improvements are much larger than with large motors. A recent survey of reported efficiency levels in manufacturers' catalogs indicates that efficiencies can range from less than 50% to over 80% for 0.50 and 0.75 hp polyphase motors (APT 2000). Other than studies of test methods in Canada in the early 1990s (McKay 1992) and a draft report prepared for DOE in 1996 (LBNL 1996), there has been little focus on efficiency in this class of product from either the efficiency community or manufacturers.

While the Canadian effort has produced a test procedure (CSA 1993), it is faulted for its lack of reproducibility and difficulty to implement. As a result, there is no widely accepted test procedure for fractional motors that compares with IEEE 112 Method B for integral motors (NEMA 1999). This is due

---

ing whether to proceed with a program, in consultation with motor manufacturers and motor program operators (EPA 2000a). Education efforts on premium motor benefits and reliability would be part of the ENERGY STAR program. Furthermore, by building demand for premium-efficiency motors, program managers hope that over time more manufacturers will introduce premium products and prices for these motors will come down.

In addition, there appears to be an emerging consensus among motor program managers that rather than promoting premium motors independently of other motor-related measures, premium-motor promotions should be tied into broader programs that seek to promote good

---

in part to the difficulty of measuring efficiency accurately. Some experts feel that the IEEE 112 Method B can be extended to fractional polyphase motors with the use of appropriate testing equipment and experience (Kellum 2001; Kline 2001). IEEE is currently attempting to develop a new test method for fractional single-phase motors, though its adoption and the testing necessary to build confidence in the test may be several years off (Stricklett 2001). One obstacle is that the cost of performing a test is many multiples of the cost of a fractional motor.

Recently, some energy efficiency programs have begun to express interest in these motors. In 2000, for example, Southern California Edison commissioned some initial market analysis and testing (Benkhart, Elliott, and Grimm 2001).

A better understanding of both product efficiency and markets will be needed if effective strategies are to be deployed to capture this efficiency potential. For fractional polyphase motors, the existing NEMA MG 1 testing, labeling, and efficiency guideline coverage of integral-horsepower motors may be able to be extended. For single-phase motors, more extensive work will probably be needed, including refinement of existing (or development of new) test procedures, collection of additional data to better understand the market, and development of program strategies.

Since a dramatic difference in efficiency exists among the different types of motors, one strategy may be to discourage use of less efficient types (such as the shaded-pole design discussed in Chapter 2) while encouraging more efficient designs (such as capacitor-start motors, also discussed in Chapter 2). However, such an approach will need to concentrate on new motors and equipment since the more efficient motor types have a different physical size that may not fit in existing equipment.

motor management practices. By packaging these promotions together, greater savings can be achieved. Also, one of the best ways to promote sales of premium motors is for motor purchasers to establish standard specifications for when premium-efficiency motors should be purchased. Therefore, we now turn our attention to motor management programs.

Motor Management and Motor Repair Practices

Good motor management practices include automated inventory of fleet motor age and efficiency, predictive/preventive maintenance practices, guidelines for repair/replace decisions, stocking guidelines for on-site replacement, and use of quality repair specifications. Increasingly, operators of motor programs are discovering that substantial energy savings can be achieved from good motor management practices. Initial efforts tended to address only one or two isolated aspects of motor management, commonly as an add-on to high-efficiency motor programs, but in the past year more integrated motor management programs have started to evolve.

Probably the largest motor management program operating today is the Drive Power Initiative operated by the Northwest Energy Efficiency Alliance. NEEA is a nonprofit organization formed by public and private officials and state government agencies in the Pacific Northwest to operate regional market transformation programs serving a four-state region (Idaho, Montana, Oregon, and Washington). The NEEA program works with large motor users and trade allies to influence customer decisions regarding motor selection and replacement. Specific objectives include

- Increasing the operating efficiency of in situ motors by assisting customers with comprehensive motor management

- Increasing the number of motors that are replaced with new, efficient motors instead of being reconditioned by helping customers with repair/replace decision-making

- Increasing quality reconditioning by educating customers, providing repair guidelines, and working to ensure an adequate supply of qualified repair shops

Among the good motor management practices being promoted are the development of economically rational replace/repair policies, purchases of premium-efficiency motors, use of quality rewind specifications and shops that can provide quality services, carrying an adequate motor inventory so the proper motor can be installed when a unit fails, proper motor sizing, and in situ motor testing. The program is designed to address several barriers to these practices, including customer unfamiliarity with them, their benefits, and how to obtain quality services; lack of time; split decision-making authority (e.g., re-place/repair decisions are made in one department and procurement rules are set in another); and, in the case of motor repair, a shortage of repair shops with the equipment, skills, and time to undertake quality repairs. In many cases the program is trying to address these barriers by educating decision-makers and trying to get them to develop stan-dard company repair and replacement policies that maintenance per-sonnel can then implement. A sample of such a policy is provided in the box on p. 292.

The initiative offers two main services: a broad customer educa-tion program and tailored one-on-one customer services to address specific motor management issues. The customer education program includes general marketing materials and a "tool-kit" of technical ma-terials that are available in printed form, a few of which are available on the Web. One-on-one services are offered by a group of five "circuit riders" with motor management experience whose job is to both re-cruit customers to participate in the program and assist current partic-ipants. The initiative also seeks to influence the practices of motor re-pair and rewind shops to support customer requests for improved services. Specific activities include preparation of a quality motor re-pair specification and supporting materials (e.g., a workbook; an eval-uation form to assist the customer in selecting a repair shop; and a sample motor inspection, repair, and test form for use by repair shops), distribution of these materials to repair shops and customers, and technical assistance from circuit riders for repair shops and cus-tomers on how to use the specification. In the longer term, the pro-gram also is considering making small grants to repair shops to help them acquire equipment needed to improve their repair services, and setting up a program to recognize "preferred providers," who follow the recommended repair specification (ELPN 1999; PEA 2000).

At this point the initiative is still too new to tell how it is achiev-ing its objectives. An initial preliminary evaluation concluded that the program is off to a strong start in hiring skilled staff and targeting potential participating customers but that the specific services to be provided by the circuit riders need to be better clarified and refined (PEA 2000).

Another program begun in 2001 is the Motor Repair Component of NEEP's Northeast Premium Efficient Motors Initiative. While this program is labeled "motor repair," the program is designed to address the repair/replace decision as well as the elements of quality repair when motors are repaired. The focus of the program is on educating

---

Sample Motor Policy Statement

- Replace all failed standard-efficiency motors that operate continuously with premium-efficiency motors.

- Repair all other failed standard-efficiency motors greater than (customer-specific threshold) and replace smaller motors with new premium-efficiency motors if annual operating hours are greater than (customer-specific threshold); otherwise, replace smaller motors with a new motor meeting EPAct standards.

- Repair all failed energy-efficient motors greater than (customer-specific threshold) and replace smaller motors with new premium-efficiency motors if annual operating hours are greater than (customer-specific threshold); otherwise, replace smaller motors with a new motor meeting EPAct standards.

- Replace any motor for which the repair cost exceeds 60% of the cost of a new motor. The new motor should be a premium-efficiency motor if annual operating hours exceed (customer-specific number).

- Assess optimal motor size for the application prior to motor replacement and specify this optimal size when ordering a new motor.

 The size thresholds above will vary depending principally on local prices for electricity, new motors, and motor repair services. The range for the repair/replace threshold varies from 40 to 75 hp among facilities that use this approach. For failed energy-efficient motors, the threshold is usu-

---

customers to make informed decisions when dealing with a motor failure. The program seeks to instruct customers in how best to choose between motor repair and motor replacement, and, if the answer is repair, how to get a proper repair. In the first year, the program is using a variety of printed materials similar to those used in the Northwest, and is also offering a series of motor repair workshops for industrial facility managers. In the second year, services will be expanded, possibly to include one-on-one technical assistance to end-users (NEEP 2000b).

 Building on these initiatives, motor manufacturers and members of CEE have begun discussions about a possible joint education strategy focusing on repair/replace decisions. Discussions are still in the initial stages but items being discussed include developing training

ally one or two sizes smaller. For new replacement motors, the annual operating hour threshold for new premium versus EPAct motors also varies with local costs but will generally be in the range of 2,000–4,000 hours.

In all cases where repair is called for, the repair shop should follow the *Motor Repair Specification for Low-Voltage Induction Motors* developed for DOE (Douglass 1999c), including

- Performing a stator core test before and after winding removal
- Avoiding overheating the stator or sandblasting the core iron
- Repairing or replacing defective or damaged stator core laminations
- Balancing the rotor
- Repairing or replacing all broken or worn parts
- Maintaining the same air gap between the rotor and stator and the same resistance of the stator windings (do not increase!)
- Having and using an ammeter, voltmeter, wattmeter, ohmmeter, megohmmeter, and a high-potential tester
- Having an appropriate power supply for running the motor at its rated voltage
- Measuring and recording winding resistance and room temperature
- Calibrating all test equipment and measuring devices at least annually
- Not making any mechanical modifications or changes to the winding design without customer approval

Sources: Douglass 1999c; ELPN undated; Suozzo et al. 2000

for distributors and utility motor program staff, developing tools to evaluate motor repair and replacement options, targeted incentives, and ways to enhance relationships with motor distributors and repair shops (CEE 2000b).

Programs that address only one aspect of motor management also have some useful lessons to teach. For example, the Advanced Energy Corp., a nonprofit based in North Carolina, has promoted a quality assurance program for motor repair facilities, the Proven Excellence Verification Program, for several years. To be certified, repair shops must follow certain procedures, as evidenced by an independent inspection of their facilities and records. Certification also includes efficiency testing on a small sample of rewound motors. However, participation in the program has been limited—four shops are now certified after several years of program operation (AE 2000). Most customers do not recognize the added value that a certified repair shop brings and thus most repair shops see little reason to make the investment of time and money in certification. In addition, the requirements for this level of certification are considered to be too complex and expensive by many repair shops.

For many years, Carolina Power & Light has provided free detailed energy audits to its industrial customers. For customers with motors that operate for long periods of time, the audit includes a motor survey. This survey includes spot metering of motor kilowatt use and focuses on units that, upon failure, should be replaced with high-efficiency motors. Auditors recommend that customers mark candidates for replacement with yellow paint and instruct maintenance staff to purchase new motors when a "yellow dot" motor fails. Follow-up surveys indicate that this system works well in practice (Johnston 1990).

Another method for readily identifying motors needing replacement is through the use of software designed to track motor maintenance needs and practices. Information on recurring maintenance tasks is put into the computer, and work orders are prepared by the computer according to a schedule set by the user. These systems can also identify motors with recurrent problems, enabling the development of replacement schedules. The *MotorMaster+*® package includes this software; other packages are available from private vendors. Evidence indicates that these systems are primarily used by large firms with at least a dozen maintenance workers on the payroll. For example, Ralston Purina reports a 2-year payback on a sophisticated maintenance software system installed at their million-square-foot headquarters complex (Sperber 1989). An evaluation of the *MotorMaster+*® package found that small firms seldom take advantage of the program's motor management components, finding that they are complicated and require substantial time to fill out (PEA 2000).

Several utilities now offer equipment loan services to their customers. For example, National Grid USA (formerly New England Electric) has been offering an equipment loan service for several years. The service provides equipment to monitor motor energy use, demand, power factor, and efficiency. Originally the utility hired private contractors to assist customers with monitoring. More recently the utility has given each of their field representatives some metering equipment and the utility staff provide most of the technical assistance (Stout 2000).

The Sacramento Municipal Utility District (SMUD) runs a Diagnostic Services program in which utility specialists run diagnostic tests on motors and other equipment. Services are offered to customers at modest costs; for example, the charge is $150 to test a motor.

When the tests indicate motor degradation, the specialists provide counseling on whether to repair or replace the motor. To date, approximately 325 motors have been tested. The major benefit of the program has been a reduction in motor failures (Coomes 2000).

A few years ago, in order to improve the capabilities of motor repair shops in the province, Manitoba Hydro offered incentives to repair shops for the purchase of core-loss testing units. The testers help shops identify motors with damaged cores that should be replaced instead of repaired (Dederer 1999).

Overall, these different programs illustrate how interest is increasing in programs that promote improved motor management practices and quality repairs. As discussed in Chapter 2, good motor management practices can result in significant energy savings, but barriers such as limited end-user knowledge about these practices and limited time to learn about and undertake them often get in the way. A possible way to overcome these barriers in many firms is to institute company-wide policies specifying specific replacement and repair policies so that, when decisions on particular motors need to be made, the policy can be quickly applied. The Northwest Drive Power program is working in this direction but in the first few months of operation had not yet accomplished this objective (Gordon 2000). Other programs have had some success in promoting specific motor management practices, illustrating the importance of emphasizing a simple approach and message to customers (most customers have not responded well to complicated motor management systems). Thus, at this point in time, while programs to promote good management practices appear promising in theory, they are largely untested in practice. Hopefully current and nascent efforts in the next few years will fully test these program approaches.

In an attempt to bring all these motor management goals together, CEE coordinated the launch of a new campaign, *Motor Decisions Matter*[SM]. This initiative (which is profiled in the "Motor Decisions Matter Campaign" box on p. 298) is intended to compliment existing regional programs with a consistent national message encouraging motor decision planning. All the major regional efficiency programs, motor manufacturers and NEMA, the repair industry (represented by EASA), and DOE have joined together to support this initiative (CEE 2001).

A related issue that also needs increased attention is how best to improve practices at motor repair shops. End-user education on good motor repair specifications and how to select a good repair shop can help for the more sophisticated end-users, but many end-users need a simpler way to recognize good repair shops. The EASA and AE certification programs are steps in this direction, but their certification requirements are considered too onerous by many companies in the repair industry, and thus participation in these programs has been limited. Simpler procedures are needed to identify and recognize good (as contrasted with superior) repair shops. Ideally these procedures would be developed within the repair industry (i.e., by EASA), but if EASA is unwilling, then independent program operators should work to develop such a program. Also, utility in-

---

Motor Challenge

The *Motor Challenge* (MC) program, operated by DOE, worked in many of the areas discussed in this chapter, including motor management and motor systems optimization. DOE's Office of Industrial Technologies (OIT) created MC in 1993 as a voluntary industry/government partnership. The primary goal of the program was to increase market penetration of efficient industrial electric motor-driven systems by helping industry adopt a systems approach to developing, buying, and managing motors, drives, and motor-driven equipment such as pumps, fans, and compressors. MC created a network of resources that supplied free motor systems information. MC's efforts included facilitating the formation of end-user groups, industry partners (e.g., equipment manufacturers), and allied partners (e.g., utilities); and providing an Information Clearinghouse, technology demonstrations that resulted in case studies, and technology tools (Scheihing 1996). The flagship technology tool was *Motor-Master+®* (which is discussed elsewhere in this chapter and in Chapter 2).

In the spring of 2000, the MC program was subsumed into the broad Industrial Best Practices program, which also covers other areas such as steam (Cockrill 2000). The program de-emphasizes a technology-specific focus, with an eye toward plantwide improvements in cost and energy efficiency. Industrial Best Practices brought together all the resources, tools, and expertise that previously made up OIT's Challenge programs (*Motor Challenge, Steam Challenge, Compressed Air Challenge,* and *Combined Heat and Power Challenge*) and the Industrial Assessment Centers (DOE 2001).

The Information Clearinghouse was the central point for accessing MC products and services, and it continues in that role with Best Practices. The clearinghouse is staffed by experts in motor systems specification, design, and maintenance. These experts are available by toll-free telephone (see Appendix D). Publications, newsletters, technical bulletins, listings of educa-

---

centives could be useful to encourage repair shops to acquire the necessary equipment and go through the certification process.

Compressed-Air Systems

As discussed in Chapters 5 and 7, there are usually opportunities to reduce the energy use of compressed-air systems by about 15–25%. Probably the biggest barrier to achieving these savings is that most

---

tion/training opportunities, and updates on program activities are available through the clearinghouse (LBNL/RDC 2001). Also, the OIT Web site (http://www.oit.doe.gov/bestpractices/motors/) provides electronic resources, including databases of motor systems components, bulletin boards, case studies, and chat services.

Industrial Best Practices: Motors' workshops, training sessions, and conferences provide various learning options, including attending regularly scheduled classes or workshops; working with one of the allied partners; or using prepared training modules, which include slides, trainer notes, and handout materials. A number of training modules are available, including Introduction to Motor-System Management; Motor Basics; Repair/Replace Decision-Making Policy; Using *MotorMaster+*® software (on-line training is also available for *MotorMaster+*® software); Adjustable Speed Applications; Pump Systems; Water and Waste Water Pump Efficiency; and Compressed Air Challenge (DOE 2001).

In 2000, an evaluation of the MC program was conducted. It concluded that the program had many positive benefits, including reducing energy use by approximately 520 gigawatt-hours (GWh) per year; saving industrial facility operators approximately $25 million annually; and stimulating nearly $75 million of private investment in energy efficiency improvements to industrial motor systems. Overall, the benefits of the program were estimated to be over four times the amount of program expenditures. However, while the program has spurred important gains, the evaluation notes that the majority of potential savings in end-user facilities has not been achieved and recommends increased focus on developing a set of easy-to-use tools and materials that will support end-users and vendors in achieving system-level savings. The evaluation also recommends devoting more resources to helping program partners convince end-users to implement projects and then assist them with the process (XENERGY 2000).

Motor Decisions Matter Campaign

While excellent information and tools exist regarding the purchase and repair of premium-efficiency motors, most industrial managers remain unaware either of this information or of how better motor management can benefit them. Motor manufacturers, the motor service industry, DOE, and CEE joined together in the fall of 2000 to launch a program packaging this information and creating a consistent market message: "motor decisions matter." The goal of the program is to increase the demand for premium motors and quality motor repair services by promoting to decision-makers the benefits of implementing a motor management plan.

The *Motor Decisions Matter*[SM] campaign is designed to

- Increase customer awareness of the benefits of better, more evaluative motor management ("customer" means the key employees involved in making key operational and policy decisions)

- Highlight the tools and resources needed to manage motors more effectively (how to tap into the savings of premium motors and high-quality motor repair)

- Help consumers develop motor management plans for their facilities

Rather than creating a new program, the campaign is designed for integration into existing programs. The hope is that, by having all major market players integrating the same simple market message into their marketing and product materials, this message will be more effectively delivered than it has been by past programs.

Campaign participants are sharing market research and program tools and materials, such as brochures and computer decision programs. A marketing firm has been retained to refine the campaign message and to develop common campaign materials that can be customized by participating programs and companies.

The campaign founders hope that in the future the *Motor Decisions Matter*[SM] campaign can serve as a vehicle for disseminating other program strategies, such as motor system energy efficiency services.

Source: CEE 2001

customers do not realize how wasteful compressed-air systems can be and how to achieve substantial savings from a variety of compressed-air system improvements. In addition, expertise in compressed air-systems is rare—there is a major need for training to improve the quality of service providers. Furthermore, many customers do not have confidence in suppliers to deliver compressed-air savings, so they don't ask for these services.

At this point, the major program operating in the United States to promote energy-saving improvements to compressed-air systems is the *Compressed Air Challenge* (CAC) program, operated by a consortium of government agencies, regional organizations, utilities, and compressed-air system component manufacturers and service providers. The mission of CAC is to develop and provide resources that educate industry regarding the opportunities to increase net profits through compressed-air system optimization. The *Compressed Air Challenge* program is intended to address the barriers discussed above, primarily by providing objective information in user-friendly formats.

Currently, the *Compressed Air Challenge* program offers several training courses and educational publications for customer maintenance and vendor sales staff. Two courses are offered: a one-day "Fundamentals of Compressed Air Systems Training" (level 1) and a two-day "Advanced Compressed Air Management Training" (level 2). Publications include the *Compressed Air Sourcebook* as well as a series of fact sheets. A computer program (*Air Master*) has also been prepared that assists end-users in evaluating the applicability and energy-saving potential of eight energy efficiency measures for compressed-air systems. Training programs have been offered in many regions of the country, often with co-sponsorship by local utilities and/or state and regional organizations (McKane 2000).

The *Compressed Air Challenge* also includes a customer awareness campaign on the benefits of effective and efficient compressed-air systems. This campaign is just getting going and is particularly focused on placing articles in magazines read by industrial facility managers, engineers, and maintenance staff (McKane 2000).

As a complement to the *Compressed Air Challenge* program, the Compressed Air and Gas Institute (CAGI) is considering developing level 3 and level 4 training courses and operating a certification program for people who pass exams based on these classes (McKane 2000). CAGI and its members have also agreed on a common set of operating conditions for rating compressors; manufacturers are starting to make efficiency ratings available for these rating conditions. A common set of rating conditions makes it possible to compare equipment from manufacturer to manufacturer.

Many utilities have used CAC materials to help develop more intensive local programs to promote compressed-air system efficiency improvements. For example, PG&E conducted market research on the compressed-air industry and offered a pilot program using a data logger to record compressor operating performance over time and *Air Master* to estimate energy and cost savings based on the data collected. The program found average potential energy savings of approximately 30%. Based on these results and the results of its market research, PG&E is developing a Compressed-Air Market Transformation Program (CAMP) that includes three main components: standardized tools for compressed-air testing; training for trade professionals and end-users; and development of case studies and other marketing materials. These components are designed to build the supply of professionals who can provide compressed-air marketing services, and to increase end-user interest in purchasing these services (Hanna and Baker 2000).

Similarly, XENERGY began a program in 2000 sponsored by NY-SERDA. The program's objective is "to encourage compressed air system distributors to look at promoting energy efficiency in the systems they sell and service as a key strategy for increasing profits and enhancing customer relationships." To achieve this objective, the program plans to provide participating distributors with tools and support so they can quickly and effectively offer these services while minimizing start-up costs. Program services include plant assessment field materials to estimate system energy and potential savings from a list of two dozen common efficiency measures, plant assessment report writing tools (e.g., templates for developing reports), project implementation tools (e.g., guidelines and scripts regarding project financing and closing project sales), hands-on project design and closing assistance provided by expert consultants, and marketing support (XENERGY 2000c).

For years, utilities in the New England states (NSTAR, National Grid, and Northeast Utilities) and others have offered technical assistance and rebates for compressed-air efficiency projects. These services, combined with utility oversight of technical quality of studies, have created a small but significant community of compressed-air optimization experts and an increasing number of sales and service and engineering firms with the ability to identify compressed-air efficiency opportunities, address the more straightforward issues, and tap into the experts for the more complex issues. Utilities in Massachusetts and New Jersey are also developing optimization case studies in a variety of industries as a complement to CAC in order to help build confidence in compressed-air system optimization among customers and vendors (Gordon 2000).

Another interesting initiative is the Sav-Air program operated by the NW Alliance. The Sav-Air program provides integrated compressed-air management systems and engineering services that effectively allow the customer to outsource management of their compressed-air systems. The Sav-Air approach includes remote monitoring and control of compressed-air systems using sensors, computers, and software. The service begins with temporary monitoring and preparation of a site-specific proposal and, if the proposal is accepted, proceeds to (1) detailed auditing and monitoring to determine baseline usage, savings potential, and system improvements; (2) installation of system improvements; (3) verification of savings; and (4) ongoing monitoring, reporting, and system optimization. These services are provided by a private company that receives some support from the alliance but is otherwise selling these services to potential customers. As of this writing, four sites have received initial monitoring and proposals, and installation of detailed monitoring is now taking place at one site. These initial sites will be used to prepare case studies on costs and benefits and it is hoped that these case studies and other marketing efforts will ultimately allow Sav-Air to be a self-sustaining business (Scott, Stout, and Gordon 2000).

A more limited approach for achieving some compressed-air energy savings is through a leak detection and repair program. For example, in the early 1990s, B.C. Hydro provided free airflow and leakage surveys of compressed-air systems. The surveys assessed leaks in the system, motor and compressor efficiency, system controls, and system pressure relative to compressed-air needs. Follow-up leakage tests were provided 3 months after the initial assessment. These identified the general location of leaks, estimated how much they were costing the customer, and suggested a leak reduction target. If the target was achieved, the company received an award and the maintenance crew got a free lunch and a door prize. In addition, to encourage regular tests of the compressor system, B.C. Hydro paid one-half the cost of a follow-up assessment 1 year after the original assessment. B.C. Hydro would also cost-share customer purchases of leak detection equipment. The hope was that customers would undertake and finance annual assessments without utility involvement. The program targeted facilities with compressor systems of 100 hp or more. After 2 years of operation, almost 200 customers participated, which was nearly 40% of eligible customers. Program staff attributed this high participation rate to the free nature of the service and to extensive marketing efforts including local seminars and personal contacts with eligible customers. However, achieving persistence of savings requires regular leak testing—more than 100 customers purchased leak detectors under the program but it is unclear how many of these customers regularly use this equipment (Ference Weicker and Company 1995; Nadel and Jordan 1994).

In 1992, in an effort to capture additional energy savings, B.C. Hydro introduced financial assistance for full-scale compressed-air system audits, and incentives and financing for implementation of audit recommendations. Approximately 50 projects were undertaken

---

Selling Compressed Air and Other Energy Services

Selling an energy service is another approach to motor systems improvement. Various forms of motor-driven equipment service are being considered in which the contractor operates (and possibly owns) the equipment and charges the customer per unit of output (e.g., cubic feet of compressed air at a specified pressure). Under this arrangement, the customer purchases the services they need on an out-source basis, and the service provider can earn higher profits by optimizing the system to reduce operating costs.

For example, several years ago Wisconsin Electric Power Company (WEPCo) operated a pilot program called the End-Use Pricing Service (EUP). Under EUP, WEPCo would design, install, own, and operate end-use systems on the customers' premises in return for a flat fee. A long-term contract (10–15 years) for the end-use service was negotiated, with the customer paying a flat fee subject to renegotiation at intervals during the contract. An option for customer purchase of the equipment was also included. The program began with pilots of HVAC, refrigeration, and compressed-air services. While these initial pilot projects were successful, the program prompted complaints of unfair competition from some trade groups and was suspended (Flanigan and Hogan 1995; Gandhi and DiGiacomo 1994). As the electric utility industry is restructured in the United States and electric service ceases to be a monopoly, these objections may no longer apply.

A number of other groups, including utilities, air compressor distributors, and energy service companies, are considering offering this type of service. In some cases, utilities are partnering with companies possessing technical expertise. One such example is the partnering of Honeywell and Duke Engineering and Services (a nonregulated subsidiary of Duke Power Company) to offer "out-sourced utility" compressed-air services (Thielemann 1997). Some companies are also looking beyond compressed air to other motor services such as pumping, cooling, or even shaft horsepower. No examples exist at this time, though this appears to be an exciting area for development.

---

under this program component. In 1995, services were scaled back and then discontinued along with other energy efficiency programs due to a change in utility priorities (Ference Weicker and Company 1995; Fleming 1995).

Overall, reasonable progress is being made in raising the ability of the compressed-air systems industry to understand and address compressed-air systems problems. Major next steps are completing the suite of training and educational materials and increasing efforts to educate distributors and end-users about opportunities to improve systems operation and reduce operating costs through compressed-air systems improvements. Local and regional program operators would do well to tie in with the Compressed Air Challenge program and also to think about complementary local activities and services along the lines of the programs in California and New York.

Fan and Pump Systems

Fan and pump systems also provide large opportunities for energy savings, and since fan and pump systems account for about 40% of motor systems energy use (see Chapter 6), they are a prime target for energy efficiency efforts. However, most of the available savings from fan and pump systems require good application-specific engineering and cannot be made in a "cookie cutter" fashion. Most end-users (and even many of the consulting engineers they hire) lack practical knowledge regarding how best to optimize systems. Furthermore, optimization can be a time-consuming process, and time is something most customers are short of.

Past Programs

Programs focusing on fans and pumps began around 1990, starting in Canada but then progressing to several regions of the United States. Early programs were offered by B.C. Hydro and Ontario Hydro and focused on identifying good applications for adjustable-speed drives. However, this focus proposed an answer before asking which technologies make the most sense for each customer. By 1993, the Canadian utilities began several pilot projects that used a systems approach to optimize the entire motor-driven system. Due to a shift in utility priorities, these programs were discontinued before they moved out of the pilot stage, but many of the people working on these programs participated in the development of a Performance Optimization Service (POS) in Wisconsin.

The Wisconsin POS program began in 1993 and was operated by the Energy Center of Wisconsin (a nonprofit organization) in partnership with the state's utilities. Under POS, utility customer-service representatives identified candidates for POS services, and a POS engineer was hired to provide the customer a quick, free engineering "walk-through" analysis of their systems. If substantial savings were projected, a feasibility study proposal was prepared to cover work needed to determine what could be done to improve efficiency and performance, and how much it would save the customer. If the proposal was accepted, a POS engineer collected system-load and operating data and prepared a feasibility study report, which recommended a design strategy and detailed technical and economic impacts of the project. As the program evolved, these steps were streamlined and combined so that following the walk-through the customer was given preliminary cost and savings estimates, along with the proposal for the detailed study.

Utilities offered a range of incentives to customers to implement POS projects: partial reimbursement of feasibility study costs; customized rebates based on projected energy savings; low-interest loans; and shared-savings contracts through an independent financing organization. A training program was developed with support materials for utility representatives, consulting engineers, trade allies, and end-users, with training tailored toward specific needs of each of these groups (Wroblewski 1996).

The POS program provided initial audits to 36 sites and detailed feasibility studies to 11 sites. Ultimately, however, only six customers decided to implement projects: the others decided against implementation or else made no decision at all. An evaluation of the program attributed the low implementation rate to several factors: (1) nothing was broken; (2) it was perceived that savings were risky or cost estimates unrealistic; (3) several plants got a second opinion from a fan vendor who told them not to do the project because of feasibility or reliability concerns; (4) reluctance on the part of plant personnel to acknowledge inefficiencies in their systems; (5) payback periods that exceeded company targets (typically 2 years); and (6) expectation/desire for financial incentives (which Wisconsin utilities were phasing out as the POS program was beginning). The evaluation also found that those customers that did implement projects did so for two main reasons: the project solved an existing problem, and/or the project was low risk and had low or no capital costs. Interviews during the evaluation found that most companies preferred to consider process changes when existing systems fail or need to be expanded. Interviewees recommended greater utility involvement in the process, including presenting the POS concept to senior management. Also, interviewees recommended educating manufacturers and design engineers about POS concepts so that these concepts can be incorporated into system designs when new equipment is installed (Bensch 1999; Sturiale 1999).

Of the projects that were implemented, four were evaluated. These projects cost an average of $48,000 and saved an average of $40,500 annually, resulting in an average simple payback period of 1.2 years. As a result of this low implementation rate, as well as the high cost of marketing and engineering, the Energy Center decided to cancel the program. Program staff felt that the concept had a lot of merit, but more work was needed to streamline procedures so that costs could be kept in check and also so that projects were more contained and easier for customers to make decisions about. For example, several of the engineers involved in the program recommended development of improved pre-screening procedures so that inappropriate sites could be better screened out prior to any on-site assessments (Bensch 1999; Meadows 2000).

Recent and Current Efforts Focusing on Particular Industries

Building on these lessons, recent efforts to capture fan and pump efficiency savings have tended to focus on particular industries and the particular fan and pumping systems that are generic to an industry. Focusing on industries allows for knowledge from one project to be applied to other projects, cutting costs. Also, word-of-mouth and case studies within an industry can be very useful in building participation. Examples of programs focused on particular industries include work in California on municipal water and wastewater systems and agricultural pumping systems; work in British Columbia, the Pacific Northwest, and North Carolina on lumber-drying kilns; and work in the Northwest on refrigerated storage warehouses.

In 1995, DOE, EASA, PG&E, and local motor distributors identified water pumping at water and wastewater treatment facilities as a major energy-saving opportunity, based on several previous demonstration projects. Using three California Energy Commission (CEC) case studies at water and wastewater sites as a foundation, they organized operations and maintenance pumping workshops for Northern California American Water Works Association (AWWA) members. The workshops focused on how to choose motors and pumps, maintenance and operation practices, and motor and pump repair. These workshops were all standing room only. As a result, AWWA partnered with CEC and utilities throughout the state to offer workshops statewide. Based on this success, DOE, CEC, and AWWA, along with the Electric Power Research Institute and the utilities, brought the Pumping System Optimization training (discussed below) to California. They conducted six sessions in both 1997 and 1998, again to standing-room-only crowds. The success of the California initiative resulted in programs in other states, including New York, Arizona, and Iowa. The strength of the program has been its focus on a narrow market segment. Also, AWWA was an essential partner. Another inducement was that the initiative worked with participating states to give Continuing Education Unit hours for the workshops, which was important to many of the water and wastewater operators who needed the hours to maintain certification (Oliver 1999).

California utilities (e.g., PG&E and SCE) have also been promoting pump system improvements to agricultural customers for decades. The foundation of these programs has been a pump-testing service that tests pumping systems to determine overall system efficiency, electrical motor performance, pump hydraulics, and water well characteristics. The result is a computerized report containing information on the testing results, and a recommendation on whether replacement or upgrading equipment is warranted. Where such changes are recommended, estimates of the capital and operating cost impacts for the upgraded system are provided. In recent years, many of these programs have added additional services such as free or subsidized engineering feasibility studies on energy-saving measures, pump system design analysis, and incentives for installing energy-saving measures (Conlon and Weisbrod 1998; SCE 2000).

Recent market research by PG&E found that the major barriers inhibiting good pumping practices are the perception that efficiency measures have many hidden costs, concerns that measures will not perform as advertised, and lack of information and the time needed to find trustworthy information. This research found that the PG&E pumping program was addressing the information barrier to a significant extent and other barriers to a lesser extent (PG&E 1999). A 1998 evaluation of SCE's pumping program found similar program impacts and further found that pumping system efficiency gradually increased in the 1990s (due in part to the SCE program), program participants saved energy relative to nonparticipants, and the market share for high-efficiency pumps was much greater in California than in a neighboring state that did not have an agricultural pumping program (Conlon and Weisbrod 1998).

However, this study also recommended developing additional intervention strategies to better address the remaining market barriers to high-efficiency pumps and good pumping system design, operation, and maintenance practices. Among the recommendations were the following: improving access to financing and other tools to reduce the first cost of more efficient equipment; developing standards for defining and distinguishing high-efficiency pumping equipment; working with municipalities to improve bidding procedures so that efficient and inefficient equipment are no longer evaluated as "comparable;" encouraging dealers to improve stocking of efficient equipment; and offering training for pumping system consultants on the value of long-term payback from investing in the acquisition of higher-efficiency and longer-lasting equipment (Conlon and Weisbrod 1998).

The N.C. Alternative Energy Corp., now Advanced Energy (AE), in cooperation with the furniture industry, North Carolina State University, and the state's electric utilities, undertook a project to demonstrate the potential of controlling airflow in hardwood lumber dry kilns. The project's goals included understanding the drying process for furniture-grade hardwood; developing a control strategy for the fans; and evaluating the potential for savings. In furniture-grade lumber, especially some hardwoods, such as oak, the control of the drying process is critical for maintaining lumber quality. The AE study developed a control strategy based on the change in the humidity in the air immediately before and after it had passed through the lumber stack to vary the fan speed. Field trials with two industrial demonstration sites confirmed that varying the airflow resulted in at least as good a lumber quality with no effect on production rates. The operators felt that the lumber quality was better with the variable air (particularly on woods difficult to dry, such as oak), though tests were not conclusive. The total energy required for a load of wood was reduced by about half. Paybacks for the fan control systems varied from 2.4 to 8.7 years, depending on motor size and kiln configuration. Following completion of the report, AE conducted a series of seminars on the topic for the Southeastern Dry Kiln Club, the industry technical association (IEL 1992).

A project to promote variable fan speed systems in softwood dry kilns was operated by B.C. Hydro for several years. The program sponsored audits of dry kiln systems in the province and recommended ASDs in applications with a 2-year simple payback or less (36% of audited kilns met this criterion). Ultimately, approximately two-thirds of the kilns in the province were audited, and of these, approximately 25% proceeded to implement the audit recommendations (Ference Weicker and Company 1995). Similarly, industrial efficiency programs offered by the Bonneville Power Administration in the 1990s provided incentives for variable fan speed systems at many plants in the Northwest (Gordon 2000).

In the Northwest, a program has recently begun to work with re-frigerated fruit storage warehouses to encourage them to install ASDs on refrigeration system evaporator fans. The program builds on several successful refrigerated storage projects in the region and seeks to use these successful projects to promote widespread applica-tion of this measure in the Northwest's large fruit industry. The initial projects resulted in significant energy savings and reduced fruit weight loss in all applications while improving the quality of the stored fruit in some applications. The program centers on educating warehouse owners—as well as vendors, contractors, and systems operators—on the benefits of ASDs in refrigerated warehouses. Pro-ject activities include a database on refrigerated warehouses and ASD installations in these warehouses, demonstration projects, and detailed reports on these case studies. Overall, the demonstration projects have reduced fan energy use by 24–78% and have a simple payback period of 1.6–21.6 years when only energy savings are considered. However, when fruit mass loss savings are also considered, simple paybacks drop to 1.1–2.9 years (Morton and McDevitt 2000; NEEA 2000).

Other Current Efforts

In addition to these industry-specific programs, a number of broader programs are being offered to promote fan and pump sys-tem energy savings. For example, DOE's Industrial Best Practices: Motors program has developed and offered a series of Pumping Sys-tem Optimization Workshops. These sessions present the fundamen-tals of optimizing pump systems and focus on the Pump System As-sessment Tool (PSAT), which helps industrial users assess the efficiency of pumping system operations. PSAT uses achievable pump performance data from Hydraulic Institute (HI) standards to calculate potential energy and associated cost savings. DOE is now developing a training program on how to use the PSAT software. Two levels of training are planned—for pump system specialists who will use the software to evaluate pumping systems, and for in-structors who will lead workshops for others on using the software (DOE 2000a).

Many utilities offer incentives for installation of ASDs in fan and pump systems. Most of these programs offer "custom" incentives for ASDs and other energy-saving measures. In these custom programs, the customer or its consultants prepare a description of the measure, its costs, and its energy and demand savings. Based on this informa-tion, the utility will provide an incentive determined according to a formula (e.g., $x per kWh saved). However, a few utilities offer pre-calculated incentives per unit horsepower controlled by the ASD, provided certain criteria are met. For example, National Grid USA offers incentives on a per horsepower basis for boiler water feed pumps; hydraulic pumps on injection molding machinery; chilled water distribution pumps employed in building HVAC systems; and supply, return, and building exhaust fans employed in variable air volume building HVAC distribution systems. For each of these applications, annual equipment operating hours must exceed eligibility levels set by the utility. National Grid offers precalculated incentives for these applications because it is confident that ASDs will provide cost-effective energy savings. For other potential applications of ASDs, National Grid accepts custom-measure applications but requires engineering calculations to verify that savings are significant and cost-effective. The National Grid ASD program has been popular with customers and ASD vendors because it is relatively simple to apply for and the amount of incentive is known in advance, making it easier for vendors to sell projects to customers. National Grid has been offering this program since the early 1990s and as of 2000 has provided incentives for nearly 500 ASDs (McAteer 2000).

Overall, it is clear that there is a lot of interest in and experimentation with approaches for promoting fan and pump system improvements. However, none of the approaches used has yet "taken off" and produced a noticeable transformation of practices in the market. However, the targeted industry approach has made significant progress in some markets and appears promising for replication in other regions. Continued work is needed to identify additional sectors and processes that lend themselves to optimization improvements, permitting development of additional industry-focused initiatives. For example, a recent analysis of motor energy use and savings opportunities in the Pacific Northwest identified four industry areas with the most promise: irrigation using groundwater pumping; the pulp and paper industry; the mining industry; and food preservation (including food processing and cold storage) (Easton Consultants and XENERGY 1999b). Broader efforts to develop good tools and training programs are a useful foundation for these more targeted efforts. They also contribute to the longer-term goal of increasing knowledge of motor and fan system optimization techniques by allowing these techniques to be employed across a wide range of applications. A particular need is to provide better training on optimization techniques for the process engineers who are designing fan and pump systems today. To the extent that these designers know how to "do things right," system efficiency at the time of construction can be significantly increased.

Standards for Pumps and Fans

Over the past two decades, there have been periodic suggestions that minimum-efficiency standards be set for pumps and/or fans. For example, in 1980, DOE investigated possible standards for pumps and concluded that "the wide range of operational requirements for a given piece of equipment would make the establishment of standards extremely complex" (ADL 1980). Likewise, in the 1990s, the Canadian Electrical Association (CEA) investigated possible standards for commercial and industrial fans and concluded that market transformation initiatives are needed for both products, and that test procedures and minimum-efficiency standards for small pumps should be pursued as well (Intek, Inc., and Libby Engineering Limited 1995; Kaminski 1994). However, work on small pumps did not make much progress before nationwide utility funding cutbacks caused CEA to scale back work on standards. On another front, the Air Movement and Control Association (AMCA) has had success with voluntary efficiency testing, certification, and labeling agricultural ventilation fans (AMCA 1995), illustrating how it may be possible to set standards for some types of fans and pumps, but only by concentrating on particular types of products. However, any efforts to set standards for particular types of pumps and fans will require substantial time and technical work to develop testing standards and efficiency levels that work across a broad range of products and applications.

Air Conditioning

Air conditioning accounts for approximately 17% of motor energy use (see Chapter 6). Air conditioning systems range from small residential-scale packaged systems to large custom-engineered chillers. Program approaches vary with the type of system.

For chillers, many programs begin with incentives for high-efficiency models. Efficiency thresholds vary with chiller size and type. Eligibility is typically stated in terms of kilowatt per ton under standard design load conditions but some programs are also specifying eligibility in terms of integrated part-load value (IPLV—a measure of part-load performance). For chillers that operate at part load most of the time, IPLV ratings are useful. For heavily loaded chillers, peak ratings are more appropriate. A sample chiller rebate eligibility schedule is provided in Table 9-3.

Table 9-3

Sample Chiller Program Incentive Schedule— Water-Cooled Units, 300 Tons Cooling Capacity and Larger

	Centrifugal		Screw	
kW/ton	Full Load ($/ton)	IPLV ($/ton)	Full Load ($/ton)	IPLV ($/ton)
0.64	—	—	$29	—
0.63	—	—	$31	—
0.62	—	—	$33	$29
0.61	—	—	$35	$31
0.60	—	—	$37	$33
0.59	$35	—	$39	$35
0.58	$37	—	$41	$37
0.57	$39	$35	$43	$39
0.56	$41	$37	$45	$41
0.55	$43	$39	$47	$43
0.54	$45	$41	$49	$45
0.53	$47	$43	$51	$47
0.52	$49	$45	$53	$49
0.51	$51	$47	$55	$51
0.50	$53	$49	$57	$53

Note: Incentives are available for meeting either the full-load or part-load (IPLV) criteria, but not both. Incentive schedules are also available for other types of chillers, including air-cooled chillers and various types and sizes of water-cooled chillers.

Source: Connectiv 2000

A 1994 review of chiller and other rebate programs found that the most successful programs typically work closely with trade allies such as manufacturer representatives and engineering consultants; use marketing that emphasizes face-to-face contact with trade allies and customers; and maintain consistency in program design from year to year (Nadel et al. 1994).

However, just as important as the efficiency of the chiller under laboratory conditions is how the system is optimized for the particular application. A system providing process cooling for a three-shift industrial operation will have very different system requirements than a system for a typical high-rise office building. System optimization is very complex and does not lend itself to simple rebates. Among other issues, the system designer needs to assess where and how to use ASDs or other techniques to match the system output to changing cooling and ventilation needs. Additionally, system optimization is most effective when based on metered data on system loads, but most modelers and designers are not accustomed to working with metered data. Here, two complementary approaches tend to prevail. First, information on, and tools to aid, good system monitoring, analysis, and optimization can be provided to design engineers. Second, technical assistance can be provided by the program sponsor, including hiring or cost-sharing an optimization expert to work with the design engineers (or at times, it just means paying the design engineers a little more in order to have them take the extra time to fully optimize the system).

A good example of the first approach is the CoolTools™ program at PG&E. The objective of the program is to develop, disseminate, and promote an integrated set of tools for the design and operation of chilled water plants. CoolTools™ products are software programs, publications, and support services that together provide an objective, analytical method for comparing alternative strategies during the design and operation of chilled water systems. The products are public domain and Internet-based (www.hvacexchange.com/cooltools). As of mid-2000, over 20 modules (software and/or written materials addressing specific topics) were up and running, with more are in preparation, including case studies of successful projects. According to the program manager, feedback on the tools from design engineers has been very positive; PG&E is now exploring options to encourage wider use of the tools (Turnbull 2000).

Leading programs using the second approach are those offered by several New England utilities, including Northeast Utilities (NU), National Grid, and N-Star. For example, National Grid has integrated their replacement chiller optimization program into the Comprehensive Design Approach (CDA) of their Design 2000 Program. CDA deals with all manner of comprehensive building opportunities. For existing chillers, National Grid uses a calibrated modeling approach, based on metered data, to credibly show building loads. This has proved important in building designers' confidence to design to actual load levels. National Grid also optimizes new chiller systems through the CDA service. The primary difference is that there is no prior load data so the design is computer model–based.

A further step beyond system optimization is to take steps to reduce the load placed on the chiller before installing a new chiller. Commonly called "integrated chiller retrofits," such projects typically involve installing efficient lighting and possibly other load-reduction measures, with the resulting load reductions allowing for purchase of a smaller chiller than the one being replaced. Other measures to improve systems operation (such as improved HVAC controls, improved pumping and airflow designs, larger pipes or ducts, or VSDs) can also be included in the project. For example, Worcester Polytechnic Institute (WPI) in Massachusetts replaced a 290-ton, 0.85 kW/ton chiller with a 170-ton, 0.62 kW/ton chiller. The chiller downsizing reduced the cost of the new chiller, in part because of reduced heat gains from installing more efficient lighting and in part because the old chiller was oversized. At the same time, WPI installed new air handling unit controls (to improve system operation), added ASDs to pumps in the system, and installed an outdoor air heat exchanger for wintertime computer room cooling. The total project reduced electricity use in buildings served by the chillers by more than 15% and had a 5.2-year payback to WPI (Gartland and Sartor 1998). Some utilities (e.g., the New England utilities discussed above) encourage such integrated retrofits by packaging technical assistance services with incentives for purchase of efficient chillers and lighting and other load reductions. For example, New England Electric provided technical assistance and incentives for the WPI project discussed above.

Existing commercial chiller systems can also be better optimized. All too often, controls and valves get out of adjustment, and in some cases they were never properly adjusted, even when the system was new. *Commissioning* is the process of checking that systems are properly installed and adjusted. For complex systems such as chillers, experienced commissioning engineers are needed to do a good job. Historically, when commissioning is done at all, it is done when systems are new. But in recent years there has been growing interest in *retrocommissioning*, meaning the commissioning of existing buildings. Several utilities and government agencies now offer programs to promote these services, based on field studies that indicate whole building energy savings from commissioning of 5–10% and retrocommissioning of 5–20%, with simple paybacks of 1–3 years. The typical retrocommissioning program includes training programs for commissioning engineers, educational materials and programs for building owners and managers on the benefits of retrocommissioning (including case studies of successful projects), and financial incentives to share the cost of commissioning and retrocommissioning services with building owners. Most of these programs address the entire HVAC and control system, and often other systems as well, although a few programs have focused just on chiller systems (Dodds, Baxter, and Nadel 2000).

The majority of commercial building cooling energy use is not attributable to chillers but instead is due to unitary (packaged) systems (E Source 1997). Units range in size from 3-phase versions of residential equipment to rooftop systems of 30 tons cooling capacity or more. For packaged commercial systems, the Consortium for Energy Efficiency (a consortium of utilities and state and nonprofit energy organizations) has developed a set of efficiency thresholds that members use to determine eligibility for incentives and other promotions. By working together on a single national specification, rather than having specifications vary from utility to utility, CEE members make it possible for manufacturers and distributors to produce and stock a common line of equipment that is eligible for programs nationwide.

The CEE specification includes two efficiency tiers—Tier 1, which is about 10% higher than current mandatory federal efficiency standards for these products, and Tier 2, which is about 10% higher than Tier 1. However, the model building code developed by the American Society of Heating, Refrigerating and Air-Conditioning Engineers, Inc. (ASHRAE), was recently upgraded to require Tier 1 levels of efficiency,

Table 9-4

CEE Eligibility Levels and NEEP Incentives for Air-Source Commercial Packaged Air Conditioners

Cooling Capacity	Required Efficiency			NEEP Incentives ($/ton)	
	Federal Standard	CEE Tier 1	CEE Tier 2	CEE Tier 1	CEE Tier 2
≤65,000 Btuh	10 SEER	12 SEER	14 SEER	$55	$85*
65,001–135,000 Btuh	8.9 EER	10.3 EER	11 EER	$38	$68
135,001–240,000 Btuh	8.5 EER	9.7 EER	10.8 EER	$43	$73
>240,000 Btuh	None	9.5 EER	10 EER	$43	$73

* For Tier 2, NEEP has slightly modified the CEE tiers, using SEER 13 for equipment under 65,000 Btuh.

Note: One ton = 12,000 Btuh. The programs also cover heat pumps and water-source equipment, with different eligibility levels and incentives. EER = energy efficiency rating.

Sources: CEE 2000a; NEEP 2000a

and the federal standard is likely to be revised soon to these levels as well. As a result, utility promotions are increasingly emphasizing Tier 2, and some programs are dropping incentives for Tier 1 equipment. Table 9-4 summarizes the different efficiency tiers and also includes information on incentives for each tier being offered in 2000 by NEEP, which probably has the largest program in the United States covering this equipment. Furthermore, EPA is considering establishing an EN-ERGY STAR labeling program for this equipment based largely on the CEE Tier 2 levels (EPA 2000b).

The NEEP program combines incentives for eligible equipment, extensive outreach and marketing to equipment distributors and vendors (via several full-time "circuit riders"), and more limited outreach to equipment purchasers. A 1999 evaluation of the program found that approximately 15% of equipment sales met the program's eligibility thresholds, split roughly between Tier 1 (60%) and Tier 2 (40%) (RLW Analytics 1999). In the first half of 2000, participation rates increased significantly, with the majority of rebates going to Tier 2 equipment (Linn 2000).

In addition to more efficient equipment, the NEEP initiative is also beginning to explore interventions to improve the quality of installation and maintenance practices for commercial packaged equipment.

In 2000, a customer education program on these issues was offered by NEEP in consultation with vendors. Also in 2000, a group of NEEP members studied the feasibility of a training and certification program on high-quality installations in commercial and industrial facilities and decided to develop and implement a pilot one-day training course for commercial HVAC technicians but to hold off on a certification program pending results of the training program. The sponsoring utilities have selected an independent organization (the Eastern Heating and Cooling Council) to run the training program, feeling that there would be more acceptance from contractors if this training was not be a direct utility activity (Neal et al. 2000).

For residential air conditioning units, there are also many opportunities to promote more efficient equipment and installation and maintenance practices. Most programs focus on central air conditioning systems since these account for more than 80% of residential cooling loads (EIA 1999). EPA has the ENERGY STAR program that offers the ENERGY STAR label for residential central air conditioners with a seasonal energy efficiency ratio (SEER) of 12 or higher. EPA also has a variety of promotional tools for this equipment that it makes available to manufacturers, distributors, utilities, and others who want to promote ENERGY STAR–qualifying equipment (EPA 2000c). Data available from the Air Conditioning and Refrigeration Institute (a manufacturers' trade association) indicate that just over 20% of residential central air conditioner and heat pump sales meet the ENERGY STAR levels (Leland 1998).

In addition to promoting ENERGY STAR, some utilities are offering incentives for equipment with SEER 12, 13, or higher. A 1997 report for CEE reviewed many of these programs and found that several of the most successful ones were achieving market shares of 40% or more for SEER 12 equipment and market shares of 10% or more for SEER 13 equipment. Particularly noteworthy was a program offered by Potomac Electric Power Company (PEPCo) that was achieving a 50% market share for SEER 13 equipment. The study found that successful programs included strong relationships with HVAC contractors, had a lot of continuity (e.g., were operating for 5 years or more), and gave contractors ample notice of program changes. The PEPCo program achieved its high SEER 13 market share by first offering incentives that covered a larger portion of the incremental cost for SEER 13 equipment than incentives offered for SEER 12 equipment, and then once the market for SEER 13 equipment was established, phasing out SEER 12 incentives entirely (CEE 1997).

These programs are important because updated federal standards for air conditioners (and heat pumps) will not take effect until 2006. In

January 2001, the Clinton Administration published final standards. Their primary effect would be to raise the minimum SEER to 13. The Bush Administration suspended that regulation and proposed a SEER 12 standard instead. As of this writing, that proposal is in its public hearing phase and the subject of litigation. The processes could have impact on the final standard level and possibly on the implementation date. Once the standards take effect, programs to promote improved-efficiency air conditioners will need to target efficiency levels above the new standards.

As with commercial equipment, some of the residential programs are beginning to target proper installation and maintenance practices. For example, utilities in New Jersey are offering training to HVAC technicians on key elements of project equipment installation, including system sizing, proper refrigerant charging, and proper airflow maintenance. The program also requires that rebate applications for new systems include a form providing information on system charge and airflow, as well as submission of load calculations, in order to encourage technicians to pay attention to these parameters on each job. This program was extended into a portion of New York State in 2001. And in California, utilities are offering incentives to customers to encourage them to hire contractors to check their air conditioning and heat pump systems for proper charge and airflow, and to check duct systems for excessive leakage. If significant duct leakage problems are found, additional incentives are available for duct sealing services (Nadel et al. 2000).

Refrigeration

Motors are also used extensively to power refrigeration systems, including built-up systems (such as those used in many supermarkets and in the food warehousing and processing industries) and packaged systems (such as refrigerators, freezers, ice-makers, and water coolers). In the past, utilities played an active role in promoting more efficient built-up refrigeration systems by promoting such measures as floating head pressure control, uneven parallel compressors, mechanical subcooling, and hot gas defrost. As a result of these efforts (plus the fact that the high operating cost of refrigeration systems provides owners with a large incentive to improve efficiency), many of these measures have become common practice.

At this point, opportunities to improve refrigeration systems are primarily promoted on a custom measure basis in which owners or their consultants propose energy-saving projects to their utility, and incentives are provided based on projected kilowatt and kilowatt-hour savings and the cost of the project. A few utilities actively promote these services. For example, Northeast Utilities has an in-house refrigeration engineer who works with local supermarkets, warehouses, and ice rinks to develop specific energy-saving projects built around a list of approximately 20 energy-saving opportunities the utility has developed. The utility engineer is well known to refrigeration system designers and developers, who inform him about projects early in the design stage. The utility engineer then meets with the owner and design team to review plans and discuss possible energy-saving modifications. Based on information provided by the designers, the utility engineer then estimates the costs and savings of energy-saving plan modifications and calculates the rebate available for the improvements, if they are implemented. Many of these recommendations are accepted. In recent years, the program has undertaken an average of 30 projects annually (Knapp 2000). According to outside observers, the key to the success of this program is the utility engineer, who has an excellent reputation in the local refrigeration community.

For packaged equipment, most promotional activities today are based on the ENERGY STAR label. DOE sponsors the ENERGY STAR program for residential refrigerators and works with manufacturers, retailers, utilities, and state governments to promote the program locally (DOE 2000b). EPA is now researching possible ENERGY STAR programs for commercial refrigerators and freezers, refrigerated vending machines, water coolers, and ice-makers (Kubo et al. 2000).

Cross-Cutting Program Approaches

The sections above discussed program approaches targeting motors and motor-driven systems. In addition, there are a wide range of cross-cutting programs that promote energy savings across a variety of end-uses, including, but not limited to, motor systems. In the sections below we summarize some of these approaches and how they have been applied to motor systems.

Performance Contracting

Performance contracts typically involve private energy service companies (ESCos), which contract with a utility or end-user to assess, finance, and install energy-saving measures. The ESCo takes as payment a share of the energy cost reduction, based on either engineering estimates or actual metered savings. In the latter case, the ESCo takes the risk that predicted savings will actually materialize.

In recent years, several states and utilities have run standard performance contracting (SPC) programs (sometimes also called standard offer programs) in which incentives are provided to ESCos (and in some cases to customers as well) for installation of energy-saving measures. SPC programs typically have two goals—developing the market for performance contracting services and acquiring energy savings. Some programs emphasize one or the other goal. SPC programs began in New Jersey in the mid-1990s and are presently under way in California, New York, and Texas.

In SPC programs, incentives are typically paid per kilowatt-hour saved, with savings pre-calculated for simple measures (e.g., efficient lighting) and metered for more complicated measures. Payments are typically made over several years and are based on expected lifetime energy savings from installed measures. For programs that emphasize developing the market for performance contracting services, eligibility is commonly limited to energy service companies, and incentives may be higher for energy-saving measures and market segments that are not typically emphasized by the ESCo industry. For programs that emphasize resource acquisition, end-users are typically eligible to obtain incentives for measures in their own facilities and there is less differentiation of incentives among measures and market segments.

Based on results to date with SPC programs, several conclusions can be drawn. First, SPC programs can achieve substantial energy savings. In 1998, an evaluation of the California program found that it resulted in annual energy savings of approximately 230 GWh. Second, most SPC programs operating today are under-subscribed (i.e., participation levels are not high enough to fully use program budgets). This means that at a minimum more time is needed to allow the market to build. These low participation rates possibly could also indicate that the size of the potential market for ESCo services is limited. Third, programs are not yet generating evidence of sustained impacts on the market for efficiency services, although this could well be because it is too soon to tell. Regarding motors, motor-related measures have been a significant part of savings in at least some programs. In New York, 31% of electricity savings to date have been from motors, mostly in the institutional and commercial sectors, as only 11% of total program savings to date have been in the industrial sector (Schiller et al. 2000).

Bidding

In the early 1990s, several utilities used bidding to select large projects to receive energy efficiency incentives. ESCos, other service providers, and customers proposed projects, and winning bids were selected on the basis of requested incentive per kilowatt-hour saved and other factors. Thus, unlike SPCs where the program administrator sets incentive levels, under bidding it is the market that determines the level of incentive, subject to a cost-cap set by the utility to ensure that the program is cost-effective to the utility and its ratepayers. However, based on experience with early bidding programs, ESCos have expressed concerns about high transaction costs for bid preparation, lengthy periods for contract negotiation, and the fact that ultimately only a few firms (the winners) had access to incentive funds. In order to address these concerns, the SPC concept was developed.

However, in the past few years, at least two utilities have resumed bidding programs—Public Service of Colorado (PSCo) and Northeast Utilities. In the case of the PSCo program, $15 million in contracts was awarded in early 2000 to 15 contractors and 8 customers. As of this writing, implementation is just beginning (Schiller, Goldman, and Henderson 2000). In the case of the NU program, the first two rounds of funding resulted in projects that are projected to save nearly 200 million kWh over the lifetime of the projects, with an average cost to the utility of less than $0.01/kWh. The program has funded a wide range of energy-saving projects; some of the projects have included ASDs and efficient motors. Most of the projects to date have been organized by ESCos or other third parties; only a few projects have been organized by customers (Odell 2000).

While experience with the latest round of bidding programs is limited, some conclusions are likely to apply. A review of these previous programs by Goldman and Hirst (1989) concluded that demand-side bidding programs can contribute most significantly to savings in existing large commercial and industrial facilities. For smaller facilities and new construction, bidding will generally not be appropriate due to high transaction costs and the lack of a solid measurement baseline.

Rebate Programs

Utilities have offered rebate programs for many years, including rebates for specific measures such as high-efficiency lighting equipment, HVAC systems, and motors, and custom incentives for measures proposed by customers. Information on rebate programs for high-efficiency motors, HVAC equipment, and ASDs was discussed above; in this section we focus on custom incentives that can cover other motor systems improvements.

In a review of custom rebate programs in the 1991 edition of this book, we found that the more successful programs generally featured all or most of the following elements

- An extensive education and technical assistance component
- Marketing that emphasizes one-on-one personal contacts with equipment dealers and large customers
- Significant rebates

We also found that customers generally prefer payments per kilowatt or kilowatt-hour saved or as a percent of measured cost because they are relatively easy for customers to understand, and to estimate the incentive for potential projects. Such payments can be set to reflect the value of energy savings to the utility. On the other hand, they encourage rapid payback measures and discourage longer payback measures.

Payments to bring the cost of a measure down to a specified payback level are more difficult for customers to understand and require considerable analysis to accurately estimate costs and savings. On the other hand, these incentives best advance a utility's long-term objective of encouraging medium- and long-term payback measures, which customers would be unlikely to implement without utility assistance. This approach is perhaps best suited to large customers that can understand and are willing to work with complex programs.

In general, these conclusions are still valid today. However, some recent experiences allow us to add several additional conclusions. First, many of the most successful programs have not just provided incentives and technical assistance to their customers but have gone a step farther and worked to develop a long-term partnership with their key customers. Second, energy savings alone will generally motivate only some customers, but when other benefits can be captured as well (e.g., production increases, productivity improvements, emissions reductions, etc.), prospective projects become attractive to many more customers.

Partnerships take many forms but tend to share certain elements (Pye et al. 1996): (1) they are developed by working one-on-one with the customer; (2) the utility meets a customer's specific needs; (3) the benefits to customers extend beyond energy efficiency; (4) the participant often offers something tangible to the utility in return (e.g., a long-term power purchase contract), as opposed to just taking incentive money; and (5) the project is often not designed with standard operating procedures but instead involves custom elements to serve particular customers' needs.

An example of a program employing these features was PSI Energy's Industrial Efficiency Improvement and Energy Awareness Program. This program relied on personal contacts by PSI account managers with an individual within a plant, usually the plant engineer. For medium-size and large industrial customers, the program provided customized energy studies and tailored incentives to encourage installation of efficient equipment that could provide both demand and energy savings. Mostly smaller/simpler projects were implemented in the early years of each partnership, but as the relationship and trust built, larger projects were considered (Pye et al. 1996).

While utility partnerships typically target large customers, several state agencies have operated programs to develop partnerships with small and medium-size industrial customers. For example, NYSERDA operates a custom-tailored technical assistance program called Flex-Tech with the goal of lowering facility operating costs, increasing productivity, and reducing air emissions. The heart of the program is engineering services provided by a FlexTech "stable" of 24 prequalified technical assistance consultants around the state. These firms have extensive experience providing consulting services to industrial facilities and a range of expertise. Specific consultants are selected that can best address each individual customer's needs, and the cost of these services is shared between NYSERDA and the customer. In most small companies, management must approve this cost-sharing arrangement, which provides management buy-in that facilitates project implementation. In addition, NYSERDA works with financial institutions to provide low-interest financing to assist customers with implementing the measures. According to a survey of FlexTech clients, more than two-thirds of the recommendations made by FlexTech contractors have been implemented. Each dollar spent on FlexTech engineering services has resulted in $17 in capital improvements and $5/yr in energy savings (NYSERDA 2000).

A 1996 review of a variety of partnerships deduced several characteristics of successful endeavors (Pye et al. 1996):

- Understanding the customer
- Providing flexibility in all aspects of the program
- Building a long-term relationship and trust
- Establishing personal contact
- Bundling value-added services
- Constructively engaging trade allies
- Having patience and persistence

Quantifying nonenergy benefits can often be the difference between a project idea that sits on the shelf and one that gets implemented. For example, Massachusetts Electric has offered an Industrial

Systems Optimization Service (ISOS) since 1998. The program focuses on industrial production or process improvements and looks beyond electrical savings to provide comprehensive project evaluations. These nonelectric benefits may include thermal savings, hazardous waste reduction, increased productivity, and/or labor or material savings. The target market is medium-size to large customers that lack the time or resources to fully evaluate the impact a process improvement may have for their business. The program offers financial incentives based on electricity savings, but by quantifying the nonelectrical benefits, it can often offer a much more attractive package to customers.

For example, the ISOS program worked with a wastewater treatment facility that used a dissolved air floatation thickener (DAFT) system to remove solids. The sludge created by the process was then trucked off-site for disposal. ISOS investigated the replacement of the DAFT with a gravity belt thickener system. The new system eliminated the need for the compressed-air system required by the DAFT and significantly reduced pump energy requirements. In addition, the treated sludge is much thicker (less water content), thereby reducing the volume of treated sludge trucked off-site. This means a savings of almost $27,000/yr in transportation costs in addition to $7,700 in annual electric cost savings. The large nonenergy benefits were an important factor in the customer's decision to implement the project (Coughlin 2000; MECo 2000).

Loan Programs

Loans can compensate for the limited access to capital that prevents many customers from investing in efficiency. Some programs, instead of offering rebates for installation of efficiency measures, use loans instead, with the interest rate commonly subsidized in order to make the loan (and hence project implementation) more attractive. For example, NYSERDA complements its FlexTech program (and other audit and technical assistance programs) with a loan program, which works with local banks and farm credit associations to buy down the interest rate 4.5% off the lender's rates for consumer and commercial loans. In the first 2 years of the program, approximately 300 loans were closed, totaling $23.5 million. Industrial projects accounted for approximately 20% of the loan dollars, primarily for HVAC and process improvements. More recently, residential sector applications have predominated as industrial customers often obtain financing through other programs and sources (Fenno 1989; Winters 2000).

A number of utilities have offered loans to their customers. One of the better-publicized programs is the PacifiCorp Energy FinAnswer program. Under Energy FinAnswer, the utility conducts comprehensive audits, provides engineering services, and can arrange financing at a favorable interest rate (the prime rate) for electric energy efficiency opportunities. The loan is paid back through an energy services charge on the customer's electricity bill, and program administrative costs are paid by the utility. The commercial new construction portion of the program has been particularly successful and is discussed in the next section. The industrial portion of the program began in 1992 and uses utility industrial energy experts to work with customers to identify areas where they can achieve the most cost-effective results. The program also offers free monitoring and verification services to help ensure that energy savings are realized and maintained over time. Marketing targets large industrial customers (demand greater than 500 kW), although other customers are also eligible. However, marketing and implementing projects are slow processes, and, as a result, the program served only ten participants in the first year, with participation building to about twenty new participants annually thereafter (Pye et al. 1996). In general, with FinAnswer the high quality of technical assistance services has been the key to program success; without these services, participation in the program would probably be much more limited (Nadel et al. 1994).

Wisconsin Electric and Puget Power and Light have offered customers a choice between a zero-interest loan and rebates with the same cost to the utility. Over 90% of participating customers selected rebates. Loans are thus less popular but are useful for the minority of customers that lack investment capital. Furthermore, both utilities found the rebates easier to administer than the loans (Clippert 1989; France 1989).

New Construction Programs

New construction embodies decisions that affect the energy use of a facility for many years to come. Incorporating efficiency measures when a facility is built is much less expensive than retrofitting it later because marginal capital, design, and installation costs are much lower for new construction. For these reasons, energy-saving opportunities in new facilities are often called "lost-opportunity resources": once the opportunity to acquire these resources inexpensively is lost, it may never come again. Due to both the unique energy efficiency opportunities during new construction and differences in the new construction and existing facility markets (discussed in Chapter 8 and earlier in this chapter), many utilities and government agencies have targeted special programs for the new construction market.

Most new construction programs concentrate on commercial buildings. Typically these programs have two tracks: a prescriptive track that provides rebates for common measures such as high-efficiency lighting and HVAC equipment; and a performance track that provides design assistance and custom rebates for comprehensive packages of efficiency measures, which are optimized through the design process. A 1994 review of these programs found that, in just a few years, several of these programs achieved participation rates of more than one-third of new commercial floor area, including a few programs that exceeded a 50% participation rate.

These high-participation programs share several common attributes including an emphasis on personal marketing and building relationships with the design and building community, and the availability of financing or financial incentives that cover most or all of the incremental cost of efficiency measures. Many of the programs emphasize ease of participation and have achieved high participation rates by stressing simple, prescriptive measures such as lighting and HVAC equipment improvements. The PacifiCorp Energy FinAnswer program has taken a different tack. The program is not simple; instead it highlights quality services, including substantial energy savings from a comprehensive package of measures, building commissioning, and a post-occupancy audit.

Savings from commercial new construction programs have rarely been empirically evaluated, but the limited available information indicates average savings relative to prevailing construction practices of about 20% for comprehensive packages of measures and about 10% for more limited packages of simple prescriptive measures (Nadel et al. 1994).

Only a few utilities have offered industrial new construction programs, and, of these, only some have actively worked to customize the program to meet the specific needs of industrial customers. Industrial new construction programs can encourage use of efficient processes in new factories and new production lines. They can also be used to encourage new plants to locate in the program sponsor's region or to encourage existing customers to expand their local operations. Because each production process and factory is different, a key component for a successful new construction program is to have consultants on retainer who are experts in particular process industries. These consultants can also advise the sponsor on which measures are the prevailing practice in a particular industry (and hence the customer should pay for) and which measures go beyond standard practice and are appropriate for incentives.

Among the industrial new construction programs that have achieved some success are United Illuminating's (UI) Energy Blueprint (which also included commercial buildings), B.C. Hydro's New Plant Design, and BPA's Energy Savings Plan programs (which dealt with both retrofits and new process lines and plants). While none of these programs are operating now due to utility-wide cutbacks in energy efficiency services, they provide some useful lessons. Each of these programs dealt with both industrial buildings and industrial processes and made extensive use of experienced process engineers to establish baselines and recommend efficiency improvements. All offered design grants to industrial customers to cover the customer's costs of analyzing alternative plant designs. Under the UI and BPA programs, incentives to industrial customers typically covered 50–80% of the incremental cost of a project and reduced energy use on average by 20–30% below baseline practice. The managers of these programs make several recommendations. UI found that it is important to staff the program with people who are experienced and well-versed in a particular industry. Several programs found that a key challenge was identifying firms early enough in the design process that there still would be time to affect design decisions. B.C. Hydro noted that several years can elapse from the time when a project is proposed and to the date the new plant is completed, and therefore patience is required on the part of the program sponsor and also a willingness to honor incentive commitments made several years earlier. The industrial component of UI's program cost the utility $0.02/kWh saved (Nadel & Jordan 1994; Nadel et al. 1994).

Dedicated Efficiency Fund for Large Customers

In some states, very large industrial customers have opposed state funding for utility energy efficiency programs. These customers argue that they have already invested in energy efficiency in their own facilities and that they should not subsidize efficiency investments by others. The typical utility energy efficiency program has a budget of roughly 1–2% of utility revenues, so for customers with annual electric bills in the millions of dollars, the cost of efficiency programs included in rates can be significant. In several states, in order to address this issue, utilities, utility regulatory commissions, and large industrial customers have negotiated an agreement in which the utility will set up special funds for large customers that do not wish to participate in the utility's normal program offerings and instead want to undertake their own efficiency investments. These programs are sometimes called "opt-out" programs. For example, in Vermont, the utility commission and the largest industrial firm in the state negotiated an agreement that allows large customers to participate in a

C&I Customer Credit Program under which 70% of that customer's payments for energy efficiency programs goes into a fund that the customer can use to fund efficiency investments in its facility. The remaining 30% is contributed to statewide efficiency programs under the rationale that all customers receive at least some indirect benefits from these programs (e.g., an improved environment and economic climate) (VDPS 1999). Somewhat similar programs have been set up in Indiana, Massachusetts, and New York.

Tax Credits

Tax credits and accelerated depreciation allowances have been used in many countries to promote energy-saving investments. For example, from 1978 to 1982, U.S. businesses could take a tax credit equal to 10% of the capital cost of energy-saving investments. Similarly, Japan allows a 7% tax credit on energy-saving investments plus a special depreciation allowance equivalent to 18% of the acquisition price of energy-saving measures (Furugaki 1988).

An analysis of the U.S. program (ASE 1983) found that the credit had little impact on energy investments made by U.S. industry; most firms that took the credit would have made the same investments if no credit had been available. This study also examined the hypothetical effects of higher tax credits, including a 40% tax credit and a 70% repayable credit, which would be repaid over a series of years. The analysis concluded that these latter two credits would have only a limited impact on project economics (i.e., the internal rate of return for a project with a 2.5-year simple payback rose from 33% in the no-credit case to approximately 45% in the high-credit cases). While the impact of high credits cannot be determined with any certainty, the authors of the study conclude that "energy tax credits are relatively ineffective in inducing industrial firms to undertake additional conservation investment."

In 1999 and again in 2000, the Clinton administration and several congressmen proposed a series of tax credits to Congress to promote the market development of several specific advanced energy-saving technologies that have very limited current market shares. These credits are limited in time duration and available funding; their purpose is not to directly acquire large amounts of energy savings but instead to kick-start the market for covered products. In the industrial sector, the proposed tax credit targets the development of combined heat and power systems and does not cover motor systems. In the residential sector, tax credits are proposed for central air conditioners with SEERs of 13.5 and 15 (Geller 1999).

Research, Development, and Demonstration Programs

Most research, development, and demonstration (RD&D) work on new motor system technologies is done by private firms, ranging from industry giants to small start-ups. This work is largely carried out by companies in the United States, Europe, and Japan and covers diverse fields such as materials, electronics, motors, and ASDs. Most RD&D work is done as a strategic effort performed by private companies to ensure the competitiveness of their products.

In addition, government agencies and electric utilities, through EPRI and the associated Power Electronics Application Center (PEAC), conduct drivepower RD&D, often in conjunction with industry. For example, DOE has worked with private manufacturers on the design of permanent-magnet motors, reluctance motors, amorphous metal motors, and copper rotor motors (these technologies are briefly discussed in Chapter 2) (Comnes & Barnes 1987; Peters 1998). DOE, through Oak Ridge National Laboratory, is also working on the development of soft-start inverters, which reduce the voltage spikes that characterize the most common types of inverters (Peng 2000). Similarly, EPRI and PEAC are conducting research on a number of motor-related technologies, including permanent-magnet motors, high-efficiency motor rewinds, manufacturability of more efficient motors, square-wave motors (motors optimized to run on the square-wave power produced by some ASDs), improved-efficiency ASDs, and clean-power ASDs (those with less harmonic distortion) (Lawrence 2000).

Much of the recent applications-related RD&D activity has involved ASDs and, to a lesser extent, improved production processes and motor efficiency in rewind and part-load applications. For example, EPRI has conducted several ASD application and demonstration projects, including one for very large motors (2,000 hp and up) used in power plants (Oliver & Samotyj 1989) and one for industrial applications of fans and pumps ranging from 5 to 1,250 hp (Poole et al. 1989). Some of these projects are discussed in Chapters 4 and 5. Advanced Energy in North Carolina has conducted drivepower RD&D in wood-dust collection systems, industrial heat pumps, compressed-air systems, variable-airflow fans for lumber dry kilns, and rewind and part-load applications (NCAEC 1989). And the Energy Center of Wisconsin has emphasized development of improved optimization strategies for fan, pump, and compressed-air systems among other projects (Pye and Nadel 1997).

Utility- and government-funded RD&D has also stimulated commercialization of advanced energy-saving products. For example, EPRI teamed up with the Carrier Corporation to produce a high-efficiency variable-speed heat pump that incorporates a permanent-magnet motor with an integral ASD—the first in the HVAC industry. Likewise, EPRI's ASD demonstrations have probably sped up market acceptance of these products, although definitive data on the effects of the demonstration projects are not available.

Summary

Most programs operating today emphasize the market transformation approach, which seeks to reduce market barriers to particular products and services, thereby resulting in long-term, sustained changes in the market.

Initial program efforts focused on energy-efficient motors, and several programs achieved substantial success, laying the groundwork for establishment of minimum-efficiency standards in the United States and Canada. Current motor-only programs focus on CEE premium-efficiency motors and have achieved some significant impacts on the market, although direct participation rates in these programs are generally low.

Increasingly, motors programs are not just addressing improved motor efficiency but are also seeking to improve motor management practices. These efforts seek to encourage customers to adopt and implement specific motor policies dealing with repair/replacement decisions and specifications for repair services. Complementary efforts are needed to encourage and assist repair shops in acquiring necessary equipment and improving the quality of their repairs.

Interest is also growing in programs to improve management and optimization practices for major types of motor-driven equipment, particularly fans, pumps, and compressors. The *Compressed Air Challenge* program has made substantial progress in developing tools and training programs to allow distributors and their customers to improve compressed-air systems. Programs to improve fan and pump systems are more limited and have evolved to focus on particular applications in particular industries that can be widely replicated.

In addition to programs focused on particular products, many efficiency programs promote energy savings more broadly while paying some attention to motor systems. These programs are particularly useful for regions that need to acquire energy savings in the short term in order to improve system reliability and meet other regional needs.

While most programs are oriented toward addressing the needs of particular markets, successful programs often share several common attributes, including

- Use of a personalized one-on-one approach for program marketing and service delivery
- Development of good working relationships with trade allies such as equipment distributors and design engineers
- Extensive education and technical assistance efforts
- Significant financial incentives
- Consideration of nonenergy benefits
- Continuity over time
- Patience

No single program or policy is likely to overcome all the barriers to motor efficiency improvements. In order to capture even a fraction of the savings opportunities identified in Chapter 7, a combination of approaches is needed. While past and current program efforts have made significant progress, much more needs to be done. Specific recommendations regarding next steps are provided in Chapter 10.

Recommendations

Full, cost-effective application of the technologies and practices discussed in this book could reduce national drivepower energy use by 28–42%. The potential savings are large, but many barriers to their capture exist. In this chapter, we recommend programs and policies to remove these barriers and advance the implementation of energy-saving measures and practices. We also discuss education and research and development needs, both to help implement existing technologies and practices and to advance the state of the art in efficient motor systems.

Significant progress was made in the decade since the first edition of this book was written. Some of the first edition's recommendations in the areas of standards, research, and education were implemented. Unfortunately, little has changed in many areas and the recommendations remain largely the same.

Programs and Policies
Testing, Labeling, and Standards

The area of motor testing, labeling, and standards has shown the greatest progress in the decade since the first edition of this book was published. Some of the first edition's key recommendations in this area have been implemented, including the passage and implementation of motor minimum-efficiency standards in the United States and Canada. However, further steps must be taken to address problems with implementation of the new standards and labeling, and to extend the scope of the current standards.

Many of the problems with the IEEE and Canadian Standards Association test procedures that were identified in the first edition have been addressed, and these harmonized test standards are now positioned to become the international standards of choice. In an increasingly global market, it will be essential to reach an international consensus on testing standards, which should be based on the IEEE/CSA methods. The European Union, which has been considering this step, should act and provide leadership in moving to a truly global testing standard.

In the United States, the Energy Policy Act of 1992, discussed in Chapter 2 and Appendix B, established testing, labeling, and minimum-efficiency requirements for most of the three-phase motors used in industry and buildings. All motors covered by EPAct ("covered motors") manufactured in or imported into the United States now must be labeled with a certified efficiency value and must meet minimum-efficiency levels. While the implementation of these regulations proved more difficult than anyone involved would have imagined in 1992, these efforts represent a major step forward in motor efficiency.

Some problems have, however, been created by the regulations. As of this writing, details of domestic enforcement are only now being clarified, and procedures for enforcement of imports have yet to be set forth. One step the U.S. Department of Energy (DOE) should take is to move quickly and aggressively to enforce all aspects of EPAct for both domestic production and imports. NEMA and domestic manufacturers have shown good faith in trying to comply with delayed and unclear rules, and should be looked upon as allies rather than adversaries by DOE.

The EPAct standard covers an important segment of the motor population. However, other segments remain in need of either mandatory or voluntary standards. As discussed in Appendix B, the EPAct standards can be applied to many noncovered three-phase motors. The NEMA definitions of "energy-efficient" motors can be applied to many three-phase induction motors, including 8-pole motors and motors above 200 hp.

Of greater significance is the other major motor population—small, single-phase motors. As discussed in Chapter 6, these motors are the most numerous in the motor population, and (as discussed in Chapter 2) they offer the greatest variation in efficiency levels among products. Due to the many types of motors that are made, this segment poses many challenges. The first challenge is the development of an accurate, cost-effective efficiency test procedure. IEEE 114-1982 (IEEE 1982) includes a method, and the CSA has developed a procedure (CSA 1995), but many experts feel that these methodologies are unworkable (Bonnett 1999; Daugherty 1999). Absent a workable test procedure, it is difficult to label or provide other procurement guidance beyond recommending the purchase of one specific design over another. Complicating the small motor issue is the fact that few of these motors are purchased as a distinct product, but instead they are mostly purchased as part of another commodity, such as an appliance. Because of these challenges, Lawrence Berkeley National Laboratory, in an assessment for DOE, did not recommend that DOE proceed with development of national standards for small motors (LBNL 1996). Even so, the government and technical associations should continue to work to develop reliable and cost-effective test methods for small motors that can serve as the basis for voluntary labeling and promotion initiatives.

Promoting Improved-Efficiency Motors

With EPAct now in effect, new motors need to at least meet the EPAct standards. However, additional savings can be achieved from use of new premium-efficiency motors and from replacement of old motors with either premium or EPAct-compliant motors. In order to promote use of these improved motors, several actions should be taken, including defining and labeling premium-efficiency motors, implementing increased education for end-users and repair shops on repair/replace decision-making criteria, and creating incentive programs to encourage stocking and sales of premium-efficiency motors.

Many regional program operators have worked through the Consortium for Energy Efficiency to set common eligibility levels for their premium-efficiency motors programs (as discussed in Chapter 2). However, no consistent labeling is used by manufacturers. NEMA has yet to develop a new definition of efficiency levels above EPAct. It should move quickly to adopt a standard definition of premium efficiency, working closely with utilities and regional organizations that are currently promoting CEE premium-efficiency motors. This new definition needs to be robust, reflecting a significant increase in efficiency over the EPAct levels. The starting point for developing these standards should be the CEE levels since program operators and manufacturers are already working with these levels. If NEMA sets a definition that is too low, the utilities and other program operators will ignore it, just as they ignored NEMA's original definition of "energy-efficient" motors in 1990. On the other hand, the CEE levels were developed in 1996 and it is appropriate to review them now in light of changes in product offerings over the past few years. By working together to develop a common definition, manufacturers and program operators can produce a definition that meets their needs and helps end-users understand exactly what is meant by "premium efficiency." If the industry is unable to move quickly on creating and disseminating a robust definition, the U.S. Environmental Protection Agency (EPA) and/or DOE should promulgate a voluntary definition based on the CEE levels.

EPA and DOE should also consider establishing a voluntary ENERGY STAR labeling or other branding program for motors, based on the definitions discussed above. The label or brand would build on other ENERGY STAR awareness promotions while addressing the market confusion about efficiency levels above EPAct. Governments are in an excellent position to provide leadership by committing to purchasing motors meeting these higher levels. The Federal Energy Management Program (FEMP) has already adopted a specification, based on the CEE levels, as a recommended purchasing guideline (FEMP 2000). This specification goes beyond the CEE standards by including motors up to 500 hp. Agencies and other governmental entities should follow FEMP's leadership.

With the implementation of the EPAct minimum-efficiency standards, substantial energy can be saved by replacing old, less-efficient motors with new motors that equal or exceed the EPAct efficiency levels. Education efforts are needed to shift the customer's repair-versus-replace decision toward replacement, and to encourage customers to purchase a premium product over an EPAct product. Regional program operators in the Northwest and Northeast are starting to operate such programs, but these efforts need to be better focused. We recommend focusing these programs on encouraging the adoption of standard motor management policies that specify which motors to repair and which to replace as a function of motor size, condition, and operating hours. A sample of such a policy is included in Chapter 9. Utilities and state governments in other regions should also develop programs along these lines.

As a complement to the above efforts, utilities and other program operators should offer incentives for the purchase of motors meeting the premium-efficiency definition. Such incentives serve several purposes. First, they encourage motor distributors and vendors to keep premium motors in stock, and to promote these motors to their customers. Second, they attract end-user attention and interest. Such incentives can increase the market share for premium-efficiency motors when new motors are purchased and will also encourage more customers to replace instead of repair old motors. Incentive programs now under way in the Northeast United States and in northern California provide possible models for other regions.

Manufacturers can and should play important roles in all of the above programs. Unfortunately, they have been involved in only a limited way because of their recent fixation on complying with the EPAct rule. Manufacturers are uniquely positioned to both reach end-users through their marketing efforts and inform and encourage dealers to promote conscious motor decision-making. To take advantage of this position and encourage information-sharing, they should

- Incorporate material on motor decisions into the educational and marketing materials that they make available to their dealers and trade allies

- Identify premium motors in their published and on-line catalogs

- Incorporate decision tools, similar to MotorMaster+ (WSU 1999), into their on-line catalogs, and make them available to distributors for use in counter and phone sales

- Provide incentives to their distributors for sales of premium-efficiency motors

Cooperation between motor manufacturers and program administrators could substantially improve the effectiveness of motor incentive programs. In turn, these programs should cooperate to coordinate on a national level, as they did in developing and deploying the CEE premium-efficiency specification.

Motor Repair and Replacement Practices

Motor repair/replacement policies are only one aspect of good motor management practices. Other aspects of good motor management include motor inventory tracking, preventive maintenance, selecting a good motor repair shop, developing good repair specifications in order to obtain quality repairs, and proper motor sizing. Education and technical assistance programs can provide information on these management measures and their benefits. The goal of these educational programs is to establish procedures for inventorying, maintaining, repairing, and purchasing motors that will reduce the cost of motor operations and minimize downtime during replacements. As discussed in Chapter 9, some efficiency programs are already promoting sound motor management practices at the utility or regional level.

Training courses and manuals for maintenance staff and for their bosses, who decide how much staff time should be allocated to maintenance, can go a long way toward improving maintenance practices in the field. If maintenance staff and their supervisors understand how costly it is to lose a percentage point or two in efficiency from inadequate O&M and appreciate how much downtime from bearing failures or other problems can be reduced by better O&M, these personnel are more likely to upgrade their maintenance practices.

One focus of this training should be to encourage users to measure operating parameters when they first get a motor and then regularly thereafter, especially before and after repairs. Motor monitoring benefits users by helping them to spot problems before they lead to failure. Regular measurements of current and voltage can identify problems with the electrical supply, as discussed in Chapter 2, and also changes in motor load that can guide future motor decisions. Software is available to compile and analyze this information. Owens-Corning Fiberglas Corporation regularly tests motors to identify units that are likely to fail and need repair, and also tests motors after repair to evaluate the repair job. The program has cut the motor failure rate, significantly improved motor repair quality, and reduced repair costs (Kochensparger 1987).

Repair shop operators, as well as customers, should be educated on the quality repair practices discussed in Chapter 2 and the benefits that these repair practices bring. Rewind shops should be encouraged to test all motors for core damage before and after rewinding. Utilities and other program operators could offer incentives for the purchase and use of the necessary test equipment, as Manitoba Hydro has done. Incentives can also be offered to repair shops for training technicians in quality repair practices. Since a quality repair is predicated on the presence of a quality-control tracking program, it may also make sense to provide shops with a small rebate for each quality-tracking form that is completed and provided to customers. The cost of completing the form is modest, and an incentive might encourage shops to offer this benefit to their customers. Likewise, since many repair shops also sell new motors, efficiency programs could provide modest incentives to shops that provide an economic analysis to customers showing the life-cycle cost of both the repair and the replacement options. While these suggestions all involve financial incentives, nonfinancial incentives, including marketing assistance, publicity, and prizes such as free trips for shops that meet specific targets, can also be used in many cases.

Several repair shop certification programs have been developed but have yet to receive broad market acceptance. Efficiency programs could encourage acceptance of certification by underwriting the cost of certification for shops in the programs' areas and promoting the use of certified shops to end-users. Current certification programs are perceived to be too complex and expensive by many repair shops. Consideration should be given to developing simpler and less expensive certification procedures.

Systems Optimization

Efficient, cost-effective performance requires careful integration of motors, controls, electric cables, drivetrains, and driven equipment. It also requires an understanding of the load that the driven equipment serves. As noted in Chapter 5, the largest source of compressed-air waste is inappropriate use. As with selecting a new motor, such optimization is essentially a lost-opportunity resource—it is easiest to implement when the system is first designed or is being modified for other purposes. While an operating system can be analyzed for efficiency opportunities, this is more complex and costly.

All too often, optimization is neglected because of inadequate time or expertise. Optimization of a system cannot be achieved piecemeal—it requires knowledge of electrical and mechanical engineering, computer optimization techniques, and practical experience with the particular systems and processes affected. Few individuals currently have the requisite skills.

Another aspect of system design that merits increased attention is the systematic addition of safety margins at each stage in the design process, resulting in dramatically oversized equipment. While safety margins have their place, engineers need improved training on equipment selection and control procedures that provide a safety margin without affecting efficiency.

Matching the speed of centrifugal equipment (i.e., pumps, fans, and blowers) to load requirements is the largest potential source of motor system energy savings. Adjustable-speed drives and other controls are crucial tools in achieving maximum system efficiencies. In the past decade we have seen increased acceptance of ASDs in the marketplace, but still the technology remains underutilized or misapplied, as is discussed in Chapter 4.

While the opportunities in systems optimization and ASDs are widely recognized, methodologies and programs that capture the potential have met with limited success. This is due in part to the complex nature of system evaluation, the specialized nature of the skills required, and the great diversity in motor-driven applications. Those areas where success has been achieved, such as pump system optimization in water and wastewater facilities, have measures that can readily be replicated among a large number of sites. We should build on these successes in our future endeavors.

Education remains an essential strategy for addressing these problems. DOE and some utility and state efficiency programs are delivering seminars and workshops on motor system optimization to both end-users and designers. In addition, innovative initiatives such as the *Compressed Air Challenge* bring together diverse market interests to develop and deliver training that can show results on the plant floor.

Because the magnitude of the savings potential from system optimization is so great, efforts in this area should be increased. Specific recommendations include the following:

- Continued implementation of the *Compressed Air Challenge* program to provide information, training, and software that can be used by individual plants as well as be a resource for regional programs.

- Development of more advanced compressed-air training courses and certification of technical staff who can demonstrate how to apply these techniques in the field.

- Development and operation of local programs to work with compressed-air equipment distributors and assist them in offering improved compressed-air management services to their customers.

- Development and implementation of regional programs to promote best practices in the areas of water supply, wastewater and irrigation pumping, and fan speed management in lumber-drying and food storage operations. These are all applications for which best practices have been identified in at least one region of the country; efforts are now needed to promote these practices elsewhere.

- Research and demonstrations to identify best practices in other specific, recurring fan and pump applications.

- Development of additional training programs, handbooks, and software on fan and pump system optimization, building on currently available tools.

- Continuation and expansion of programs to promote more efficient air conditioning equipment, including system installation and maintenance.

- Expansion of education and technical assistance programs to encourage better optimization of large HVAC systems, including the design of new systems and "retrocommissioning" of existing systems.

Other Programs

Technical assistance and generic rebates can be offered for kilowatt-hours or kilowatts saved. These incentives can target end-users and, in some cases, energy service companies as well. Among the measures to include are the following:

- Rebates per horsepower for ASD applications meeting specific criteria that have been shown to result in effective installations.

- New construction incentives tied to the incremental cost of improved designs and equipment. These efforts need to focus more on industrial process improvements in order to complement the existing focus on commercial buildings.

- Incentives for refrigeration system efficiency improvements that exceed standard practice.

Another useful set of programs would involve implementing ENERGY STAR labeling for packaged commercial refrigeration equipment, including reach-in refrigerators and freezers, ice-makers, vending machines, and water coolers. These programs would help customers consider efficiency in their purchase decisions, thereby motivating equipment manufacturers to pay more attention to efficiency when they design equipment and take advantage of many low-cost opportunities for improving equipment efficiency.

Other ideas that merit exploration include the following:

- A utility-subsidized motor maintenance service in which outside contractors would perform ongoing preventive maintenance and efficiency tune-ups in customer facilities.

- Leasing of efficient equipment, whereby lease payments are less than monthly energy savings and the lease payments are included on customers' electric bills.

- Internal shared savings programs in which departments within a company or agency can keep a portion of the money they save and use it for funding further improvements and staff bonuses. For example, the state of Washington had a program under which employees involved in energy conservation programs shared 25% of the savings resulting from the measures they implemented, and the other 75% went to the state (Lannoye 1988).

Finally, more attention is needed to specifically develop programs and services targeting small commercial and industrial customers. Past experience shows that, with many programs, large customers tend to participate much more often than small customers because large customers have more staff and also because program marketing efforts generally target large savings opportunities (Nadel 1990). Also, experience to date with electric utility industry restructuring shows that power and service marketers are primarily targeting large customers (Flaim 2000). From an equity point of view, additional programs for smaller customers are needed. Frequently governments operate programs to assist small businesses. For example, DOE's Industrial Assessment Centers provide audits and technical assistance to small industrial firms on saving energy and reducing waste. Utilities and other program providers can work with these programs to expand services or add an efficiency component. The Energy Center of Wisconsin (ECW), for example, has set up a cooperative relationship with the Wisconsin Manufacturing Extension Partnership (WMEP), which has a cadre of manufacturing specialists who work with firms on ways to modernize and improve operations. Under the WMEP/ECW partnership, ECW supports the cost for energy experts to accompany the WMEP specialists on visits to the factory. The two experts then work together on developing recommendations (Shipley, Hinge, and Elliott 2000). Likewise, Massachusetts Electric has teamed with the Massachusetts Division of Energy Resources (the state energy office) to conduct assessments of electricity, gas, and nonenergy savings opportunities. The utility pays for the electric portion of the analysis, the state for the other portions (Elliott, Nadel, and Pye 1996). Another useful model is the FlexTech program in New York State (discussed in Chapter 9). Programs such as these should be replicated in other states and new creative program ideas developed on ways to best assist small businesses to implement motors and other efficiency improvements.

Program and Policy Evaluation

Programs and policies too often fail to include plans for assessing how well they work. While this situation has improved in the decade since the first edition of this book was published, the admonition needs to be repeated that programs and policies from the beginning should contain an evaluation component. With programs increasingly based on the market transformation approach to program design, evaluation needs to follow from each program's market transformation objectives. Thus, if programs aim to improve user knowledge and vendor stocking of some type of efficient equipment, evaluations should track these parameters over time as well as interview market participants to determine which aspects of the program appear to be working, which do not, and what program modifications are needed. Evaluation results need to be circulated so that others can benefit, encouraging implementation of good ideas and cautioning against problematical strategies.

Education

Throughout the sections above, dozens of areas are pointed out where more education is needed for end-users, equipment vendors, and service providers. In addition to these very specific education needs, there is a need for broader efforts as well. Motor system efficiency should be incorporated into engineering curricula, junior engineers need one-on-one field training with experienced engineers, and practicing professionals should have ready access to continuing education programs. In recent years, electric motor training has been eliminated from most engineering curricula at technical universities. These courses were originally in the electrical engineering curriculum, but with the shift to electronic and computer engineering in these departments, the courses have often been eliminated. Mechanical engineering departments should be encouraged to include motor system courses in their programs. The one notable example where motor system efficiency is being incorporated into engineering curricula has been the Industrial Assessment Centers discussed in Chapter 9. Students participating in the program are provided a background in motor systems in order to allow them to identify opportunities when they perform the assessments. The students carry this awareness of motor system opportunities with them into industry.

Universities should not be the only venue for motor efficiency education. Increasingly, technical schools and community colleges have assumed responsibility for educating the technical workforce, as well as providing continuing education for many engineers. The biggest challenge in all these sectors has been the absence of curriculum materials and educators familiar with motor systems. Groups such as NEMA, HI, and AMCA should undertake curriculum development, as has CAGI in concert with the *Compressed Air Challenge*. Each motor program should encourage educational institutions in its region to add motor system courses to their course offerings, and provide them assistance in procuring expert instructors who can make these courses meaningful.

Research, Development, and Demonstration

Our recommendations are grouped in three areas: equipment, tools, and data.

Equipment

With the introduction of EPAct and lines of premium-efficiency motors, we may have achieved most of the efficiency improvements that are economically viable for the basic three-phase, integral-horsepower, general-purpose induction motor. As discussed in Chapter 2, to move to a higher level in this design will involve a shift to a technology such as the cooper rotor. Improving the efficiency of other induction motors, including single-phase units and additional polyphase models, should now be given priority because these motors represent more than 25% of U.S. motor electricity use. Attention should also be directed toward other emerging motor designs, such as permanent-magnet, switched-reluctance, and written-pole motors, which offer high levels of efficiency as well as precise speed control (Nadel et al. 1998). Further work is also needed to optimize motors for use with ASDs.

ASDs also require continued attention. The past decade has seen significant progress in improving the reliability of drives and reducing their tendency to both cause power quality problems and be affected by them. More work is needed to refine existing designs and develop new designs featuring improved torque and speed characteristics, higher efficiency, improved power factor, increased applicability, reduced harmonics and interference problems, and reduced cost. In addition, efforts need to focus on improving the efficiency of the motor/drive system. The recent trend (discussed in Chapter 2) of marketing an integrated motor/drive package offers promise in this area. Government and manufacturers should continue their support of efforts to bring these improved products to the market.

Tools

An important part of education and technical assistance programs is written materials, training courses, and software tools that teach practitioners proper techniques and aid them in applying these techniques. The *Industrial Best Practices: Motors* and *Compressed Air Challenge* programs have developed a useful set of tools on motors and compressed-air systems, but additional tools are needed. First, a model motor management policy is needed that industries can modify for their needs. The material in Chapter 9 in this book is a start, but expansion and elaboration is needed. Second, there is a need for additional training programs on fan and pump system optimization. *The Industrial Best Practices: Motors* program has an introductory pumping workshop, and ECW has an introductory fan systems workshop, but more advanced training programs are needed to complement these introductory programs. Likewise, while DOE has developed an excellent Pumping System Assessment Tool, there is a need to develop an analogous tool for fan systems. Similarly, improved software would be helpful for chiller system optimization, and streamlined commissioning procedures for HVAC (and other) systems would also be useful.

Data

Data on motor operating hours, load profiles, actual efficiencies, and applications are limited. While this situation has improved significantly since the publication of the first edition of this book, much work remains to be done. The issuance of the *United States Industrial Electric Motor Systems Market Opportunities Assessment* (XENERGY 1998), funded by DOE, has significantly enhanced our understanding of motors and their use in industry. In addition, the motor-testing programs undertaken by Advanced Energy and the Canadian Standards Association have contributed significantly to our understanding of motor performance and the impact of repair.

The XENERGY study has provided valuable insights. However, additional work focusing on specific motor-intensive industries is needed to refine our understanding of motors in the industrial sector. The study examined most industries at the two-digit SIC level. For many industry groups, significant variation exists among the individual industries. There is a need for greater resolution in order to provide a better understanding of the opportunities. The XENERGY report also dealt mainly with the manufacturing sector. More information is needed on motor use in nonmanufacturing sectors of the economy, including water treatment, agricultural water pumping, and mining.

A study similar to the XENERGY report was recently completed for DOE on motor use in commercial and residential buildings (ADL 1999). This study was a good start toward developing a better understanding of motor energy use in these sectors. However, Arthur D. Little, Inc.'s estimates of opportunities to reduce energy use were limited to two measures (improved-efficiency motors and use of variable-speed drives); further analysis is needed on a much broader array of measures to reduce motor energy use in these sectors.

In addition, plans should be made to periodically update data so trends can be tracked over time. These data are needed to estimate the potential savings across a wide range of measures, from high-efficiency motors and ASDs to better equipment sizing and maintenance in a variety of end-use sectors. Such information, preferably stored in a database, will allow engineers, utilities, and other motor efficiency practitioners to better focus their efforts. The data will allow them to develop rules of thumb and expert systems, as has already been done in the industrial sector using the XENERGY study (Rosenberg 2000).

We still need more data on actual motor performance. The Canadian government has made a commitment to testing as part of its motor standards program, and with implementation of the EPAct standards, manufacturers in the United States are making similar commitments. Still, additional independent motor-test facilities are needed in this country to provide verification of the performance of motors and to undertake research into performance issues such as the impact of ASDs on motor life.

Our call for publication of sales data by motor size class for high-efficiency motors was heeded by the U.S. Census Bureau; this information is now included in their *Current Industrial Reports* series (U.S. Census Bureau 2000). Unfortunately, with the implementation of the EPAct minimum standards, the term *energy efficient* is no longer meaningful. Once a new "premium-efficiency" level is defined, the U.S. Census Bureau should begin to collect data on sales of these motors.

Similar information on ASDs is not now readily available. Information on the total number of these devices sold by size class would be very useful to program designers and policymakers. The U.S. Census Bureau currently combines ASDs with other types of electronic equipment in their tracking system. NEMA and the bureau should collaborate to gather and disseminate separate information on ASD sales.

Conclusions

Tapping the riches of the drivepower gold mine will not be easy. While new and better hardware is welcome, we have only begun to take advantage of the advanced motors; the controls; the drivetrains; and the monitoring, maintenance, repair, and system optimization systems already available. We need interdisciplinary education and training, involving the entire market channel: those who make, sell, specify, buy, and use drivepower technologies. We need to better communicate the importance of planning economic motor decisions. This includes clearly conveying the efficiency choices for products in motor catalogs and sales and educational literature, and on the Internet, an increasingly important part of the information channel. We also need our engineers to approach design decisions from a system optimization perspective, considering all components from the wire to the load. Utility or government rebates and technical assistance are vital, as are internal incentives that reward employees for saving energy. Finally, continued research is needed on new hardware, analysis tools, and program and policy options.

With roughly 60% of U.S. electricity use at stake, and a similar share in most other industrialized countries, motor systems can be fairly characterized as the mother lode of energy savings. We have made significant progress in the United States in realizing the savings, but a tremendous opportunity remains. While the challenges are large, the potential rewards are vast, in terms of energy and financial savings, economic competitiveness, and environmental protection. We hope this book helps motor users everywhere to reap these important benefits.

References

Abbate, G. 1988. "Technology Developments in Home Appliances." In *Demand-Side Management and Electricity End-Use Efficiency*, 435–448. Edited by A.T. de Anibal and A. Rosenfeld. NATO Advanced Science Institutes Series E, vol. 149, Applied Sciences. Norwell, Mass.: Kluwer Academic Publishers.

[ADL] Arthur D. Little, Inc. 1980. *Classification and Evaluation of Electric Motors and Pumps*. Report DOE/CS-1047. Prepared for the U.S. Department of Energy, Office of Industrial Programs. Springfield, Va.: National Technical Information Service.

————. 1996. *Energy Savings Potential for Commercial Refrigeration Equipment*. Prepared for the U.S. Department of Energy, Office of Building Technologies. ADL Reference No. 46230. Cambridge, Mass.: Arthur D. Little, Inc.

————. 1999. *Opportunities for Energy Savings in the Residential and Commercial Sectors with High-Efficiency Electric Motors, Final Report*. Prepared for the U.S. Department of Energy. Washington, D.C.: Arthur D. Little, Inc.

[AE] Advanced Energy. 2000. *The Proven Excellence Verification Program*. http://www.advancedenergy.org/root/industrial/consulting/pev.html. Raleigh, N.C.: Advanced Energy.

Aegerter, R. 1999. "Compressed Air System Optimization." In *Proceedings: 21st National Industrial Energy Technology Conference (IETC)*, 179–181. Houston, Tex.: Texas A&M University, Energy Systems Laboratory.

Albers, T. (Electric Motors Corporation). 1998. Personal communication to Sam Wheeler. May. Quoted in S. Nadel, L. Rainer, M. Shepard, M. Suozzo, and J. Thorne, *Emerging Energy-Saving Technologies and Practices for the Buildings Sector*. Washington, D.C.: American Council for an Energy-Efficient Economy.

[AMCA] Air Movement and Control Association International, Inc. 1995. *AMCA Directory of Agricultural Products with Certified Ratings*. Arlington Heights, Ill.: Air Movement and Control Association International, Inc.

Andreas, John. 1982. *Energy-Efficient Electric Motors: Selection and Application*. New York, N.Y.: Marcel Dekker.

[ANSI] American National Standards Institute. 1995. Electric Power Systems and Equipment—Voltage Ratings (60 hz). Report #ANSI C84.1. Washington, D.C.: American National Standards Institute.

[APT] Benkhart, B. 2000. *Fractional Motor Analysis and Evaluation Report (Draft)*. Springfield, Mass.: Applied Proactive Technologies.

Argonne National Laboratory. 1980. *Classification and Evaluation of Electric Motors and Pumps*. Report DOE/TIC-11339. Prepared for the U.S. Department of Energy. Springfield, Va.: National Technical Information Service.

[ARI] Air Conditioning and Refrigeration Institute. 1993. *Statistical Profile of the Air-Conditioning, Refrigeration, and Heating Industry*. Arlington, Va.: Air Conditioning and Refrigeration Institute.

[ASE] Alliance to Save Energy. 1983. *Industrial Investment in Energy Efficiency: Opportunities, Management Practices, and Tax Incentives*. Washington, D.C.: Alliance to Save Energy.

[Baldor] Baldor Motors and Drives. 2001. *Super-E Premium Efficient Motors*. Fort Smith, Ariz.: Baldor Electric Company.

Baldwin, S. 1989. "Energy-Efficient Electric Motor Drive Systems." In *Electricity: Efficient End-Use and New Generation Technologies, and Their Planning Implications*. Edited by T.B. Johansson, B. Bodlund, and R.H. Williams. Lund, Sweden: Lund University Press.

Bannerjee, B. (Electric Power Research Institute). 1998. Personal communication to Sam Wheeler. June. Quoted in S. Nadel, L. Rainer, M. Shepard, M. Suozzo, and J. Thorne, *Emerging Energy-Saving Technologies and Practices for the Buildings Sector*. Washington, D.C.: American Council for an Energy-Efficient Economy.

Barbour, J. (Energy Solutions). 2000. Personal communication to Steven Nadel. July.

Barbour, J., S. Kulakowski, and A. Harwick. 2000. "We're Cranking Now: A Motors Program Success Story." In *Proceedings of the ACEEE 2000 Summer Study on Energy Efficiency in Buildings*, 6:11–20. Washington, D.C.: American Council for an Energy-Efficient Economy.

[B.C. Hydro] British Columbia Hydro. 1988. *High-Efficiency Motors*. Vancouver, B.C., Canada: B.C. Hydro.

Benkhart, B., R.N. Elliott, and W. Grimm. 2001. "Evaluations of Fractional Polyphase Motor Efficiency and Market Intervention Strategies." In *Proceedings of the ACEEE 2001 Summer Study on*

Energy Efficiency in Industry. Washington, D.C.: American Council for an Energy-Efficient Economy.

Bensch, Ingo. 1999. *POS Evaluation: Looking Back on the Performance Optimization Service Program, Report Summary.* Madison, Wis.: Energy Center of Wisconsin.

Bodine, C., ed. 1978. *Small Motor, Gearmotor, and Control Handbook.* 4th ed. Chicago, Ill.: Bodine Electric Company.

Bonnett, A. (U.S. Electric Motors Corporation). 1999. Personal communication to Neal Elliott. November.

Bose, B. 1986. *Power Electronics and AC Drives.* Englewood Cliffs, N.J.: Prentice-Hall.

Boteler, R. (U.S. Electric Motors). 1999. Personal communication to Neal Elliott. March.

Brithinee, W. (Brithinee Electric). 1999. Personal communication to Neal Elliott. March.

[CEC] California Energy Commission. 1984. *Energy-Savings Potential in California's Existing Office and Retail Buildings.* Analysis by P. Gertner and T. Tanton. Sacramento, Calif.: California Energy Commission, Technology Assessments Project Office.

[CEE] Consortium for Energy Efficiency. 1996. *Premium Efficiency Motor Initiative.* Boston, Mass.: Consortium for Energy Efficiency.

———. 1997. *Leading Residential HVAC Programs: Lessons Learned.* Boston, Mass.: Consortium for Energy Efficiency.

———. 1998. Minutes from the Quality Motor Repair Roundtable, Chicago, Ill., June 4. Boston, Mass.: Consortium for Energy Efficiency.

———. 2000a. "High-Efficiency Commercial Air Conditioning and Heat Pumps Initiative." http://www.ceeforMT.org/com/hecac-main.php3. Boston, Mass.: Consortium for Energy Efficiency.

———. 2000b. Minutes from the Energy-Efficient Motor Program Summit: Expanding the Market for Energy-Efficient Motors, Chicago, Ill., June 30. Boston, Mass.: Consortium for Energy Efficiency.

———. 2000c. *National Report: Premium Efficient Motor Programs and Market Summary.* Boston, Mass.: Consortium for Energy Efficiency.

———. 2001. *Motor Decisions Matter^{SM} Business Plan.* Boston, Mass.: Consortium for Energy Efficiency.

Clarkson, J. (Southwire Company). 1990. Personal communications to Steven Nadel and Michael Shepard. September.

Clippert, P. 1989. *Commercial/Industrial/Farm—Smart Money Energy Program, Total Completed Status: 17 March 1989.* Milwaukee, Wis.: Wisconsin Electric Power Company.

Cockrill, C. (United States Department of Energy). Personal communication to Neal Elliott. March.

Colby, R., and D. Flora. 1990. "Measured Efficiency of High Efficiency and Standard Induction Motors." Paper presented at the IEEE Industry Applications Society Annual Meeting, Seattle, Wash., October.

Comnes, G.A., and R. Barnes. 1987. *Efficient Alternatives for Electric Drives.* Report ORNL/TM-10415. Oak Ridge, Tenn.: Oak Ridge National Laboratory.

Control Engineering. 1998. "Efficiency to the Masses—of Electric Motors, That Is." http://www.controleng.com/archives/1998/ctl0701.98/07g701.htm. Oak Brook, Ill.: Control Engineering.

Coomes, H. (Sacramento Municipal Utility District). 2000. Personal communication to Steven Nadel. July.

Cornell Pump Company. 1987. "Model 6NHP-Various Speed." Pump Curve Number 770-130. Portland, Oreg.: Cornell Pump Company.

Coughlin, T. (National Grid USA). 2000. Personal communication to Steven Nadel. August.

[CSA] Canadian Standards Association. 1993. *Energy Efficiency Test Methods for Three-Phase Induction Motors.* Standard C-390-93. Toronto, Ontario, Canada: Canadian Standards Association.

———. 1995. *Energy Efficiency for Single- and Three-Phase Small Motors.* C-747. Toronto, Ontario, Canada: Canadian Standards Association.

Darby, S. (Darby Electric Company). 1997. Personal communication to Steven Nadel. October.

Daugherty, R. (Rockwell Automation). 1999. Personal communication to Neal Elliott. March.

de Almeida, A.T. (University of Coimbra, Portugal). 1999. Personal communication to Neal Elliott. October.

de Almeida, A.T., and S. Greenberg. 1994. *Technology Assessment: Energy-Efficient Belt Transmissions.* Berkeley, Calif.: Lawrence Berkeley National Laboratory.

Dederer, D.H. 1991. "Rewound Motor Efficiency." *Ontario Hydro Technology Profile.* Kingston, Ontario, Canada: Ontario Hydro.

———. 1999. *Report on the Canadian Power Smart Motor Program for the Project Management Office of the World Bank.* Toronto, Ontario, Canada: Canadian Power Smart Motor Program.

Dodds, D., E. Baxter, and S. Nadel. 2000. "Retrocommissioning Programs: Current Efforts and Next Steps." In *Proceedings of the ACEEE 2000 Summer Study on Energy Efficiency in Buildings,* 4:79–95. Washington, D.C.: American Council for an Energy-Efficient Economy.

[DOE] U.S. Department of Energy. 1998. *Improving Compressed Air System Performance: A Source Book for Industry.* Washington, D.C.: U.S. Department of Energy.

———. 1999. *Technical Support Document: Energy Efficiency Standards for*

Consumer Products: Residential Central Air Conditioners and Heat Pumps. Washington, D.C.: U.S. Department of Energy.

———. 2000a. http://www.oit.doe.gov/bestpractices/compressed_air. Washington, D.C.: U.S. Department of Energy.

———. 2000b. http://www.energystar.gov. Washington, D.C.: U.S. Department of Energy.

———. 2001. *U.S. Department of Energy FY 2002 Congressional Budget Request, Energy Efficiency and Renewable Energy, Energy Conservation, Industry Sector.* Washington, D.C.: U.S. Department of Energy.

Douglass, J. 1999a. *Model Repair Specifications for Low Voltage Induction Motors.* Olympia, Wash.: Washington State University Energy Program.

———. 1999b. *Motor Repair Tech Brief.* Olympia, Wash.: Washington State University Energy Program.

———. 1999c. *Service Center Evaluation Guide.* Olympia, Wash.: Washington State University Energy Program.

——— (Washington State University Energy Program). 2000. Personal communication to Neal Elliott. March.

Dreisilker, Henry. 1987. "Modern Rewind Methods Assure Better Rebuilt Motors." *Electrical Construction and Maintenance* 86 (August): 30–36.

E Source. 1997. *Commercial Space Cooling and Air Handling Technology Atlas.* TA-SC-97. Boulder, Colo.: E Source, Inc.

———. 1999. *Drivepower Technology Atlas Series, Volume IV.* Prepared by B. Howe, A. Lovins, D. Houghton, M. Shepard, and B. Stickney. Boulder, Colo.: E Source, Inc.

[EASA] Electrical Apparatus Service Association. 1985. *Core Iron Study.* St. Louis, Mo.: Electrical Apparatus Service Association.

———. 1998. *EASA-Q Quality Management System for Motor Repair.* St. Louis, Mo.: Electrical Apparatus Service Association.

Easton Consultants. 1996. *National Market Transformation Strategies for Industrial Electric Motor Systems: Volume II, Market Assessment.* DOE/PO-0044. Washington, D.C.: U.S. Department of Energy.

———. 2000. *Market Research Report: Variable Frequency Drives.* Report #00-054. Portland, Oreg.: Northwest Energy Efficiency Alliance.

Easton Consultants and XENERGY. 1999a. *Northeast Premium Motor Initiative: Market Baseline and Transformation Assessment.* Stamford, Conn.: Easton Consultants.

———. 1999b. *Opportunities for Industrial Motor Systems in the Pacific Northwest.* Portland, Oreg.: Northwest Energy Efficiency Alliance.

Eaton Corporation. 1988. *Dynamatic Industrial Drives Catalog.* Kenosha, Wis.: Eaton Corporation.

[EIA] Energy Information Administration. 1995. *Household Energy Consumption and Expenditures 1993.* DOE/EIA-0321(93). Washington, D.C.: U.S. Department of Energy.

———. 1997. *Manufacturing Consumption of Energy 1994.* DOE/EIA-0512(94). Washington, D.C.: U.S. Department of Energy.

———. 1999. *Residential Energy Consumption Survey 1997.* DOE/EIA-0632(97). Washington, D.C.: U.S. Department of Energy.

Elliott, N. 1995. *Energy Efficiency in Electric Motor Systems.* Washington, D.C.: American Council for an Energy-Efficient Economy.

Elliott, R.N., S. Nadel, and M. Pye. 1996. *Partnerships: A Path for the Design of Utility/Industrial Energy Efficiency Programs.* Washington, D.C.: American Council for an Energy-Efficient Economy.

[ELPN] Electric League of the Pacific Northwest. Undated. "Your Motors, Your Money: Motor Repair and Replacement Decision Making." Training Brochure. Bellevue, Wash.: Electric League of the Pacific Northwest.

———. 1999. *Drive Power Initiative I: Work Plan and Schedule (Task I Deliverable).* Prepared for the Northwest Energy Efficiency Alliance. Bellevue, Wash.: Electric League of the Pacific Northwest.

Emerson Motors. 2001. Exploded Graphic of TEFC Motor. St. Louis, Mo.: Emerson Motors.

Energy Center of Wisconsin. 1997. *Forced-Air Furnace and Central Air Conditioner Markets: Tracking Sales through Wisconsin HVAC Contractors.* Madison, Wis.: Energy Center of Wisconsin.

Englander S., and L. Norford. 1988. "Fan Energy Savings: Analysis of a Variable-Speed Drive Retrofit." In *Proceedings of the ACEEE 1988 Summer Study on Energy Efficiency in Buildings.* 3:51–64. Washington, D.C.: American Council for an Energy-Efficient Economy.

[EPA] U.S. Environmental Protection Agency. 2000a. http://yosemite1.epa.gov/estar/consumers.nsf/content/motors.htm. Washington, D.C.: U.S. Environmental Protection Agency.

———. 2000b. http://yosemite1.epa.gov/estar/consumers.nsf/content/lightvac.html. Washington, D.C.: U.S. Environmental Protection Agency.

———. 2000c. http://yosemite1.epa.gov/estar/consumers.nsf/content/cac.html. Washington, D.C.: U.S. Environmental Protection Agency.

[EPRI] Electric Power Research Institute. 1982. *Evaluation of Electrical Interference to the Induction Watthour Meter.* Report EL-2315. Palo Alto, Calif.: Electric Power Research Institute.

———. 1985. "Electronic Adjustable-Speed Drives for Boiler Feedpumps." *First Use* Document FS5414B/D/E, Results Series. Palo Alto, Calif.: Electric Power Research Institute.

———. 1989. *Proceedings: Advanced Adjustable Speed Drive R&D Planning Forum.* Report CU-6279. Palo Alto, Calif.: Electric Power Research Institute.

———. 1992. *Electric Motors: Markets, Trends and Applications.* TR-100423. Palo, Alto, Calif.: Electric Power Research Institute.

———. 1994. *Market Assessment Study for the Single-Phase Written Pole TM Motor.* Prepared by CRS Sirrine Engineers, Inc. TR-104072. Palo Alto, Caflif.: Electric Power Research Institute.

[ERM] ERM-Siam Co., Ltd. 1999. *Energy Efficiency Standards Regime Study.* Bangkok, Thailand: ERM-Siam Co., Ltd.

European Union/CEMEP. 1999. *Agreement on a Motor Classification Scheme.* Brussels, Belgium: CEMEP.

Faruqui, A., C.W. Gellings, M. Mauldin, S. Schick, K. Seiden, and G. Wikler. 1990. *Efficient Electricity Use: Estimates of Maximum Energy Savings.* Report CU-6746. Palo Alto, Calif.: Electric Power Research Institute.

Federal Register. 1999. "Energy Conservation Program for Consumer Products: Energy Conservation Standards for Central Air Conditioner and Heat Pumps; Proposed Rule." *Federal Register.* Part III, 10 CFR Part 430, 64: 54,142–51. Prepared by the U.S. Department of Energy, Office of Energy Efficiency and Renewable Energy. Washington, D.C.: U.S. Government Printing Office.

[FEMP] Federal Energy Management Program. 1998. *How to Buy an Energy-Efficient Electric Motor.* http://www.eren.doe.gov/femp/procurement/pdfs/motor.pdf. Washington, D.C.: U.S. Department of Energy.

Fenno, S. (New York State Energy Office). 1989. Personal communication to Steven Nadel. August.

Ference Weicker & Company. 1995. *Process Evaluation of B.C. Hydro's Fans, Pumps, and Compressed Air Programs.* Vancouver, B.C., Canada: Ference Weicker & Company.

Fitzgerald, A., C. Kingsley, and S. Umans. 1983. *Electric Machinery.* New York, N.Y.: McGraw-Hill.

Flaim, T. 2000. "The Big Retail 'Bust': What Will It Take to Get True Competition?" *The Electricity Journal* 13 (March): 41–51.

Flanigan, Ted, and Barb Hogan. 1995. *Industrial Energy Efficiency Programs: Building Lasting Partnerships for Mutual Benefit.* Basalt, Colo.: The Results Center.

Fleming, A. (British Columbia Hydro). 1995. Personal communication to Steven Nadel. June.

France, S. (Puget Power). 1989. Personal communication to Steven Nadel. August.

Friedman, R., C. Burrell, J. DeKorte, N. Elliott, and B. Meberg. 1996. *Electric Motor System Market Transformation.* Washington, D.C.: American Council for an Energy-Efficient Economy.

Furugaki, I. 1988. "The Energy Conservation Policy System in Japan." *Energy Efficiency Strategies for Thailand.* Edited by D. Bleviss and V. Lide. Lantham, Md.: University Press of America.

Futryk, R., and J. Kaman. 1987. "Variable-Speed Control of Lorain Assembly Plant Boiler Fans." Paper presented at the 1987 Ford International Energy Conference, Cologne, Germany, October 5–7.

Gandhi, N., and M. DiGiacomo. 1994. "Compressed Air Efficiency—Moving beyond Custom Programs." In *Proceedings of the ACEEE 1994 Summer Study on Energy Efficiency in Buildings,* 10:35–48. Washington, D.C.: American Council for an Energy-Efficient Economy.

Garay, P. 1990. *Pump Application Desk Book.* Lilburn, Ga.: Fairmont Press.

Gartland, L., and D. Sartor. 1998. "The Benefits of Integrated Chiller Retrofits: Excerpts from Case Studies." In *Proceedings of the ACEEE 1998 Summer Study on Energy Efficiency in Buildings,* 4:157–168. Washington, D.C.: American Council for an Energy-Efficient Economy.

Geller, H. (American Council for an Energy-Efficient Economy). 1990. Personal communication to Steven Nadel. Data from PROCEL, the Brazilian national electricity conservation program. June.

———. 1999. *Tax Incentives for Energy-Efficient Technologies.* Washington, D.C.: American Council for an Energy-Efficient Economy.

———. 2000. *Transforming End-Use Energy Efficiency in Brazil.* Washington, D.C.: American Council for an Energy-Efficient Economy.

Gilmore, W. (Walco Electric Company). 1989. Personal communications to Steven Nadel and Michael Shepard. September.

———. 1990. Personal communication to Steven Nadel. October.

Goldman, C., and E. Hirst. 1989. *Key Issues in Developing Demand-Side Bidding Programs.* Report LBL-27748. Berkeley, Calif.: Lawrence Berkeley Laboratory.

Gordon, Fred (Pacific Energy Associates). 2000. Personal communication to Steven Nadel. July.

Greenberg, S. 1996. *Electric Motor and Belt Retrofits: Measured Savings and Lessons Learned.* Berkeley, Calif.: Lawrence Berkeley National Laboratory.

Greenberg, S., J. Harris, H. Akbari, and A.T. de Almeida. 1988. *Technology Assessment: Adjustable-Speed Motors and Motor Drives (Residential and Commercial Sectors).* Report LBL-25080. Berkeley, Calif.: Lawrence Berkeley Laboratory.

Greenheck Fan Corporation. 1986. "Curve and Table for Model 30BISW." *Backward-Inclined Centrifugal Fans.* Publication No. BISW/BIDW-3-86 R. Schofield, Wisc.: Greenheck Fan Corporation.

Guttman, M., and A. Stotter. 1984. "The Influence of Oil Additives on Engine Friction and Fuel Consumption." In *Proceedings of the 39th Annual Meeting of the American Society of Lubrication Engineers.* Park Ridge, Ill.: American Society of Tribologists and Lubrication Engineers.

Hamer, P. (Chevron Research and Technology). 1999. Personal communication to Neal Elliott. October.

Hanna, Jim, and Michael Baker. 2000. "Making Performance Analysis Business-as-Usual in the Industrial Compressed Air Market." In *Proceedings of the ACEEE 2000 Summer Study on Energy Efficiency in Buildings,* 6:145–155. Washington, D.C.: American Council for an Energy-Efficient Economy.

Hanson, M. (Energy Center of Wisconsin). 1997. Personal communication to Neal Elliott.

Houghton, D., R. Bishop, A. Lovins, B. Stickney, J. Newcomb, M. Shepard, and B. Davids. 1992. *The State of the Art: Space Cooling and Air Handling.* Boulder, Colo.: Competitek.

Hudson, W. (Penn Dower Petroleum Co.). 1989. Personal communication to Steven Nadel. October.

Ibanez, P. 1978. "Electromechanical Energy." In *Efficient Electricity Use,* 369–409. Edited by C. Smith, et al. Elmsford, N.Y.: Pergamon Press.

[IEEE] Institute of Electrical and Electronics Engineers. 1981. *IEEE Guide for Harmonic Control and Reactive Compensation of Static Power Converters.* IEEE Standard 519. New York, N.Y.: Institute of Electrical and Electronics Engineers.

———.1982. IEEE 114-1982: *IEEE Standard Test Procedure for Single-Phase Induction Motors.* New York, N.Y.: Institute of Electrical and Electronics Engineers.

———. 1996. *Standard Test Procedure for Polyphase Induction Motors and Generators.* Standard 112-1996. New York, N.Y.: Institute of Electrical and Electronics Engineers.

———. 1998. *Guide for Distribution Transformer Loss Evaluation.* Standard PC57.12.33. Piscataway, N.J.: Institute of Electrical and Electronics Engineers.

[IEL] Industrial Electrotechnology Laboratory. 1992. *ENERGY SAVERS: Variable Airflow Control to Cut Lumber Drying Costs.* Raleigh, N.C.: Industrial Electrotechnology Laboratory.

Ingersoll-Rand. 2000a. Single-Acting Reciprocating Air Compressor and Helical-Screw Air Compressor Graphic. Davidson, N.C.: Ingersoll-Rand.

————. 2000b. Centac Model C950 Centrifugal Air Compressor Photo. Davidson, N.C.: Ingersoll-Rand, Air Solutions Group.

Intek Inc. and Libby Engineering Limited. 1995. *Technology Profile Summary: Commercial and Industrial Fans*. Montreal, Quebec, Canada: Canadian Electrical Association.

Jaccard, M., A. Fogwill, and J. Nyboer. 1993. "How Big Is the Electricity Conservation Potential in Industry?" *The Energy Journal* 14 (2): 139–156.

Jackson, J. 1987. *New England Power Pool Commercial Energy Demand Model Systems*. Sandwich, Mass.: Jerry Jackson and Associates.

Jacobs Engineering Group, Inc. 1996. *ASDMaster User's Guide*. Palo Alto, Calif.: Electric Power Research Institute.

Johnston, W. (North Carolina Industrial Extension Service). 1990. Personal communication to Steven Nadel. September.

Kaminski, Tony. 1995. *Technology Profile Report: Small Pumps*. Montreal, Quebec, Canada: Canadian Electrical Association.

Katz, G. (Momentum Engineering). 1990. Personal communication to Steven Greenberg. June.

Kellum, Z. (Advanced Energy). 2001. Personal communication to Neal Elliott. February.

Kent, J. 1989. (Kent Oil Co.). Personal communication to Steven Nadel and Michael Shepard. October.

Kline, J. (Emerson Motors). 2001. Personal communication to Neal Elliott. February.

Knapp, Bill (Northeast Utilities). 2000. Personal communication to Steven Nadel. November.

Kochensparger, J. 1987. "Applying Predictive Maintenance Testing to Minimize Motor Failure Downtime." *Plant Engineering* (March 12).

[Krupp-Widia] Krupp-Widia Magnet Engineering. 1987. Information extracted from an advertisement. Essen, Germany: Krupp-Widia Magnet Engineering.

Kubo, T., S. Nadel, and M. Suozzo. 2000. "Commercial Packaged Refrigeration: An Untapped Lode for Energy Efficiency." In *Proceedings of the ACEEE 2000 Summer Study on Energy Efficiency in Buildings*, 3:203–214. Washington, D.C.: American Council for an Energy-Efficient Economy.

Lannoye, M. (Washington State Energy Office). 1988. Letter to A. Lovins of the Rocky Mountain Institute. September 22.

Lawne, R.J., ed. 1987. *Electric Motor Manual*. New York, N.Y.: McGraw-Hill.

Lawrence, Roger (EPRI ASD Center). 2000. Personal communication to Neal Elliott. July.

[LBNL] Lawrence Berkeley National Laboratory. 1996. *Draft Report on*

Energy Conservation Potential for Small Electric Motors. Prepared for the U.S. Department of Energy. Berkeley, Calif.: Lawrence Berkeley National Laboratory.

[LBNL/RDC] Lawrence Berkeley National Laboratory and Resource Dynamics Corporation. 2001. *Improving Motor and Drive System Performance: A Sourcebook for Industry*. Washington, D.C.: U.S. Department of Energy.

Leland, Ted (Air Conditioning and Refrigeration Institute). 1998. Personal communication to Steven Nadel. April.

Leonard, W. 1984. *Control of Electrical Drives*. New York, N.Y.: Springer-Verlag.

———. 1986. "Microcomputer Control of High Dynamic Performance AC Drives—A Survey." *Automatica* 22 (1): 1–19.

Linn, C. 1987. "Calculating Daylight for Successful Retail Design." *Architectural Lighting* (January): 29–34.

Linn, J. (NEEP). 2000. Personal communication to Steven Nadel. April 21.

Liu, P. (International Institute for Energy Conservation). 2000. Personal communication to Steven Nadel. January.

Lloyd, T.C. 1969. *Electric Motors and Their Applications*. New York, N.Y.: Wiley Interscience.

Lovins, A., B. Bancroft, T. Flanigan, P. Kiernan, J. Neymark, and M. Shepard. 1989. *The State of the Art: Drivepower*. Snowmass, Colo.: Rocky Mountain Institute (Competitek).

Magnusson, D. 1984. "Energy Economics for Equipment Replacement." *IEEE Transactions on Industry Applications* IA-20 (March/April): 402–406.

Marbek Resource Consultants, Ltd. 1987. *Energy-Efficient Motors in Canada: Technologies, Market Factors, and Penetration Rates*. Ottawa, Ontario, Canada: Marbek Resource Consultants.

McAteer, Michael (National Grid). 2000. Personal communication to Steven Nadel. April.

McCoy, G., and J. Douglass. 1997. *Energy Management Guide for Motor-Driven Systems*. Olympia, Wash.: Washington State University Energy Program.

McGovern, W. 1984. "High-Efficiency Motors for Upgrading Plant Performance." *Electric Forum* 10 (2): 4–6.

McKane, A. (Lawrence Berkeley National Laboratory). 2000. Personal communication to Steven Nadel. July.

McKay, S. 1992. *Development of Test Standards for Single- and Three-Phase AC Induction Fractional HP Motor Efficiency*. Toronto, Ontario, Canada: Ontario Hydro Research Division.

Meadows, K. (Energy Center of Wisconsin). 2000. Personal communication to Steven Nadel. August.

[MECo] Massachusetts Electric Company. 2000. *Industrial Systems Optimization Service (ISOS), Primary Objective.* Northborough, Mass.: Massachusetts Electric Company.

Milton, B., and E. Carter. 1982. "Fuel Consumption and Emission Testing of an Engine Oil Additive Containing PTFE Colloids." *American Society of Lubrication Engineers Transactions* 39 (2): 105–110.

Mohan, N. 1981. *Techniques for Energy Conservation in AC Motor-Driven Systems.* EPRI Report EM-2037. Palo Alto, Calif.: Electric Power Research Institute.

Montgomery, D. 1989. "Testing Rewinds to Avoid Motor Efficiency Degradation." *Energy Engineering* 86 (3): 24–40.

Morash, R. (Precise Power Corporation). 1998. Personal communication to Sam Wheeler. April. Quoted in S. Nadel, L. Rainer, M. Shepard, M. Suozzo, and J. Thorne, *Emerging Energy-Saving Technologies and Practices for the Buildings Sector.* Washington, D.C.: American Council for an Energy-Efficient Economy.

Morton, R., and M. McDevitt. 2000. *Evaporator Fan VFD Effects on Energy and Fruit Quality.* Walla Walla, Wash.: Cascade Energy Engineering.

Nadel, S. 1990. *Lessons Learned: A Review of Utility Experience with Conservation and Load Management Programs for Commercial and Industrial Customers.* Washington, D.C.: American Council for an Energy-Efficient Economy.

———. 1996. *Providing Utility Energy Efficiency Services in an Era of Tight Budgets: Maximizing Long-Term Energy Savings While Minimizing Utility Costs.* Washington, D.C.: American Council for an Energy-Efficient Economy.

Nadel, S., F. Gordon, and C. Neme. 2000. *Using Targeted Energy Efficiency Programs to Reduce Peak Electrical System Reliability Problems.* Washington, D.C.: American Council for an Energy-Efficient Economy.

Nadel, S., and J. Jordan. 1994. *Designing Industrial DSM Programs That Work.* Washington, D.C.: American Council for an Energy-Efficient Economy.

Nadel, S., and L. Latham. 1998. *The Role of Market Transformation Strategies in Achieving a More Sustainable Energy Future.* Washington, D.C.: American Council for an Energy-Efficient Economy.

Nadel, S., M. Pye, and J. Jordan. 1994. *Achieving High Participation Rates: Lessons Taught by Successful DSM Programs.* Washington, D.C.: American Council for an Energy-Efficient Economy.

Nadel, S., L. Rainer, M. Shepard, M. Suozzo, and J. Thorne. 1998. *Emerging Energy-Saving Technologies and Practices for the Buildings Sector.* Washington, D.C.: American Council for an Energy-Efficient Economy.

Nadel, S., M. Shepard, S. Greenberg, G. Katz, and A.T. de Almeida. 1991. *Energy-Efficient Motor Systems.* Washington, D.C.: American Council for an Energy-Efficient Economy.

Nailen, Robert. 1987. *Motors, Volume 6: Power Plant Electrical Reference Series.* Palo Alto, Calif.: Electric Power Research Institute.

[NCAEC] North Carolina Alternative Energy Corporation. 1989. "Energy-Efficient Electric Motor Systems." Project Proposal. Research Triangle Park, N.C.: North Carolina Alternative Energy Corporation.

Neal, L., K. Grabner, and F. Gordon. 2000. *HVAC Installation Practices: Draft Final Report.* Newark, N.J.: Public Service Electric and Gas.

[NEEA] Northwest Energy Efficiency Alliance. http://www.nwalliance. org/ projects/current/warehouse.html. Portland, Oreg.: Northwest Energy Efficiency Alliance.

[NEEP] Northeast Energy Efficiency Partnership. 2000a. "Commercial and Industrial HVAC Equipment Application." Form #CCI-01.01.00. Lexington, Mass.: Northeast Energy Efficiency Partnership.

———. 2000b. *Northeast Premium Efficiency Motors Initiative, Motor Repair Component (Draft).* Lexington, Mass.: Northeast Energy Efficiency Partnership.

[NEMA] National Electrical Manufacturers Association. 1978. *Motors and Generators.* NEMA Standards Publication No. MG 1-1978. Rosslyn, Va.: National Electrical Manufacturers Association.

———.1989. *Motors and Generators.* NEMA Standards Publication No. MG 1-1989. Rosslyn, Va.: National Electrical Manufacturers Association.

———. 1993. *Motors and Generators.* NEMA Standards Publication No. MG 1-1993. Rosslyn, Va.: National Electrical Manufacturers Association.

———. 1994. *Energy Management Guide for Selection and Use of Polyphase Motors.* NEMA Standards Publication No. MG 10-1994. Washington, D.C.: National Electrical Manufacturers Association.

———. 1996. *Guide for Determining Energy Efficiency for Distribution Transformers.* NEMA Standards No. TP-1-1996. Rosslyn, Va.: National Electrical Manufacturers Association.

———. 1999. *Motors and Generators.* NEMA Standards Publication No. MG 1-1998. Rosslyn, Va.: National Electrical Manufacturers Association.

———. 2000. "NEMA Premium Energy Efficiency Motor Program." Fact Sheet. http://www.nema.org/premiummotors. Rosslyn, Va.: National Electrical Manufacturers Association.

Neme, C., S. Nadel, and J. Proctor. 1999. *Energy Savings Potential from*

Addressing Residential Air Conditioner and Heat Pump Installation Problems. Washington, D.C.: American Council for an Energy-Efficient Economy.

[NEPSCo] New England Power Service Company. 1989. *Rhode Island Motor Survey*. Westborough, Mass.: New England Power Service Co.

Niagara Mohawk. 1987. N*iagara Mohawk Power Corporation Motor Retrofit Program: An Industrial Customer Rebate Demonstration Program*. Project No. C1P-17. Syracuse, N.Y.: Niagara Mohawk.

Nugent, D. 1993. "High Efficiency Chillers—Why Stop at .42 kW/Ton?: System Integration and Close-Approach Design Maximize Energy Benefits." In *Proceedings, Second National New Construction Programs for Demand-Side Management Conference*. Sacramento, Calif.: ADM Association.

[NYSERDA] New York State Energy Research & Development Authority. 2000. *New York State Energy Smart Program Evaluation and Status Report—December*. Albany, N.Y.: New York State Energy Research & Development Authority.

Odell, C. (Northeast Utilities). Personal communication to Steven Nadel. August.

[OIT] Office of Industrial Technologies. 1997. *The Challenge: Improving Ventilation System Energy Efficiency in a Textile Plant*. Motor Challenge Showcase Demonstration Case Study Series. Washington, D.C.: U.S. Department of Energy, Office of Industrial Technologies.

———. 1998. *The Challenge: Reduce BOF Hood Scrubber Energy Costs at a Steel Mill*. Motor Challenge Showcase Demonstration Case Study Series. Washington, D.C.: U.S. Department of Energy, Office of Industrial Technologies.

———. 2000. *Pumping System Assessment Tool*. Washington, D.C.: U.S. Department of Energy, Office of Industrial Technologies.

Oliver, J. (U.S. Department of Energy, Motor Challenge Program). 1999. Personal communication to Steven Nadel. January.

Oliver, J., and M. Samotyj. 1989. *Lessons Learned from Field Tests of Large Induction Motor Adjustable Speed Drives, 1984–1989*. Palo Alto, Calif.: Electric Power Research Institute.

Ontario Hydro. 1988. *Marketing High-Efficiency Motors*. Toronto, Ontario, Canada: Ontario Hydro.

———. 1992. *Hydro Motor Efficiency Levels*. Toronto, Ontario, Canada: Ontario Hydro.

Ostertag, K. 1999. *Transaction Costs of Raising Energy Efficiency*. Paris, France: Centre International de Recherche sur l'Environnement et le Développement.

Paco Pumps. 1983. "Pump Curve Number RC-2010." Oakland, Calif.: Paco Pumps Company.

————. 1985. "A1b End Suction Centrifugal Pumps Type L; Pump Selection Charts." Oakland, Calif.: Paco Pumps Company.

Payton, R. (Reliance Electric Company). 1988. Personal communication to Steve Greenberg. February.

[PEA] Pacific Energy Associates, Inc. 1998. "Premium Efficiency Motors Program." *Market Progress Evaluation Report.* NEEA Premium Efficiency Motors Program. Portland, Oreg.: Northwest Energy Efficiency Alliance.

————. 2000. *Market Progress Evaluation Report: Alliance Drive Power Initiative.* Portland, Oreg.: Northwest Energy Efficiency Alliance.

[PEAC] Power Electronics Applications Center. 1987. ASD Directory. 2nd ed. Knoxville, Tenn.: Power Electronics Applications Center.

Peddie, R. 1988. "Smart Meters." In *Demand-Side Management and Electricity End-Use Efficiency,* 171–180. Edited by A.T. de Anibal and A. Rosenfeld. NATO Advanced Science Institutes Series E, vol. 149, Applied Sciences. Norwell, Mass.: Kluwer Academic Publishers.

Peng, F. (Oak Ridge National Laboratory). 2000. Personal communication to Neal Elliott. August.

Perkins, R. (Compaq Computer). 1989. Personal communication to Steven Nadel. October.

Peters, Dale (Copper Development Association, Inc). 1998. Personal communication to Neal Elliott. June.

[PG&E] Pacific Gas and Electric Company. 1999. *PY97 Agricultural Energy Efficiency Incentives: Pumping and Related Market Effects Study.* Study ID #335A. San Francisco, Calif.: Pacific Gas and Electric Company.

Poole, M., J. Moran, D. Seitzinger, T. Johnson, G. Stengl, S. Salib, and D. Wangerin. 1989. *Commercial and Industrial Applications of Adjustable-Speed Drives.* CU-6883. Palo Alto, Calif.: Electric Power Research Institute.

Precise Power. 1998. Product catalog for written pole motors. Tel. 941-746-3515. Bradenton, Fla.: Precise Power Corporation.

Pye, M. 1998. *Making Business Sense of Energy Efficiency and Pollution Prevention.* Washington, D.C.: American Council for an Energy-Efficient Economy.

Pye, M., R.N. Elliott, and S. Nadel. 1996. *Partnerships: A Path for the Design of Utility/Industrial Energy Efficiency Programs.* Washington, D.C.: American Council for an Energy-Efficient Economy.

Pye, M., and S. Nadel. 1997. *Energy Technology Innovation at the State Level: Review of State Energy Research, Development, and Demonstration (RD&D) Programs.* Washington, D.C.: American Council for an Energy-Efficient Economy.

RLW Analytics. 1998. *Southern California Edison Hydraulic Services*

Program Market Effects Study: Final Report. Study ID #3507. Prepared for Southern California Edison. Sonoma, Calif.: RLW Analytics.

———. 1999. *Massachussetts Commercial HVAC Study: Final Report.* Middletown, Conn.: RLW Analytics.

Rosenberg, M. 1996. *The United States Motor Systems Baseline: Inventory and Trends.* Burlington, Mass.: XENERGY.

——— (XENERGY). 1999. Personal communication to Steven Nadel. November.

——— (XENERGY). 2000. Personal communication to Neal Elliott. June.

[SCE] Southern California Edison. 2000. "SCE Business Advisor: Pump Tests." http://www.scebiz.com/solutionscc/services/pumptest.htm. Rancho Cucamonga, Calif.: Southern California Edison.

Scheihing, Paul E. 1996. "United States Department of Energy's Motor Challenge Program: A National Strategy for Energy-Efficient Industrial Motor-Driven Systems." Paper presented at the European Commission Conference: "Energy Efficiency Improvements in Motors and Drives." Lisbon, Portugal, October 29–31.

Schiller, S., C. Goldman, and B. Henderson. 2000. "Public Benefit Charge Funded Performance Contracting Programs—Survey and Guidelines." In *Proceedings of the ACEEE 2000 Summer Study on Energy Efficiency in Buildings*, 5:299–317. Washington, D.C.: American Council for an Energy-Efficient Economy.

Schueler, V., P. Leistner, and J. Douglass. 1994. *Industrial Motor Repair in the United States.* Portland, Ore.: Bonneville Power Administration.

Scott, S., J. Stout, and F. Gordon. 2000. *Market Progress Evaluation Report: Sav-Air Market Transformation Initiative.* Portland, Oreg.: Northwest Energy Efficiency Alliance.

Seton, Johnson, & Odell, Inc. 1983. *Summary Data from proprietary Industrial Motor Drive Study.* Seattle, Wash.: Seattle City Light.

———. 1987a. *Energy Efficiency and Motor Repair Practices in the Pacific Northwest.* Portland, Oreg.: Bonneville Power Administration.

———. 1987b. *Lost Conservation Opportunities in the Industrial Sector.* Portland, Oreg.: Bonneville Power Administration.

Shipley, A.M., A. Hinge, and R.N. Elliott. 2000. *Energy-Efficient Programs for Small- and Medium-Sized Industries—Draft.* Washington, D.C.: American Council for an Energy-Efficient Economy.

Smeaton, R. 1987. *Motor Application and Maintenance Handbook.* New York, N.Y.: McGraw-Hill.

Sperber, R. 1989. "Maintenance Software System Yields Two-Year Payback." *Food Processing* (October): 72–78.

[Standards Australia] Joint Standards Australia/Standards New Zealand Committee EL/46. 1999. *Three-Phase Cage Induction Electric Motors—Draft.* Canberra, ACT, Australia: Standards Australia.

Stout, T. (New England Electric Service). 1990. Personal communication to Steven Nadel. June.

———— (National Grid). 2000. Personal communication to Steven Nadel. July.

Stricklett, K. (National Institute for Standards and Technology). 2001. Personal communication to Neal Elliott. February.

Strohs, R. 1987. "Application of Variable-Speed Drive Pumping Systems for Energy Savings." Paper presented at the 1987 Ford International Energy Conference, Cologne, Germany, October.

Sturiale, Jo Anne. 1999. *POS Evaluation*. Prepared for the Energy Center of Wisconsin. Madison, Wis.: Sturiale & Company.

Suozzo, M., J. Benya, P. DuPont, N. Elliott, M. Hydeman, and S. Nadel. 2000. *Guide to Energy-Efficient Commercial Equipment*. Washington, D.C.: American Council for an Energy-Efficient Economy.

Suozzo, M., and S. Nadel. 1998. *Selecting Targets for Market Transformation Programs: A National Analysis*. Washington, D.C.: American Council for an Energy-Efficient Economy.

Sutcliffe, Lynn (Onsite Sycom Energy Corp.). 1999. Personal communication to Neal Elliott. July.

Thielemann, Kurt (Duke Energy). 1997. Personal communication to Neal Elliott. July.

Treadle, S. 1987. "The Interaction of Lighting and HVAC Systems." *Lighting Design and Application*. May.

Turnbull, Peter (Pacific Gas & Electric). 2000. Personal communication to Steven Nadel. November.

U.S. Census Bureau. 1988. *1986 Annual Survey of Manufacturers*. M86(AS)-1. Washington, D.C.: U.S. Department of Commerce, Economics and Statistical Division.

————. 1989. "Motors and Generators—1988." *Current Industrial Reports*. MA36H. Washington, D.C.: U.S. Department of Commerce, Economics and Statistical Division.

————. 1994. "Motors and Generators—1993." *Current Industrial Reports*. MA36H. Washington, D.C.: U.S. Department of Commerce, Economics and Statistical Division.

————. 1995. "Motors and Generators—1994." *Current Industrial Reports*. MA36H. Washington, D.C.: U.S. Department of Commerce, Economics and Statistical Division.

————. 1996a. "Motors and Generators—1995." *Current Industrial Reports*. MA36H(95)-1. Washington, D.C.: U.S. Department of Commerce, Economics and Statistical Division.

————. 1996b. *1994 Annual Survey of Manufacturers: Statistics for Industry Groups and Industries*. Report M94(AS)-1. Washington,

D.C.: U.S. Department of Commerce, Economics and Statistical Division.

———. 1997. "Motors and Generators—1996." *Current Industrial Reports.* MA36H(96)-1. Washington, D.C.: U.S. Department of Commerce, Economics and Statistical Division.

———. 1998a. *1996 Annual Survey of Manufacturers: Statistics for Industry Groups and Industries.* MA96(AS)-1. Washington, D.C.: U.S. Department of Commerce, Economics and Statistical Division.

———. 1998b. "Motors and Generators—1997." *Current Industrial Reports.* MA36H(97)-1. Washington, D.C.: U.S. Department of Commerce, Economics and Statistical Division.

———. 2000. "Motors and Generators—1998." *Current Industrial Reports.* MA335H(98)-1. Washington, D.C.: U.S. Department of Commerce, Economics and Statistical Division.

U.S. Congress. 1992. *Energy Policy Act of 1992.* Washington, D.C.: U.S. Government Printing Office.

Van Son, D. (Baldor Motors). 1989. Personal communication and letter to Michael Shepard. August.

———. 1994. "Looking Forward to What Will Happen to Efficiency through the Decade." In *Proceedings of the World Energy Engineering Congress.* Atlanta, Ga.: Association of Energy Engineers.

[VDPS] Vermont Department of Public Service. 1999. *Bilateral Agreement between the Department of Public Service and International Business Machines Corporation.* Montpelier, Vt.: Vermont Department of Public Service.

Wallace, R. (Reliance Corporation). 1998. Personal communication to Sam Wheeler. April. Quoted in S. Nadel, L. Rainer, M. Shepard, M. Suozzo, and J. Thorne, *Emerging Energy-Saving Technologies and Practices for the Buildings Sector.* Washington, D.C.: American Council for an Energy-Efficient Economy.

Wilke, K., and T. Ikuenohe. 1987. "Guidelines for Implementing an Energy-Efficient Motor Retrofit Program." In *Proceedings of the 10th World Energy Engineering Congress,* 399–406. Atlanta, Ga.: Association of Energy Engineers.

Winters, Rachel (New York State Energy Research and Development Authority). 2000. Personal communication to Steven Nadel. November.

Wroblewski, Ron (Energy Center of Wisconsin). 1996. Personal communication to Steven Nadel.

[WSEO] Washington State Energy Office. 1994. *Industrial Motor Repair in the United States: Current Practice and Opportunities for Improving Customer Productivity and Energy Efficiency.* DOE/BP-2749. Prepared for the Electric Power Research Institute, Bonneville Power

Administration, and U.S. Department of Energy. Portland, Oreg.: Washington State Energy Office.

[WSU] Washington State University Energy Program. 1999. *MotorMaster+®*. Version 3.01. Olympia, Wash.: Washington State University.

XENERGY. 1989. "Motor Inventory for Wisconsin Electric." Burlington, Mass.: XENERGY.

————. 1998. *United States Industrial Electric Motor Systems Market Opportunities Assessment*. Prepared for the U.S. Department of Energy, Office of Industrial Technologies and Oak Ridge National Laboratory. Washington, D.C.: U.S. Department of Energy, Office of Energy Efficiency and Renewable Energy.

————. 2000a. *Assessment of the Market for Compressed Air Efficiency Services*. Prepared for the U.S. Department of Energy and the Compressed Air Challenge. Burlington, Mass.: XENERGY.

————. 2000b. *Final Report of the Motor Challenge Program*. Prepared for the Oak Ridge National Laboratory. Burlington, Mass.: XENERGY.

————. 2000c. *Program Description: The Compressed Air and Pump Efficiency Program*. Burlington, Mass.: XENERGY.

Zeller, M. 1992. *Rewound High-Efficiency Motor Performance*. Vancouver, B.C., Canada: B.C. Hydro in association with Powertech Labs, Inc.

About the Authors

ANIBAL T. DE ALMEIDA, professor of electrical engineering at the University of Coimbra, Portugal, holds a Ph.D. in power systems from the University of London. He was the co-chairman of two international conferences, "Energy Efficiency Improvements in Electric Motors and Drives" in Lisbon in 1996 and "Energy Efficiency in Motor-Driven Systems" in London in 1999. He has conducted research on drivepower systems in Europe and the United States, coordinated several international projects on motor market transformation in Europe, and participated in similar projects in developing countries.

NEAL ELLIOTT, senior sssociate and industry program director, joined the American Council for an Energy-Efficient Economy in 1993. Elliott received B.S. and M.S. degrees in mechanical engineering from North Carolina State University and a Ph.D. from Duke University and is a registered professional engineer in North Carolina. He became involved with electric motors in his previous position at the North Carolina Alternative Energy Corp. (now Advanced Energy), where he oversaw the establishment of the joint Industrial Electrotechnology Laboratory, which included the first independent motor test facility in the United States. Since joining ACEEE, he has become a national leader in motor efficiency activities, including advising DOE on the development of the *Motor Challenge* program (now *Industrial Best Practices: Motors*), being involved in the development of the EPAct motor rule, chairing the Consortium for Energy Efficiency's Motor System Committee (which developed the first premium motor specification), and conducting leading-edge research on the development of motor market transformation initiatives, including the *Compressed Air Challenge*.

STEVE GREENBERG holds a bachelor's degree in mechanical engineering and a master's degree in energy and resources, both from the University of California at Berkeley, and is a licensed professional engineer. Now a senior energy management engineer at the Lawrence Berkeley National Laboratory, since 1980 he has researched and applied energy-efficient motor systems for a variety of clients on three continents and has authored numerous publications on the topic. He has worked with manufacturers, distributors, system designers, contractors, and test-facility and building maintenance personnel on a wide variety of motor system applications, going from concept to implementation and performance verification.

He has also assisted in developing test procedures and standards for motors and motor systems.

GAIL KATZ held degrees in electrical and mechanical engineering from Portland State University. She spent 7 years at the engineering firm of Seton, Johnson, & Odell, developing energy conservation plans for Pacific Northwest industries, before founding her own firm, Momentum Engineering, in 1988. Her consulting practice grew to include extensive work with industrial facilities and utilities throughout the country, national laboratories, and other groups. In 1990 she was invited to join a National Academy of Sciences delegation to help Poland address its environmental problems. She died at her home of heart and kidney failure on August 14, 1990.

STEVEN NADEL is the executive director of the American Council for an Energy-Efficient Economy, a position he assumed after serving as deputy director for many years. He holds an M.A. in environmental studies from Wesleyan University and an M.S. in energy management from the New York Institute of Technology and has over 20 years of experience in the energy efficiency field. He is the author of dozens of papers on the results of and lessons learned from energy efficiency programs, particularly programs operated by utilities and regional market transformation organizations. He also played a key role in the development of the motor minimum-efficiency standards adopted by the U.S. Congress in 1992. Before joining ACEEE, he planned and evaluated energy efficiency and load-management programs for the New England Electric System, including programs to improve the efficiency of motor systems.

MICHAEL SHEPARD is senior vice president at E Source Inc. in Boulder, Colorado, which provides information on energy-efficient technologies and strategies to energy providers, end-users, and other clients in 35 countries. He holds a bachelor's degree in natural resource conservation from Cornell University and a master's degree from the Energy and Resources Group at the University of California at Berkeley. He has written and consulted extensively on energy issues and was a co-author of E Source's *Drivepower Technology Atlas Series*. Shepard formerly was director of the Energy Program at Rocky Mountain Institute, a senior feature writer for *EPRI Journal*, and publications director of the New Mexico Solar Energy Association.

Abbreviations / Acronyms

AC	alternating current
AEC	Advanced Energy Corp.
AEDM	alternative efficiency determination method
AMCA	Air Movement and Control Association International, Inc.
ANSI	American National Standards Institute
ASD	adjustable-speed drive
ASHRAE	American Society of Heating, Refrigerating and Air-Conditioning Engineers, Inc
AWWA	American Water Works Association
BHP	brake horsepower
BOF	basic oxygen furnace
BPA	Bonneville Power Administration
BS	British Standard
BSC	Bethlehem Steel Corporation
Btuh	British thermal unit per hour
CAC	*Compressed Air Challenge*
CAGI	Compressed Air and Gas Institute
CC	Compliance Certification
CDA	Comprehensive Design Approach
CEA	Canadian Electrical Association

CEC	California Energy Commission
CEE	Consortium for Energy Efficiency
CEMEP	Committee of European Manufacturers of Electrical Machines and Power Electronics
cfm	cubic feet/minute
CSA	Canadian Standards Association
CSE	cost of saved energy
CSI	current-source inverters
DAFT	dissolved air flotation thickener
DC	direct current
DOE	U.S. Department of Energy
DSM	demand-side management
EASA	Electric Apparatus Service Association
ECPMs	electronically commutated permanent-magnet motors
EER	energy efficiency rating
EMI	electromagnetic interference
EMS	energy management system
EPA	U.S. Environmental Protection Agency
EPAct	Energy Policy Act of 1992
EPRI	Electric Power Research Institute
ESCo	energy service company
EU	European Union
EUP	End-Use Pricing Service
EXP	explosion-proof motor enclosure
EXPFC	explosion-proof fan-cooled
EXPNV	explosion-proof nonventilated
FCC	Federal Communications Commission
FEMP	Federal Energy Management Program
GCC	General Conservation Corporation
gpm	gallons per minute
GW	gigawatts

GWh	gigawatt-hours
HI	Hydraulic Institute
hp	horsepower
HVAC	heating, ventilating, and air conditioning
Hz	Hertz
I^2R	current squared times resistance
ID	induced draft
IEC	International Electrotechnical Commission
IEEE	Institute of Electrical and Electronics Engineers
IGT	insulated gate transistor
IPLV	integrated part load value
ISOS	Industrial Systems Optimization Service
JEC	Japanese Electrotechnical Commission
kVA	kilovolt-amperes
kVAR	kilovolt-ampere-reactive
kW	kilowatts
kWh	kilowatt-hours
kW-yr	kilowatt-years
LBNL	Lawrence Berkeley National Laboratory
MC	*Motor Challenge*
M-G	motor-generator
MW	megawatts
MWh	megawatt-hours
NEEA	Northwest Energy Efficiency Alliance
NEEP	Northeast Energy Efficiency Partnerships
NEMA	National Electrical Manufacturers Association
NIST	National Institute of Standards and Technology
NPSH	net positive suction head
NU	Northeast Utilities

| NVLAP | National Voluntary Laboratory Accreditation Program |
| NYSERDA | New York State Energy Research & Development Authority |

O&M	operation and maintenance
ODP	open drip-proof motor enclosure
OEMs	original equipment manufacturers
OIT	Office of Industrial Technologies

PAM	pole amplitude modulation
PC	personal computer
PCBs	polychlorinated biphenyls
PEAC	Power Electronics Application Center
PEMs	premium-efficiency motors
PEPCo	Potomac Electric Power Company
PFC	power-factor controller
PG&E	Pacific Gas & Electric Co.
PIC	power-integrated circuit
PM	permanent-magnet
POS	Performance Optimization Service
PSAT	Pump System Assessment Tool
PSC	permanent split-capacitor
PSCo	Public Service of Colorado
psi	pounds per square inch
psig	pounds per square inch gage
PSP	PowerSaving Partner
PWM	pulse-width modulation

RD&D	research, development, and demonstration
RMS	root mean square
rpm	revolutions per minute

| SCR | silicon-controlled rectifier |
| SEER | Seasonal Energy Efficiency Ratio |

SMUD	Sacramento Municipal Utility District
SPC	standard performance contracting
SR	switched-reluctance
TEFC	totally enclosed fan-cooled motor enclosure
TDH	total dynamic head
TWh	terawatt-hours
UI	United Illuminating
V	volt
VAV	variable-air-volume
VIV	variable-inlet-vane
VSI	voltage-source inverter
w.c.	water column
WEPCo	Wisconsin Electric Power Company
WP	written pole
WPI	Worcester Polytechnic Institute
WSU	Washington State University

Economics

The end-user and the electric supply system both benefit from demand-side distributed resources such as efficient motors. Motor users and the electric supply system each have reasons for installing efficient motors and drives: the motor users will save money on operating costs, and the electric supply system can obtain a demand-side resource that can be less expensive than new transmission and generating assets. As a result of recent restructuring of the electric power industry, the electric system is no longer a single vertically integrated company but now may be made up of several entities variously engaged in electricity generation, transmission, distribution, and related services. Any one or a combination of these entities may offer demand-side management programs that encourage energy efficiency. This appendix presents some general economic concepts used by each party, followed by technology-specific discussions and tables and worksheets for evaluating the economics of drivepower investments.

Economic Concepts

Motor users and DSM programs use very different criteria when making decisions about investments. The perspective of the motor user is discussed first.

Motor User Perspective

Firms have choices about where to spend money to produce a return on their investment. Each option, including the purchase of energy-efficient equipment, must compete for scarce capital with other potential investments. Therefore, the economic analysis of efficiency investments should be formatted in the same way as the analysis of other capital investments in order that all options can be compared on an equal basis.

The economic return needed to persuade a company to purchase energy-efficient equipment varies among firms. Many companies select products or make capital investments solely on the basis of least first cost. However, the most common method used by equipment buyers to evaluate conservation investments is the simple payback, or the time that it will take for the savings to pay back the cost of the investment. The simple payback is calculated by dividing the incremental cost of the efficient equipment by the value of the expected annual energy savings. For example, if an efficient motor costs $500 more than a standard motor and is expected to save $400/yr, the simple payback will be 1.25 years.

The use of the simple payback introduces some errors into the calculation by assuming that inflation is zero and utility rates are constant. It also ignores the life of the measure. A device with a 6-month payback may seem like a good investment, but it's not if it lasts only 8 months. Because of the short payback requirements of most motor users, however, and the relatively low cost of installing efficient motors and drives, the errors in simple payback analysis are generally minor.

Some motor users use a related analysis, called the return on investment (ROI). This method looks at the percent of the investment returned annually. For example, if an efficient motor costs $200 more than a standard motor and is expected to save $100/yr, it returns 50% of the investment annually. In general, this method produces results equivalent to the simple payback analysis and, additionally, is capable of evaluating a payment stream.

Within the format of a simple payback or ROI analysis, the cutoff value for investments varies from company to company, and with the nature and risk of the investment. However, survey data and discussions with motor users suggest that very few companies will invest in conservation improvements with paybacks exceeding 3 years, and many companies need to see a payback of 2 years or less (Marbek Resource Consultants, Ltd. 1987).

Energy service companies are willing to accept a much longer payback for an investment if it has an asset life appropriately relative to return and certainty of the savings. As noted in several of the case studies in Chapter 4, many ESCos are willing to take on an investment with a payback of 7 years (a ROI of about 14%) or longer if the terms are attractive. In addition, some motor suppliers are now beginning to offer service contracts in which they own and maintain the motor for a set annual fee, which can be offset by the customer savings. This rate of return, 14%, is attractive to most financial institutions (Sutcliffe 1999).

Another method, called life-cycle cost analysis, calculates the total present cost of owning and operating the equipment over the life of the equipment, assuming that there is a time value of money. In other words, future costs and savings are discounted back to the present so that the cost of different options over the life of the equipment can be compared on an equal basis. This method presents the most accurate picture of investment options over the long term. In practice, however, very few companies use this method to make decisions about efficient equipment unless the project has a very large cost because most companies are typically looking only at short-term gains. A more detailed discussion of this analysis approach can be found in *Making Business Sense of Energy Efficiency and Pollution Prevention* (Pye 1998).

It is interesting to compare the results of different techniques for evaluating the economics of conservation measures. For example, assume that an energy-efficient motor costs $1,200 more than a standard motor and is expected to save the owner approximately $600/yr. The simple payback for this investment is $1,200/$600, or 2 years. The return on investment is $600/$1,200, or 50%/yr. The motor is expected to be in service for 15 years. Over that period, it will save the owner $9,000. The present value of the energy savings depends on the discount rate used for the analysis. Using a real discount rate of 10%, the present value of these savings is $4,563, or almost four times the incremental cost of the efficient motor. If a real discount rate of 6% is used, the present value is $5,827, or almost five times the incremental cost. Discount rates are user-specific. The higher the discount rate, the more the user values money in hand today over a stream of future savings.

Since simple payback is the most prevalent method used by companies, it will be used here in the tables and examples regarding the economics of efficient equipment. It is also important to note that we will consider only the energy savings. As noted in the main text, many efficient motor projects produce significant nonenergy benefits. These benefits could significantly increase the attractiveness of a motor system efficiency project (Pye 1998).

Utility Rates

Of course, the economics of efficiency investments hinges on utility rates. Most utilities charge commercial and industrial customers for energy use (by kilowatt-hour) and peak power demand requested by the customer from the utility (by kilowatt). The ratio of these two charges is highly utility-specific and can result in charges such that the demand component accounts for up to 45% of the total electric bill.

Many utilities also levy an additional charge if the power factor falls below a certain level (typically 0.85 to 0.95 depending on the utility). Again, there is a range of charges for power factor, although this charge rarely exceeds 5% of the total bill. A few utilities charge based on kilovolt-amperes rather than power (kilowatt-hour), which has a similar impact, as discussed in Chapter 2.

Electric Supply System Perspective

Entities that make up the electric supply system routinely evaluate investments based on life-cycle costs. These entities are accustomed to purchasing power plants and transmission assets that pay back over 15 years or more. Depending upon the regulatory environment, these entities may apply the same kind of long-term analysis to demand-side resources. This policy stands in strong contrast to the average energy user's requirement that conservation investments pay back in under 3 years. This difference in perspective is referred to as the payback gap, as discussed in Chapter 8.

In evaluating conservation investments, the electric supply system entity compares the costs it avoids by making the investment (fuel, operation and maintenance, and the avoided cost of new generating facilities) to the amount it has to pay for the energy savings plus program implementation and administrative costs (typically 20–30% of direct costs). If the present value of the expenses avoided over the life of the conservation investment exceeds the cost of the investment, the measure is attractive for the entity.

Avoided cost figures will often vary with off-peak and on-peak periods. Consider an entity that has an avoided cost of $.08/kWh for peak periods and $.03/kWh for all other times. It is considering giving a rebate for an efficient 10 hp, 1,800 rpm, TEFC motor in a new installation. The specifics of the project include:

Rebate or incentive level:	$100
Program cost:	$30
Total utility cost:	$130
Total motor operation:	4,000 hrs/yr
On-peak motor operation:	1,000 hrs/yr
Off-peak motor operation:	3,000 hrs/yr
Motor load:	75% of rated load
Engineering motor life:	15 yrs
Expected operational life:	10 yrs

Utility real discount rate:	7%
Power savings (Column D, Table A-l):	0.053 kW
Energy savings on-peak:	53 kWh/yr
Energy savings off-peak:	159 kWh/yr
Annual dollar savings:	$9.01

The net present value of the energy savings over the 10-year period, using a real discount rate of 7%, is $63.28. Since this does not exceed the amount that the utility would pay to capture the savings ($130), the rebate is not cost-effective for the utility.

On a more complex level, utilities running conservation programs assume that some of their customers would have implemented conservation measures without an incentive. These customers, known as "free riders," receive rebates but their savings cannot be credited toward the conservation amount resulting from the utility program. Also, as discussed in the main text, many of these investments would have been cost-effective without the conservation program from the end-user's perspective, or attractive to an ESCo.

Using the above example, if 20% of the end-users of efficient 10 hp motors who received a rebate would have purchased these motors in the absence of the rebate, the effective cost to the utility would be the same but the energy savings would be 20% less. The new avoided cost for the investment would be $306, which is still greater than the $130 spent by the entity and therefore still cost-effective.

The above factors, along with the cost of rebates and implementation problems, have encouraged the shift from demand-side programs to market transformation programs, which seek to permanently shift market behavior from existing patterns to more efficient decisions. In the above example, the entity may choose to provide the rebate or incentive to a motor distributor rather than an end-user in order to encourage the distributor to stock the more efficient motor. See *The Role of Market Transformation Strategies in Achieving a More Sustainable Energy Future* (Nadel and Latham 1998) for a more detailed discussion of this topic.

Another popular analytic approach is to compare the cost to the utility per kilowatt-hour saved (often called the cost of saved energy, or CSE) to the utility's avoided cost per kilowatt-hour. The CSE can be calculated using the formula

$$CSE = \frac{\text{Present value of utility costs to achieve savings}}{\text{Annual kilowatt-hour savings} \times \text{present-value factor}}$$

where the present-value factor is the present value of annual payments of $1, made for the life of the measure, assuming a specific discount rate. Present-value-factor tables can be found in many economics and business textbooks, or one can use the "present value" function in most computer spreadsheet programs. In the example above, the CSE is as follows:

$$CSE = \frac{\$130 \text{ utility cost}}{1,280 \text{ kWh saved/yr} \times 7.024} = \$.013/kWh$$

The cost of saved energy ($.013/kWh) is well below the electric supply system's avoided costs of $.03/kWh (off-peak) and $.08 kWh (on-peak), so the investment is cost-effective for the entity.

Having covered some general economic principles and analytic methods, we present in the following sections tables and worksheets to illustrate how to evaluate specific installations.

Efficient Motors

An efficient motor can be installed when a new motor is purchased (in lieu of rewinding an existing motor that has failed) or as a retrofit replacing an operating standard-efficiency unit.

The relevant cost for financial analysis depends on the type of installation. When a new motor is purchased, the incremental cost of a premium-efficiency model over an EPAct unit is the value to be used in calculations. When a new efficient motor is installed instead of rewinding a burned-out motor, the actual cost for the efficiency improvement is the cost difference between rewinding the old motor and purchasing a new efficient motor. When an efficient motor is installed as a retrofit, the costs of the efficiency gains include the full purchase price of the new efficient motor plus the labor to remove the old motor and install the new one. We ignore the salvage value of the retiring motor because it should be scrapped, not re-used, and the scrap value of the metal is small.

Motor Costs

Tables A-1 and A-2 present the costs for new efficient motors, new standard motors, and motor rewinds. These tables are based on some generic assumptions about the type of motor, the location, and the duty factor and can be used in the worksheets to follow on motor economics. However, actual motor costs for specific applications should be used in calculations to more accurately reflect the economics of a specific project. The *MotorMaster+*® software, discussed in Chapters 2 and 9, is perhaps the best source of motor-specific data currently available, incorporating economic evaluation tools that aid in comparing motor options.

Table A-1

Costs and Performance of TEFC CEE Premium-Efficiency Motors, EPAct Motors, and Rewinds

A	B	C	D	E	F	G	H	I	J	K	L	M	N
Horsepower	Avg. EPAct Efficiency (@ 75% load)	Avg. Premium Efficiency (@ 75% load)	New Motor kW Savings (@ 75% load)	Avg. Cost EPAct	Avg. Cost Premium	Avg. Price Premium	Avg. Eff. Stock & Rewind	Premium Retrofit and Rewind kW Savings (@ 75% load)	Rewind Cost	Labor Cost Retrofit	Baseplate Adapter Cost U to T Frame	Baseplate Adapter Cost Pre-Stnd. to T	Adjustable Sheave Cost
1	82.4	85.2	0.022	213	256	120%	73.0	0.11	220	139	13	39	11
2	84.6	87.3	0.041	268	340	127%	78.3	0.15	270	139	13	39	11
3	87.9	89.9	0.042	326	372	114%	79.5	0.24	300	139	19	64	26
5	88.5	90.5	0.070	342	409	120%	82.0	0.32	330	139	19	59	26
7.5	90.5	91.7	0.061	485	552	114%	83.9	0.43	380	139	22	74	47
10	91.1	91.9	0.053	579	633	109%	84.7	0.52	500	139	22	72	47
15	91.7	92.8	0.11	654	733	112%	85.5	0.77	550	139	29	105	68
20	92.3	93.5	0.16	931	1,009	108%	87.3	0.85	600	232	29	133	68
25	93.1	93.7	0.096	1,112	1,260	113%	87.9	0.99	660	232	41	133	80
30	93.2	94.0	0.15	1,300	1,456	112%	88.2	1.2	760	232	41	133	80
40	93.5	94.4	0.23	1,613	2,023	125%	88.4	1.6	880	381	53	172	103
50	93.9	94.8	0.28	2,114	2,419	114%	90.6	1.4	980	381	75	172	103
60	94.3	95.2	0.34	3,088	3,409	110%	90.9	1.7	1,100	381	97	186	121
75	94.5	95.7	0.56	3,915	4,274	109%	91.0	2.3	1,320	381	97	287	121
100	94.9	95.7	0.49	4,543	5,296	117%	91.2	2.9	1,650	899	107	custom	162
125	94.6	95.6	0.77	5,419	5,930	109%	90.6	4.0	2,200	899	custom	custom	custom
150	95.1	95.9	0.74	6,651	8,256	124%	91.8	3.9	2,400	899	custom	custom	custom
200	95.3	96.3	1.2	8,751	9,939	114%	92.3	5.0	2,650	899	custom	custom	custom

Note: Columns B and C are the average 75% load efficiency values obtained from the *MotorMaster+*® database version 3.0 (WSU 1999). Column D is calculated from B and C. Columns E and F are 65% of the average full list price reported in the *MotorMaster+*® database. G is calculated from Columns E and F and is the average price premium for a CEE premium motor over an EPAct motor. Column H is the average stock motor efficiency reported in the *MotorMaster+*® database. Column I is calculated from Columns C and H and reflects the average efficiency difference between a stock motor and a CEE premium motor. Column J is the average rewind cost reported in the *MotorMaster+*® database. Columns J, K, L, and M are values from the previous edition of this book inflated by 5%.

Table A-2
Costs and Performance of ODP CEE Premium-Efficiency Motors, EPAct Motors, and Rewinds

A	B	C	D	E	F	G	H	I	J	K	L	M	N
Horsepower	Avg. EPAct Efficiency (@ 75% load)	Avg. Premium Efficiency (@ 75% load)	New Motor kW Savings (@ 75% load)	Avg. Cost EPAct	Avg. Cost Premium	Avg. Price Premium	Avg. Eff. Stock & Rewind	Premium Retrofit and Rewind kW Savings (@ 75% load)	Rewind Cost	Labor Cost Retrofit	Baseplate Adapter Cost U to T Frame	Baseplate Adapter Cost Pre-Stnd. to T	Adjustable Sheave Cost
1	83.1	85.8	0.021	148	189	128%	74.9	0.09	200	139	13	39	11
2	84.8	87.4	0.039	189	219	116%	78.9	0.14	240	139	13	39	11
3	87.3	89.9	0.056	216	209	97%	81.2	0.20	270	139	19	64	26
5	88.6	90.2	0.056	255	264	104%	82.3	0.30	300	139	19	59	26
7.5	89.5	91.6	0.107	346	383	111%	84.0	0.41	350	139	22	74	47
10	90.8	91.9	0.074	428	456	107%	85.4	0.46	450	139	22	72	47
15	91.9	92.9	0.098	584	585	100%	86.8	0.63	500	139	29	105	68
20	92.1	93.3	0.16	683	767	112%	87.4	0.81	550	232	29	133	68
25	92.7	94.0	0.21	794	903	114%	88.0	1.0	600	232	41	133	80
30	93.2	94.2	0.19	937	1,023	109%	88.6	1.1	700	232	41	133	80
40	93.5	94.7	0.30	1,098	1,287	117%	88.3	1.7	800	381	53	172	103
50	93.8	95.0	0.38	1,258	1,450	115%	90.2	1.6	900	381	75	172	103
60	94.1	95.7	0.60	1,613	1,949	121%	90.7	1.9	1,000	381	97	186	121
75	94.4	95.3	0.42	1,981	2,108	106%	91.3	1.9	1,200	381	97	287	121
100	94.4	96.0	0.99	2,397	2,916	122%	91.7	2.7	1,500	899	107	custom	162
125	94.9	96.0	0.84	2,780	3,327	120%	91.8	3.3	2,000	899	custom	custom	custom
150	95.1	96.0	0.83	3,913	4,291	110%	92.1	3.7	2,200	899	custom	custom	custom
200	95.4	96.3	1.1	5,146	5,271	102%	92.4	4.9	2,400	899	custom	custom	custom

Note: Columns B and C are the average 75% load efficiency values obtained from the *MotorMaster+*® database version 3.0 (WSU 1999). Column D is calculated from B and C. Columns E and F are 65% of the average full list price reported in the *MotorMaster+*® database. G is calculated from Columns E and F and is the average price premium for a CEE premium motor over an EPAct motor. Column H is the average stock motor efficiency reported in the *MotorMaster+*® database. Column I is calculated from Columns C and H and reflects the average efficiency difference between a stock motor and a CEE premium motor. Column J is the average rewind cost reported in the *MotorMaster+*® database. Columns J, K, L, and M are values from the previous edition of this book inflated by 5%.

The critical factors influencing the purchase price of motors are (in order of importance): size, discount structure, speed, enclosure type, and efficiency. For most electrical equipment, manufacturers have a published "list" price that is typically 30–50% above the trade price actually paid by industry. High-volume customers can often negotiate discounts of up to 50% off the trade price. In this appendix, we have made the simplifying assumption that all motors are purchased at the industry trade price, which is approximately 10–30% higher for CEE premium-efficiency motors than for EPAct motors.

The slower a motor, the more material it requires and the higher its capital cost. Because more than 50% of the motors sold are rated at 1,800 rpm, the tables in this section are based on this operating speed.

Enclosure type also influences the cost of a motor. Totally enclosed fan-cooled motors require more material and must operate in more severe environments and therefore cost more than open drip-proof models.

Other Costs for Motor Replacement

U-frame motors can usually be replaced with T-frame motors of the same rating. However, the mounting holes do not line up for the two frame types, so a conversion baseplate must be used. Depending on the motor size, the adapter plate can cost between $10 and $100, with $25 as the most typical value. Using an adapter also adds an hour of labor to the installation.

As stated in Chapter 2, many motors now in service are oversized for the application. One way to save money on an efficient replacement is to install a new motor that is smaller than the original unit. In this case, however, both the mounting system and the shaft may be different sizes than for the original equipment. This change in the physical size of the equipment generally will force the user to install an adapter plate for mounting the motor and to change pulleys or put in a sleeve for attaching the existing pulleys to the new shaft.

Motor starters typically include protective devices designed to disconnect the motor if it is overloaded. Most starters use a temperature-activated switch that is warmed by a set of protective devices known as thermal overload elements or heaters. The heaters are sized according to the motor's full-load input current rating and will typically need to be changed if the motor is downsized. Also, as discussed in Chapter 2, some premium-efficiency motors have higher inrush currents than do older standard motors. This situation may require the replacement of the heaters or the entire starter in order to avoid nuisance trips.

Repair Costs

The cost of repair depends on the extent of the repair and local labor rates. In the simplest version of a repair, new bearings are installed and the motor is cleaned, as described in Chapter 2. In many repairs, the motor windings are also replaced—the tables in this appendix include both winding and bearing costs, based on stripping in a burnout oven.

Energy Savings

The energy an efficient motor will save depends on a number of factors. The potential is defined by the difference in efficiency between the efficient motor and the standard motor it replaces (including any degradation to the efficiency of the standard motor from aging and past repairs). The magnitude of the savings results from the load and number of hours of operation. The efficiency values listed in Tables A-1 and A-2 are for 1,800 rpm TEFC and ODP standard- and high-efficiency motors, operating at 75% of full load, after averaging across the offerings of the major manufacturers serving the U.S. market. The difference in energy use for these two sets of products was used to calculate the savings for new or retrofit applications. The rewind savings values in the tables assume that the motor being replaced lost one percentage point in efficiency from previous rewinds. Of course, if more specific values are available, they should be used in the calculations instead of relying on the average efficiencies listed in Tables A-1 and A-2. Note also that 3,600 rpm motors tend to be one or two points more efficient, and 1,200 rpm motors one or two points less efficient, than 1,800 rpm models (see Figure 2-10 [top]).

The savings values in Tables A-1 and A-2 assume that the motors run at 75% of the rated load. In most cases, the motor user is unlikely to know the actual load on the motor. However, if this load is known, the worksheet allows the user to adjust the calculation accordingly.

Estimating Costs and Benefits

Calculating Savings from Efficient Motors

The following completed worksheets illustrate how to estimate the costs and benefits of installing a CEE premium-efficiency motor in the following situations: (1) as a new application; (2) as an alternative to rewinding an old U-frame motor (an installation that requires an adapter baseplate); (3) as a retrofit for an in-service standard motor; and (4) as a downsizing application. Immediately following the examples are two blank worksheets: one (Worksheet A-1) is to be used for the first three cases listed above, and the other (Worksheet A-2) is to be used for downsizing applications. These blank forms may be copied or used to develop spreadsheets for evaluating specific applications.

Example #1 involves buying a new 20 hp TEFC motor that will operate at 85% load and 4,000 hrs/yr with utility rates of $.063/kWh and $88/kW-yr. Because the motor is a new installation, the relevant cost for analysis is the difference in price between a standard- and a high-efficiency motor.

Example #2 analyzes the choice between rewinding a 50 hp, U-frame ODP motor and buying a new high-efficiency motor. The motor will operate 4,000 hrs/yr at utility rates of $.06/kWh and $70/kW-yr. Because the operating load is not known, the calculation assumes 75% loading. The conversion from a U-frame to a T-frame model requires an adapter baseplate and, in this case, new heaters.

Example #3 calculates the economics of replacing an operating standard-efficiency 30 hp TEFC motor with a high-efficiency unit. The motor runs 4,000 hrs/yr at unknown loading, with utility rates of $.07/kWh and $50/kW-yr. No baseplate adapter is required, but the cost of labor to remove the old motor and install the new one must be counted. We assume no heater replacement is needed, but a new pulley must be installed.

Calculating Savings from Downsizing

As a general rule, downsizing a motor can be cost-effective if the existing motor is operating at less than 40% of rated load. When a motor is running this lightly loaded, the combination of the energy savings that accrue from eliminating the reduced efficiency at low loads plus the potential capital cost savings due to using a smaller motor (in new installations and, in some cases, instead of rewinds) will provide a good return on investment for downsizing. In addition, moving to a smaller motor can help to raise the facility's power factors, as discussed in Chapter 3. (Columns F and J in Tables A-l and A-2 show that it is cheaper to buy motors below approximately 15 hp than it is to rewind them. Rewinding is less costly for motors larger than approximately 10 hp.) However, downsizing a motor that operates above 40% load is often not cost-effective. The cost of installing a new motor (including adapter plates, new pulleys, and heaters) plus the efficiency loss from operating a smaller, less efficient motor outweigh the savings from running a new, smaller motor at higher loading. Because large motors maintain their efficiency better at low loads than do small motors, the 40% cutoff is only a general guideline. Downsizing decisions should be evaluated for each specific application.

Example #4 evaluates the economics of replacing a standard-efficiency 40 hp TEFC motor with a high-efficiency 15 hp unit when the 40 hp motor runs at only 11 hp (28% of load).

Example #1:
Installing an efficient motor in a new application

Application: new motor __✔__ rewind _____ retrofit _____
1. Motor size (hp) __20__
2.[a] Operating hours per year __4,000__
3.[b] Operating load (if known) __85__ %
4.[a] Utility rate (energy) (a) $__0.063__ per kWh
             (demand) (b) $__88__ per kW-yr
5. Motor enclosure: ODP (Table A-2) _____ TEFC (Table A-1) __✔__
6. Enter kW savings @ 75% load from Table A-1 or A-2 __0.16__
    (Column D for new installations or Column I for retrofit or rewind)
7. Adjust kW savings to actual operating load, if known
    __0.16__ × __85__ % / 75% = __0.181__ kW[c]
    from Line 6   Line 3
8. Calculate kWh savings
    __0.181__ × __4,000__ = __724__ kWh
    from Line 7   Line 2
9. __724__ × __0.063__ + __0.181__ × __88__ = $ __61.54__
    from Line 8   Line 4(a)   Line 7   Line 4(b)
10. Costs (from Table A-1 or A-2)
    (a) $__1,009__ new efficient motor (Column F)
    (b) $__0__ motor baseplate (U-frame, Column L; pre-NEMA, Column M)
    (c) $__0__ labor (retrofit only, Column K)
    (d) $__0__ heaters ($20 if needed)
    (e) $__0__ pulley (if needed, Column N)
   subtotal $__1,009__
    (f) $__931__ cost of alternative (standard motor, Column E; rewind, Column J)
11. Calculate net cost:
    Add Lines 10(a) through 10(e), then subtract Line 10(f) = $__78__
12. Simple payback __78__ / __61.54__ = __1.27__ years
               from Line 11   Line 9

[a] If possible, break out operating hours, load profile, and rates by on- and off-peak, using a separate worksheet for each segment.
[b] If load is under 40%, consider downsizing and refer to Worksheet A-2. If load is unknown, use 75%.
[c] If actual load is not known, use Line 6 value.

Example #2:
Installing an efficient motor as an alternative to rewinding

Application: new motor _____ rewind __✔__ retrofit _____
1. Motor size (hp) __50__
2.[a] Operating hours per year __4,000__
3.[b] Operating load (if known) __N/A__ %
4.[a] Utility rate (energy) (a) $__0.06__ per kWh
                       (demand) (b) $__70__ per kW-yr
5. Motor enclosure: ODP (Table A-2) __✔__ TEFC (Table A-1) _____
6. Enter kW savings @ 75% load from Table A-1 or A-2 __1.6__
   (Column D for new installations or Column I for retrofit or rewind)
7. Adjust kW savings to actual operating load, if known
   __0.16__ × __75__ % / 75% = __1.6__ kW[c]
   from Line 6   Line 3
8. Calculate kWh savings
   __1.6__ × __4,000__ = __6,400__ kWh
   from Line 7   Line 2
9. __6,400__ × __0.06__ + __1.6__ × __70__ = $ __496__
   from Line 8   Line 4(a)   Line 7   Line 4(b)
10. Costs (from Table A-1 or A-2)
    (a) $ __1,450__ new efficient motor (Column F)
    (b) $ __75__ motor baseplate (U-frame, Column L; pre-NEMA,
              Column M)
    (c) $ __0__ labor (retrofit only, Column K)
    (d) $ __20__ heaters ($20 if needed)
    (e) $ __0__ pulley (if needed, Column N)
    subtotal $ __1,545__
    (f) $ __900__ cost of alternative (standard motor, Column E;
              rewind, Column J)
11. Calculate net cost:
    Add Lines 10(a) through 10(e), then subtract Line 10(f) = $ __645__
12. Simple payback __645__ / __496__ = __1.3__ years
                   from Line 11   Line 9

[a] If possible, break out operating hours, load profile, and rates by on-
   and off-peak, using a separate worksheet for each segment.
[b] If load is under 40%, consider downsizing and refer to Worksheet A-2.
   If load is unknown, use 75%.
[c] If actual load is not known, use Line 6 value.

Example #3:
Installing an efficient motor as a retrofit for an in-service motor

Application: new motor _____ rewind _____ retrofit __✔__

1.   Motor size (hp) __30__
2.ᵃ Operating hours per year __4,000__
3.ᵇ Operating load (if known) __N/A__ %
4.ᵃ Utility rate (energy)   (a) $__0.07__ per kWh
                (demand) (b) $__50__ per kW-yr
5.   Motor enclosure: ODP (Table A-2) _____   TEFC (Table A-1) __✔__
6.   Enter kW savings @ 75% load from Table A-1 or A-2 __1.2__
     (Column D for new installations or Column I for retrofit or rewind)
7.   Adjust kW savings to actual operating load, if known
     __1.2__ × __75__ % / 75% = __1.2__ kWᶜ
     from Line 6   Line 3
8.   Calculate kWh savings
     __1.2__ × __4,000__ = __4,800__ kWh
     from Line 7   Line 2
9.   __4,800__ × __0.07__ + __1.2__ × __50__ = $__396__
     from Line 8   Line 4(a)   Line 7   Line 4(b)
10. Costs (from Table A-1 or A-2)
     (a) $__1,456__ new efficient motor (Column F)
     (b) $__0__ motor baseplate (U-frame, Column L; pre-NEMA,
               Column M)
     (c) $__232__ labor (retrofit only, Column K)
     (d) $__0__ heaters ($20 if needed)
     (e) $__80__ pulley (if needed, Column N)
     subtotal $__1,768__
     (f) $__900__ cost of alternative (standard motor, Column E;
               rewind, Column J)
11. Calculate net cost:
     Add Lines 10(a) through 10(e), then subtract Line 10(f) = $__868__
12. Simple payback __868__ / __396__ = __2.2__ years
                   from Line 11   Line 9

ᵃ If possible, break out operating hours, load profile, and rates by on-
  and off-peak, using a separate worksheet for each segment.
ᵇ If load is under 40%, consider downsizing and refer to Worksheet A-2.
  If load is unknown, use 75%.
ᶜ If actual load is not known, use Line 6 value.

Example #4:
Downsizing as a retrofit for an in-service motor

1. Current motor size (hp) __40__
2. Operating load (a) __11__ hp; (b) __28__ %[a]
3. Proposed motor size __15__ hp
4. Estimated efficiency of current motor running at current percent of rated load, using Figure 3-5 and, if speed is not 1 rpm, Figure 2-10 (top) __83__ %
5. Motor enclosure: ODP (Table A-2) _____ TEFC (Table A-1) __✔__
6. Efficiency of the proposed smaller motor operating at calculated percent of rated load, from Column C of Table A-1 or A-2 or from Table 2-8 for specific 1,800 rpm models __94.4__ %
7.[a] Operating hours per year __4,000__
8.[a] Utility rate (energy) (a) $ __0.06__ per kWh
    (demand) (b) $ __70__ per kW-yr
9. Calculate change in power requirements

$$0.746 \text{ kW/hp} \times \underset{\text{from Line 2(a)}}{11} \times \left[\frac{1}{\underset{\text{Line 4}}{0.83}} - \frac{1}{\underset{\text{Line 6}}{0.944}} \right] = \underline{1.19} \text{ kW}$$

10. Calculate energy savings
    $\underset{\text{from Line 9}}{1.19} \times \underset{\text{Line 7}}{4,000} = \underline{4,760} \text{ kWh/yr}$

11. Cost savings
    $\underset{\text{from Line 10}}{4,760} \times \underset{\text{Line 8(a)}}{0.06} + \underset{\text{Line 9}}{1.19} \times \underset{\text{Line 8(b)}}{70} = \$ \underline{369}$

12. Costs (from Table A-1 or A-2)
    (a) $ __2,023__ new efficient motor (Column F)
    (b) $ __172__ motor baseplate[b]
    (c) $ __381__ labor (retrofit only, Column K)
    (d) $ __20__ heaters ($20 if needed)
    (e) $ __103__ pulley (if needed, Column N)
    total cost $ __2,699__
13. Simple payback $\underset{\text{from Line 12}}{2,699} / \underset{\text{Line 11}}{369} = \underline{7.3} \text{ years}$

[a] If possible, break out operating hours, load profile, and rates by on- and off-peak, using a separate worksheet for each segment.
[b] Baseplate costs are application-specific in downsizing installations.

Worksheet A-1 (for new applications, as an alernative to rewinding old U-frame motors, and as a retrofit for in-service standard motors)

Application: new motor _____ rewind _____ retrofit _____

1. Motor size (hp) _____

2.[a] Operating hours per year _____

3.[b] Operating load (if known) _____ %

4.[a] Utility rate (energy) (a) $_____ per kWh
  (demand) (b) $_____ per kW-yr

5. Motor enclosure: ODP (Table A-2) _____ TEFC (Table A-1)_____

6. Enter kW savings @ 75% load from Table A-1 or A-2 _____
   (Column D for new installations or Column I for retrofit or rewind)

7. Adjust kW savings to actual operating load, if known
   _____ × _____ % / 75% = _____ kW[c]
   from Line 6   Line 3

8. Calculate kWh savings
   _____ × _____ = _____ kWh
   from Line 7   Line 2

9. _____ × _____ + _____ × _____ = $_____
   from Line 8   Line 4(a)   Line 7   Line 4(b)

10. Costs (from Table A-1 or A-2)
    (a) $_____ new efficient motor (Column F)
    (b) $_____ motor baseplate (U-frame, Column L; pre-NEMA, Column M)
    (c) $_____ labor (retrofit only, Column K)
    (d) $_____ heaters ($20 if needed)
    (e) $_____ pulley (if needed, Column N)
    subtotal $_____
    (f) $_____ cost of alternative (standard motor, Column E; rewind, Column J)

11. Calculate net cost:
    Add Lines 10(a) through 10(e), then subtract Line 10(f) = $_____

12. Simple payback _____ / _____ = _____ years
    from Line 11   Line 9

[a] If possible, break out operating hours, load profile, and rates by on- and off-peak, using a separate worksheet for each segment.

[b] If load is under 40%, consider downsizing and refer to Worksheet A-2. If load is unknown, use 75%.

[c] If actual load is not known, use Line 6 value.

Worksheet A-2 (downsizing existing motors)

1. Current motor size (hp) _____

2. Operating load (a) _____ hp; (b) _____ %[a]

3. Proposed motor size _____ hp

4. Estimated efficiency of current motor running at current percent of rated load, using Figure 3-5 and, if speed is not 1 rpm, Figure 2-10 (top) _____ %

5. Motor enclosure _____ ODP (Table A-2) _____ TEFC (Table A-1)

6. Efficiency of the proposed smaller motor operating at calculated percent of rated load, from Column C of Table A-1 or A-2 or from Table 2-8 for specific 1,800 rpm models _____ %

7.[a] Operating hours per year _____

8.[a] Utility rate (energy)   (a) $ _____ per kWh
            (demand) (b) $ _____ per kW-yr

9. Calculate change in power requirements

$$0.746 \text{ kW/hp} \times \underset{\text{from Line 2(a)}}{\underline{\quad 11 \quad}} \times \left[\dfrac{1}{\underset{\text{Line 4}}{\underline{\quad}}} - \dfrac{1}{\underset{\text{Line 6}}{\underline{\quad}}} \right] = \underline{\quad} \text{ kW}$$

10. Calculate energy savings

$$\underset{\text{from Line 9}}{\underline{\quad\quad}} \times \underset{\text{Line 7}}{\underline{\quad\quad}} = \underline{\quad\quad} \text{ kWh/yr}$$

11. Cost savings

$$\underset{\text{from Line 10}}{\underline{\quad\quad}} \times \underset{\text{Line 8(a)}}{\underline{\quad\quad}} + \underset{\text{Line 9}}{\underline{\quad\quad}} \times \underset{\text{Line 8(b)}}{\underline{\quad\quad}} = \$ \underline{\quad\quad}$$

12. Costs (from Table A-1 or A-2)

    (a) $ _____ new efficient motor (Column F)
    (b) $ _____ motor baseplate[b]
    (c) $ _____ labor (retrofit only, Column K)
    (d) $ _____ heaters ($20 if needed)
    (e) $ _____ pulley (if needed, Column N)

total cost $ _____

13. Simple payback $\underset{\text{from Line 12}}{\underline{\quad\quad}}$ / $\underset{\text{Line 11}}{\underline{\quad\quad}}$ = _____ years

[a] If possible, break out operating hours, load profile, and rates by on- and off-peak, using a separate worksheet for each segment.
[b] Baseplate costs are application-specific in downsizing installations.

Calculating the Economics of Efficient Motors: Electric Supply System Perspective

Many motor purchase decisions are made in a hurry after a motor fails. When this occurs, it is difficult to evaluate the exact economics of improving motor efficiency since the efficiency of the original, failed, motor can no longer be measured. Thus, the transaction cost of procuring an efficient motor must be kept to a minimum or a less desirable motor decision will be made. As a result, DSM programs often provide fixed rebates for efficient motors based solely on the size of the motor in order to expedite the purchase of an efficient motor when an existing unit needs to be replaced or rewound. To ensure the cost-effectiveness of its investment in rebates, the program will sometimes set minimum bounds on the operation of the motor (such as a minimum number of operating hours per year).

These DSM programs use avoided-cost data along with estimates of energy savings to set rebate amounts and define applications where incentives should be offered. In general, the steps include the following:

1. Determine the avoided cost (sometimes differentiated for different seasons and/or time periods such as on- and off-peak).

2. Determine the average operating life of the measure, allowing for equipment that is removed before the end of its rated life because of production line remodeling or other reasons.

3. Look at the cost of efficient motors in the three cases: new motor purchases; installing efficient motors instead of rewinds; and retrofitting with efficient motors.

4. Estimate the utility cost savings for each application for different operating hours and loads. This calculation involves multiplying the energy or demand reductions by the utility's avoided cost per kilowatt-hour and/or kilowatt. Alternatively, the cost of saved energy (in $/kWh) can be calculated for each application.

5. Compare the avoided cost to the utility with the incremental cost or cost of saved energy for the application. When costs for the applications are expressed in terms of $/kWh, avoided costs per kilowatt-hour need to be adjusted to incorporate avoided capacity costs. This is generally done by taking avoided capacity

costs (expressed in $/kW-yr) and dividing by the average annual operating hours of the capacity in question (approximately 8,000 hours for base load capacity). The resulting value is added to the avoided costs per kilowatt-hour, which yields an estimate of capacity-adjusted costs per kilowatt-hour.

An example of these calculations is provided in the first section of this appendix. A discussion of motor program strategies was provided in Chapter 9.

Adjustable-Speed Drives

As discussed in Chapter 4, ASDs for variable-torque loads (such as centrifugal pumps and fans) cost 10–20% less than ASDs that drive constant-torque loads (such as conveyors) because the latter require heavier-duty electronics that can withstand the full motor inrush current.

ASDs are similar to motors in that typically the list price has little meaning in the marketplace, and the trade price is the actual purchase price for a low-volume user. Typical installed ASD prices are presented in Figure 4-10.

Most of the potential for energy savings from ASDs is in centrifugal fan and pump applications. Therefore, most of the costs used in this section are based on variable-torque controllers suitable for centrifugal equipment. One example outlines converting an older DC drive system to an AC adjustable-frequency drive that requires constant-torque equipment.

Two major factors besides the ASD affect its installed cost: the other equipment required to make the ASD a usable part of the system, and the options ordered with the unit.

Consider the example of a pump being installed to control the pH of the fluid in a basin by adding caustic. The base installation (without the ASD) includes a pH sensor in the basin; a feedback controller that takes the reading from the sensor, compares it to the desired pH, and sends out a signal to a control valve; and the control valve itself, which changes the flow to the basin. All of the components required to control the flow of caustic by varying the speed of the pump are already planned for this system, so there is no additional cost for using the ASD except for the ASD itself.

A second application example involves a factory that has three well pumps. When the plant is operating, each pump runs continuously, producing water pressure that varies between 50 and 100 psi, depending on the water use at any given time. The minimum pressure required to keep the equipment at the factory supplied with water is 50 psi.

In this application, there currently is no sensor that sees the complete picture of the water flow needed by the plant. Each pump rides up and down its own pump curve so that the pressure in the system increases as the flow decreases. As a result, the costs for installing an ASD must include installing a pressure sensor in the plant at a central or critical location; providing a feedback controller that takes the reading from the sensor, compares it to the desired pressure, and sends out a signal to the ASD; and the ASD itself on one of the pumps.

The costs for adding a feedback control loop are specific to the project since they depend on the type of sensor and controller that are required. A pressure sensor for an industrial process application will cost between $150 and $1,000, depending on the location, the environment, the range, and the brand. A pressure sensor used in an HVAC duct might cost only $30. Dual sensors (a primary unit and a backup unit) are sometimes used for reliability in critical industrial applications.

An inexpensive, stand-alone controller for a process application will cost between $250 and $400, while a stand-alone, self-tuning process controller can cost up to $1,200. In a plant where there is a central programmable logic controller, controlling an additional application will have no incremental cost if there are extra channels available to run the control circuit without adding any input/output hardware. Again, components that control HVAC equipment are far cheaper than industrial process control systems.

Most ASDs include options for control panels with different enclosures and different features such as switches, safeties, overloads, and pilot lights. In addition, most manufacturers can either include the control panel in the same physical location as the controller or wire the ASD so that it can be controlled from a remote station.

Historically, most manufacturers recommended that ASD users purchase an isolation transformer to keep harmonics generated by the ASD from entering the electric distribution system. Modern ASD technology has reduced the harmonic components emitted (particularly for units under 150 hp) to the point where an isolation transformer is not necessarily recommended. However, these transformers are still frequently installed if there is any question of lower power quality due to either the harmonics emitted by the ASD or the possibility of damage to the ASD by transients in the electric distribution system. An isolation transformer adds approximately 10–20% to the cost of an ASD.

Other less frequently seen options are as follows:

- ASDs sometimes emit radio frequency noise, which can be suppressed.

- Automatic restart after a power failure or a motor trip is offered by some manufacturers.

- Some ASDs have circuits that substitute for a process controller.

- Some ASDs have signal outputs so that more than one ASD can be set to track together for applications such as conveyors.

The capital costs for ASDs used in this section include the costs of a basic unit operating at 480 V and an attached control panel. These data are shown in Figure 4-10. The total costs used in the following examples include materials, labor, and whatever else is needed to integrate the ASD into the system. Such costs are too application-specific to be listed in a "cookbook" table.

Calculating ASD Energy Savings

The amount of savings from the use of ASDs on centrifugal machines such as pumps and fans is dictated by both the variation in flow for the system and the way the system is currently controlled. As a result, the actual savings are site-specific. Nevertheless, there are some general conditions that offer some clues as to whether a specific application is likely to be cost-effective. Several case studies have been used to help the reader screen for applications where some potential exists. The case studies include

1. A pump that provides variable flow into a long pipeline so that most of the energy of the liquid as it exits the pump goes into overcoming the frictional losses in the pipeline.

2. A pump that provides intermittent flow into a long pipeline so that most of the energy of the liquid as it exits the pump goes into overcoming the frictional losses in the pipeline.

3. A pump that feeds into a header that supplies a number of faucets, where the pressure in the pipe is allowed to vary with the demand.

4. A fan that supplies air to a variable-air-volume system in a commercial building.

5. The replacement of an older DC drive system, used for speed control, by an AC system with ASDs.

Example #1:
Pump with variable flow into a long pipeline

A pump is used to transfer wastewater from an intermediate catch basin in a factory to a sewage treatment plant. Under current operation, dirty water enters the basin at a varying rate depending on how the plant is operating. There is a throttle valve on the outlet from the pump that controls the flow to maintain a constant level in the catch basin. The system operates continuously at varying flow rates and has the following characteristics:

Pump: Cornell 6NHP

Rated Pump Flow: 1,200 gpm @ 55 feet of pressure, known as the total dynamic head (TDH)

Actual Pump Flow:

Percentage of Rated Flow	Percentage of Time at Each Flow
51–60%	20%
61–70%	20%
71–80%	20%
81–90%	20%
91–100%	20%

To calculate the energy savings from adding an ASD, complete the following steps:

1. Calculate the system curve that shows the pressure drop in the piping system. A table or a formula can be used to establish the pressure drop through the piping at different flow rates. In this example, the pipe is 6,800 feet long and 10 inches in diameter. The losses in the pipe are as follows:

Flow Rate (gpm)	Losses per 100 feet	Total Losses
200	.028 foot	1.9 feet
400	.099 foot	6.7 feet
600	.213 foot	14 feet
800	.370 foot	25 feet
1,000	.569 foot	39 feet
1,200	.811 foot	55 feet

2. Draw the system curve on the pump curve by plotting the pressure drop at each given flow rate calculated in Step 1 and connecting the points (see Figure A-1).

Figure A-1

Variable-Speed Pump Curve with a Variable-Pressure System Curve Superimposed

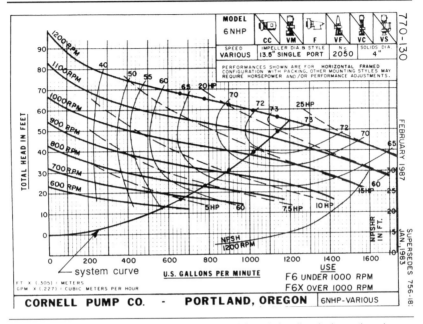

Note: The upper set of heavy dots represents flow control through throttling; the lower, through speed control. The upper and lower sets of operating points in this figure apply to Example #1, while the lower set only applies also to Example #2.

3. Estimate the power needed for each flow band from the pump curve for the throttled system. In this case, the power needed for the given flow rate is equal to the power required by the base pump curve (or the pump curve when the pump is operating at full speed). For example, looking at the pump curve, when the actual flow is 85% of the design flow (1,200 gpm × 0.85 = 1,020 gpm), the power required at the pump shaft is 21 hp (since this flow rate falls about one-fifth of the way from the 20 to 25 hp lines). The electrical power required to drive the shaft is 21 hp times the conversion factor from horsepower to kilowatt (0.746) divided by the motor efficiency at the estimated motor load. For a standard 25 hp motor, this would be approximately 21 hp × (0.746 kW/hp)/0.897 = 17 kW. The number 0.897 is from Table A-1. Note that the pressure produced by the pump exceeds the pressure required by the system at most flow rates. This extra pressure represents wasted

energy that is released across the valve to control the flow to the required level.

This method can be used to estimate the power needed at each flow rate as follows:

Flow Rate	Midpoint (gpm)	Throttled Power (hp)	Throttled Power (kW)
51–60%	660	18	14.9
61–70%	780	19	15.7
71–80%	900	20	16.6
81–90%	1,020	21	17.4
91–100%	1,140	22	18.2

4. Estimate the power needed for the system using an ASD by estimating the energy needed for each flow range to follow the system curve and adding the losses for the ASD. For example, if the pump was only turning fast enough to overcome the frictional losses in the piping at 85% of rated flow, the pump shaft would be rotating at about 1,020 rpm (based on the point where the system curve crosses a line of equal motor speed at 1,020 gpm). Looking at the pump curve, the power needed at the pump shaft at this speed and flow would be about 15 hp. The electrical power required to drive the shaft is 15 hp times the conversion factor from horsepower to kilowatt (0.746) divided by the efficiency of the motor and the ASD at the estimated motor load and ASD speed. (See Figure 4-9 for typical ASD efficiencies as a function of speed; the present example is a variable-torque load.) For a 25 hp motor, the required electrical power would be approximately 15 hp × (0.746 kW/hp)/0.897/0.95 = 13 kW.

The above method can be used to estimate the power needed at each flow rate as follows:

Flow Rate	Midpoint (gpm)	Horsepower with Speed Control	ASD Efficiency	Input Power w/ASD (kW)
51–60%	660	5	0.89	4.7
61–70%	780	7	0.92	6.3
71–80%	900	10	0.94	8.8
81–90%	1,020	15	0.95	13.1
91–100%	1,140	20	0.96	17.2

5. Estimate the total energy savings due to the ASD by calculating the

power savings for each flow range (the value derived in Step 3 minus the value derived in Step 4) and multiplying by the number of hours in that flow range (20% of full-time operation, or 1,753.2 hrs/yr at each flow range):

Flow Range	Power Reduction	Energy Savings per Year
51–60%	10.2 kW	17,883 kWh
61–70%	9.4 kW	16,480 kWh
71–80%	7.8 kW	13,675 kWh
81–90%	4.3 kW	7,539 kWh
91–100%	1.0 kW	1,753 kWh
Total		57,330 kWh

Note that the energy used at full flow increases slightly (compared to the original system) due to the inefficiencies of the ASD—from 18.6 kW with a wide-open throttle to 19.4 kW with the ASD at 100% speed. At 95% flow (used in this example for the range of flow from 91 to 100%), the ASD still saves energy. If a system demands full flow for a significant fraction of its operating time, operating the ASD in bypass mode (which requires the installation of a bypass switch) can eliminate the ASD losses.

6. Calculate the dollar savings for the above at $.07/kWh equals $4,013.

7. Estimate the installed costs at $5,100 for a retrofit and $4,300 for new construction.

8. Calculate the simple paybacks at 1.3 years for retrofit and 1.1 years for new construction.

In this example, there is no change in the flow rate when the control valve is changed to an ASD. The energy savings come from the reduction in pressure from the pump for all of the flow ranges. This pressure reduction is substantial since the pump output can follow the system curve, which has a low pressure requirement at low flows.

Because the base system is already designed with a level sensor, a feedback controller, and a control valve, the cost for the installation includes the cost of the ASD plus some labor time for wiring the unit.

The project is cost-effective because of the relatively low cost and the fact that there is a sufficient period when the system operates at low flows to produce significant savings and justify the investment.

Example #2:
Pump with intermittent flow into a long pipeline

A pump is used to transfer wastewater from an intermediate catch basin in a factory to a sewage treatment plant. Under current operation, dirty water enters the basin at a varying rate depending on how the plant is operating. The basin has a control that turns the pump on and off to maintain the level between the high and low set-points. When all of the lines in the plant are operating, the pump runs 95% of the time. When only one line is running, the pump runs 50% of the time. The system operates continuously with the pump cycling to meet the varying flow rates. The system has the following characteristics:

Pump: Cornell 6NHP

Rated Pump Flow: 1,200 gpm @ 55 feet TDH

Actual Pump Flow:

Percentage of Rated Flow	Percentage of Time at Each Flow
51–60%	20%
61–70%	20%
71–80%	20%
81–90%	20%
91–100%	20%

To calculate the energy savings, complete the following steps:

1. Calculate the system curve that shows the pressure drop in the piping system using the same calculation as in Example #1.

2. Draw the system curve on the pump curve (see Figure A-1).

3. In the base case (with the existing equipment), the pump either operates at the rated flow or is shut off. As a result, the power needed for each flow band can be estimated by taking the energy use at full flow and multiplying it by the percent of time the pump is operating to meet the flow band. For example, the power at full flow is

$$22.5 \text{ hp} \times \frac{0.746 \text{ kW/hp}}{0.897} + 18.7 \text{ kW}$$

The average power needed at 85% of full flow is

$$18.7 \text{ kW} \times 0.85 = 15.9 \text{ kW}$$

This method can be used to estimate the power needed by the existing system at each flow rate as follows:

Flow Rate	Power
51–60%	10.3 kW
61–70%	12.2 kW
71–80%	14.0 kW
81–90%	15.9 kW
91–100%	17.8 kW

4. Estimate the power needed for the system with an ASD by estimating the energy needed for each flow range to follow the system curve and adding the losses for the ASD. The same calculation was done for Example #1 and yielded the following estimates of the power needed at each flow rate:

Flow Rate	Power
51–60%	4.7 kW
61–70%	6.3 kW
71–80%	8.8 kW
81–90%	13.1 kW
91–100%	17.2 kW

5. Estimate the total energy savings from the ASD by calculating the power savings for each flow range (Line 3 minus Line 4) and multiplying by the number of hours in that flow range. Again, the continuously operating system runs 20% of the time (1,753.2 hrs/yr) at each flow range:

Flow Range	Power Reduction	Energy Savings per Year
51–60%	5.6 kW	9,818 kWh
61–70%	5.9 kW	10,343 kWh
71–80%	5.2 kW	9,117 kWh
81–90%	2.8 kW	4,909 kWh
91–100%	0.6 kW	1,052 kWh
Total		35,239 kWh

6. The dollar savings for the above at $.07/kWh equals $2,467.

7. The installed costs are $6,800 for a retrofit and $5,700 for new construction. The costs are higher for this application than for the system in Example #1 since these costs include a flow sensor and a feedback controller, which were not needed for the base system.

8. The simple paybacks are 2.8 years for retrofit and 2.3 years for new construction.

In this example, there is a tradeoff between running the pump continuously at reduced pressure and flow with the ASD and running the pump for limited periods at full flow. In the base case, the pump cycles and operates at a relatively high efficiency when it runs. There are some savings from allowing the water to flow at lower rates, producing a lower pressure drop, but these savings are not as large as in Example #1 since there is no deliberately wasted energy to control the valve in this system. Note that running the pump continuously at reduced speed instead of cycling will reduce wear on the pump, motor, and associated electrical equipment.

The paybacks may not be attractive enough for many motor users but will be for some users and for most utilities.

Example #3:
Pump with variable flow into a header

A pump is used to supply water to a factory with multiple water uses. Under current operation, the water is stored in a tank and pumped into the plant by a single pump, which pressurizes the system. Since the water demand varies, the pump will ride up and down the pump curve so that the pressure in the system varies from 55 to 70 feet of head and thus requires a relatively high minimum pressure. The system operates continuously at varying flow rates and has the following characteristics:

Pump: Cornell 6NHP

Rated Pump Flow: 1,200 gpm @ 55 feet TDH

Actual Pump Flow:

Percentage of Rated Flow	Percentage of Time at Each Flow
51–60%	20%
61–70%	20%
71–80%	20%
81–90%	20%
91–100%	20%

This system has very large pipes feeding the plant. As a result, the pressure drop is minimal at all flow rates, and the pressure required by the system is constant regardless of the flow. This pressure is 55 feet of head. To calculate the energy savings, complete the following steps:

1. Draw the pressure requirements on the pump curve (see Figure A-2). Note that this system curve is a straight line at 55 feet of head instead of the more typical system curve (as in Figure A-1).

Figure A-2

Variable-Speed Pump Curve with a Constant-Pressure System Curve Superimposed

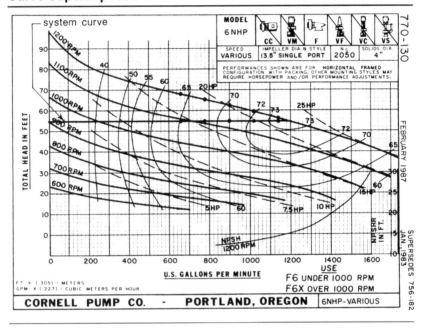

Note: The upper set of operating points (along the 1,200 rpm pump curve) shows the system (same pump as in Figure A-1) with no pressure regulation. The lower set (constant pressure at 55 feet) shows the operation using speed control to maintain constant pressure at varying flow.

2. From the pump curve, estimate the power needed for each flow band for the existing, throttled, system. In this case, the power needed for a given flow rate is equal to the power required by the base pump curve (that is, the pump curve when the pump is operating at full speed). Note that this is the same calculation for the base system in Example #1.

Flow Rate	Power
51–60%	14.9 kW
61–70%	15.7 kW
71–80%	16.6 kW
81–90%	17.4 kW
91–100%	18.2 kW

3. Estimate the power needed for the same system using an ASD by

estimating the power needed for each flow range to maintain a constant pressure of 55 feet and adjusting for the losses in the motor and ASD:

Flow Rate	Midpoint (gpm)	Horsepower with ASD	ASD Efficiency	Power with ASD (kW)
51–60%	660	14.0	0.95	12.2
61–70%	780	16.0	0.95	13.9
71–80%	900	17.5	0.95	15.3
81–90%	1,020	20.0	0.96	17.2
91–100%	1,140	21.5	0.96	18.5

4. Estimate the total energy savings due to the ASD by calculating the power savings for each flow range and multiplying by the number of hours in that flow range:

Flow Range	Power Reduction	Energy Savings per Year
51–60%	2.7 kW	4,734 kWh
61–70%	1.8 kW	3,156 kWh
71–80%	1.3 kW	2,279 kWh
81–90%	0.2 kW	350 kWh
91–100%	.3 kW	−526 kWh
Total		9,993 kWh

5. Calculate the annual dollar savings for the above at $.07 equals $700.

6. Estimate the costs at $6,800 for a retrofit and $5,700 for new construction. The costs are higher for this application than for the system in Example #1 since they include a flow sensor and a feedback controller, which were not needed for the base system.

7. Calculate the simple paybacks at 9.7 years for retrofits and 8.1 years for new construction.

In this example, the energy savings are produced by controlling the pressure at the minimum acceptable level for the application instead of letting the pressure increase at low flow rates. In other words, the wasted energy is the extra pressure between the pump curve and the system requirement of 55 feet of head generated by the pump at low flow rates. The savings for this application are lower than in Examples #1 and #2 because there is no reduction in pressure requirements at low flow rates, and the pressure generated by the pump only narrowly exceeds the pressure needed by the system. In general, it will not be cost-effective to install an ASD on systems that require a high minimum pressure.

As noted in Example #1, if the system is expected to operate much of the time at close to full flow, the ASD can be operated in bypass mode when close to full load (at additional installed cost and control complexity).

Example #4:
ASD on a variable-air-volume fan in a commercial building

Many commercial buildings have air-handling systems in which the air volume is varied to meet the cooling demand in the building. These systems, known as VAV systems, have boxes that serve each thermal zone in the building, with the airflow to that zone adjusted by a local thermostat. Inlet vanes have historically been used on the supply fan to match the airflow to the output of the boxes. While the power requirement of a fan with inlet vanes decreases as the flow decreases, the power does not fall off as fast as the flow because the inlet vanes reduce the efficiency of the fan. For example, the power that is required by a typical inlet vane system at 50% flow is about 75% of full power.

There are also many commercial buildings with constant-volume systems that may be good candidates for conversion to VAVs using ASDs. Such systems include terminal-reheat and dual-duct configurations. Note that converting such simultaneous heating and cooling systems to VAVs saves significant heating and cooling energy as well as ventilating energy.

In general, it is difficult to do a hand calculation of the impact of installing an ASD on a constant-volume system, or in place of an inlet vane or discharge damper on a VAV system, because the impact depends on the building's thermal characteristics, which change with time. For example, in the morning, waste heat from lights may be absorbed by the cool mass inside the building, which helps to make the space more comfortable. Later in the day, as the building's mass charges up, that same waste heat may serve to overwarm the space. Even with a constant outside daytime temperature of, say, 50°F, a building might thus need heating in the morning and cooling in the afternoon. Because of such issues, the easiest way to evaluate the savings from installing an ASD on an air handler would be to run a thermal model (such as DOE2, Trak-Load, or ASEAM—see Appendix D) of the building and look at the relative energy use for ventilation using different assumptions for controlling the air flow. Such simulations would also estimate the heating and cooling savings made possible by converting constant-volume systems to VAVs.

Assume that running this type of a model produces the following for a 100,000-square-foot office building:

Base fan energy use:	160,000 kWh/yr
Fan energy use with an ASD:	96,000 kWh/yr
Energy savings:	64,000 kWh/yr
Dollar savings @ $.07/kWh:	$4,480

The economics of the project to the motor user will depend on the building. If the building is a three-story suburban office building with two 50 hp supply fans, the installed cost will be about $14,000, and the simple payback will be 3.1 years.

If the building is a 12-story office building in a downtown core area with a floor-by-floor air handler system and twelve 7.5 hp motors, the installed cost will be $35,000, and the simple payback will be 7.8 years.

Note that the economics of the project to the utility may be very different from that of a project with the same customer economics in the industrial sector because there are differences in the value of energy savings during different time periods. Specifically, the projects in industry in the first three examples can be assumed to be driven by process requirements, which are typically consistent during the day and the year. However, the use of an ASD on a VAV system would save energy when the building cooling load is low, typically in the winter and in the early morning hours in the summer. If a utility has a summer peak, the use of an ASD on a VAV system may not yield any energy savings during the system peak unless the system is oversized (which is common).

Example #5:
DC drive system to be replaced with AC variable-speed drives

An older plant that manufactures metal widgets uses a 5 hp DC motor to drive the shaft on each of 20 milling machines with 5 hp DC motors. The plant currently has a motor-generator (M-G) set where an AC motor drives a DC generator to supply DC power to the machines. The motor on the M-G set runs for 4,000 hrs/yr and draws 110 kW. The M-G set needs to see a constant load in order to regulate the voltage, so it is designed so that the DC generator sees an artificial resistive load if all of the DC motors are not in operation. The plant uses all 20 machines on the day shift (2,000 hrs/yr) but only 5 machines on the swing (evening) shift.

Use the following steps to calculate the energy savings:

1. The current energy use is 110 kW × 4,000 hrs = 440,000 kWh. Note that there are no savings during the swing shift when fewer machines are running since the M-G set must see a constant load to properly regulate the DC voltage.

2. Calculate the expected energy use based on 15 motors of 5 hp running 2,000 hrs/yr and 5 motors of 5 hp running 4,000 hrs/yr. Assuming the motors operate at 50% load, and the overall efficiency of the motor and ASD together is 70%, the total energy use is

$$0.5 \times 15 \times 5 \text{ hp} \times 0.746 \text{ kW/hp} \times (1/0.70) \times 2,000 \text{ hours}$$
$$+ 0.5 \times 5 \times 5 \text{ hp} \times 0.746 \text{ kW/hp} \times (1/0.70) \times 4,000 \text{ hours}$$
$$= 130,000 \text{ kWh/yr}$$

The total use for the proposed system will be 130,000 kWh/yr.

3. Calculate the energy savings at 310,000 kWh/yr.

4. Calculate the dollar savings at $.07/kWh is $21,700.

5. Calculate the cost for the retrofit (assuming that constant-torque ASDs are needed for the application) at about $70,000, which includes the cost for new motors.

6. Calculate the simple payback at 3.2 years.

In general, AC drives provide speed control at a lower energy premium than DC drives, particularly when the DC drives are the older-style systems that use M-G sets. In addition, AC motors have much lower ongoing maintenance costs for rewinding and repair.

Oversized Wiring as a Conservation Measure

As discussed in Chapter 3, it is often cost-effective (in new installations and remodels but not retrofits) to install cable larger than required by code because larger wire has lower losses. In general, it is cost-effective to substitute larger wire for motors that operate for long periods of time at close to full load, particularly if larger wire can be installed in the same-size conduit. Two case studies will be used to illustrate when using oversized wiring is cost-effective.

Case Study #1

A 100 hp, three-phase, 480 V motor is being installed at a distance of approximately 500 feet from the motor control center (MCC). The motor is expected to run for two shifts, 5 days/week (approximately 4,000 hrs/yr). Using the tables in the National Electric Code (NEC):

Full-load current:	124 amps
Base wire size:	1/0 XHHW copper[a]
Base conduit:	1.5 inches
Estimated loss @ 100% load:[b]	2.8 kW or 11,000 kWh
Estimated loss @ 75% load:	1.6 kW or 6,400 kWh
Base cost, wire:	$1,980
Base cost, conduit:	$2,185
Base cost, total:	$4,165

[a]American Wire Gauge copper wire sizes #8, 6, 4, 3, 1/0, 2/0, 3/0, and 4/0 have respective diameters of 0.129, 0.162, 0.204, 0.229, 0.325, 0.365, 0.410, and 0.460 inches, or 3.642, 4.115, 5.189, 5.827, 8.252, 9.266, 10.40, and 11.68 mm. These diameters are for solid conductors; standard wire has larger overall diameters to yield the same net cross-sectional area of copper.

[b]Based on loss = $3I^2R$ (because each of the three phases has the same loss). Resistance (R) in ohms per 1,000 feet of wire at 75°C (167°F) for the various wire diameters is as follows: 0.12 for 1/0 copper; 0.10 for 2/0 copper; 0.077 for 3/0 copper; and 0.062 for 4/0 copper. Because these values are for 1,000 feet of wire, R equals one-half these values for the 500-foot runs used in this example. To adjust these resistance values for other temperatures, use the formula

$$R_2 = R_1 \times (1 + 0.00323 \, [T_2 - 75])$$

where R_2 is the new resistance, R_1 is the given resistance, and T_2 is the new temperature in degrees Celsius.

It is proposed that the wire size be increased to 2/0 XHHW, which would require 2" conduit:

Estimated loss @ 100% load:	2.3 kW or 9,200 kWh
Estimated loss @ 75% load:	1.3 kW or 5,200 kWh
Proposed cost, wire:	$2,200
Proposed cost, conduit:	$2,525
Proposed cost, total:	$4,725
Incremental cost:	$560
Simple payback @ $.07/kWh, 100% load:	4.4 years
Simple payback @ $.07/kWh, 75% load:	6.7 years

Because using larger wire in this case requires larger conduit, the paybacks are longer than many consumers are willing to accept, although such paybacks might be acceptable to a utility.

Case Study #2

A 125 hp, three-phase, 480 V motor is being installed at a distance of approximately 500 feet from the MCC. The motor is expected to run constantly except during 3 weeks of maintenance downtime (approximately 8,200 hrs/yr). Using the tables in the NEC:

Full-load current:	156 amps
Base wire size:	3/0 XHHW
Base conduit:	2 inches
Estimated loss @ 100% load:	2.8 kW or 23,000 kWh
Estimated loss @ 75% load:	1.6 kW or 13,000 kWh
Base cost, wire:	$2,500
Base cost, conduit:	$2,525
Base cost, total:	$5,025

It is proposed that the wire size be increased to 4/0 XHHW, which can still use the 2-inch conduit:

Estimated loss @ 100% load:	2.3 kW or 19,000 kWh
Estimated loss @ 75% load:	1.3 kW or 11,000 kWh
Proposed cost, wire:	$2,800
Proposed cost, conduit:	$2,525
Proposed cost, total:	$5,325
Incremental cost:	$300
Simple payback @ $.07/kWh, 100% load:	1.1 years
Simple payback @ $.07/kWh, 75% load:	2.1 years

Case Study #1 has a much better payback than Case Study #2 because of the combination of longer operating hours (which produce larger savings) and the lower incremental cost, since the conduit size does not change when the wire size is increased. In addition to receiving a fast payback, the motor user would save $140/yr for the life of the installation if the motor ran at 75% load.

Drivetrains

As discussed in Chapter 3, synchronous belts can be considerably more efficient than V-belts. Synchronous belts can be used effectively in both new and retrofit applications. They cost more than V-belts and require more costly cogged pulleys, but they last longer. V-belts are sometimes used for safety reasons, because they will slip if the equipment jams. Because synchronous belts do not slip, some applications might require safety equipment such as clutches or shear pins, which will add to the cost of the system.

There is no simple correlation between motor size and the cost of synchronous belts. The cost of the belts and pulleys depends on the gear ratio between the motor and the equipment, the amount of torque that the belts will see, and the distance between the centers of the pulleys. In new construction, a conventional V-belt system will cost about 65–75% less than a synchronous belt system.

Although belt costs do increase with motor size, the increase is nonlinear, so retrofitting a belt on a larger motor is more cost-effective than doing the same retrofit on a smaller motor. Two examples follow:

Example #1

A synchronous belt drive is being considered for a fan in an air handler that is driven by a 5 hp motor via a conventional V-belt. The fan operates 10 hrs/day, 5 days/week (2,600 hrs/yr). The system parameters are as follows:

Motor efficiency:	88.7% (efficient motor)
V-belt efficiency:	92%
Synchronous belt efficiency:	97%
Energy cost:	$.07/kWh
Motor load:	75%

Energy use for the V-belt system is

$$5 \text{ hp} \times 0.746 \text{ kW/hp} \times 0.75 \text{ load} \times \frac{2{,}600 \text{ hrs/yr}}{0.92 \times 0.887} = 8{,}900 \text{ kWh/yr}$$

Operating cost is

$$8{,}900 \text{ kWh/yr} \times \$.07/\text{kWh} = \$623/\text{yr}$$

The energy use for the synchronous belt system is

$$5 \text{ hp} \times 0.746 \text{ kW/hp} \times 0.75 \text{ load} \times \frac{2{,}600 \text{ hrs/yr}}{0.92 \times 0.887} = 8{,}450 \text{ kWh/yr}$$

Operating cost is

$$8{,}450 \text{ kWh/yr} \times \$.07/\text{kWh} = \$592/\text{yr}$$

The dollar savings due to the use of the synchronous belt is

$31/yr. The cost as a retrofit is $300, and the incremental cost in new construction is $170. The simple paybacks are 9.7 years for retrofits and 5.5 years for new construction.

Example #2

A synchronous belt is being considered for a fan in an air handler driven by a 75 hp motor. The fan operates 24 hrs/day, 5 days/week (6,240 hrs/yr). The system parameters are as follows:

Motor efficiency:	95.1% (efficient motor)
V-belt efficiency:	92%
Synchronous belt efficiency:	97%
Energy cost:	$.07/kWh
Motor load:	75%

The energy use for the V-belt system is

$$75 \text{ hp} \times 0.746 \text{ kW/hp} \times 0.75 \text{ load} \times \frac{6{,}240 \text{ hr/yr}}{0.92 \times 0.951} = 299{,}000 \text{ kWh/yr}$$

Operating cost is

$$299{,}000 \text{ kWh/yr} \times \$.07/\text{kWh} = \$20{,}900/\text{yr}$$

The energy use for the synchronous belt system is

$$75 \text{ hp} \times 0.746 \text{ kW/hp} \times 0.75 \text{ load} \times \frac{6{,}240 \text{ hrs/yr}}{0.97 \times 0.951} = 284{,}000 \text{ kWh/yr}$$

Operating cost is

$$284{,}000 \text{ kWh/yr} \times \$.07/\text{kWh} = \$19{,}900/\text{yr}$$

The dollar savings due to the use of the synchronous belt is $1,000/yr. The cost as a retrofit is $1,375 and the incremental cost in new construction is $780. The simple paybacks are 1.4 years for retrofits and 0.8 year for new construction.

The paybacks in Example #2 are far more attractive because of the larger equipment size and longer operating hours. In cases where synchronous belts are not economical, cogged V-belts should be considered. They fall between conventional V-belts and synchronous belts in cost and efficiency.

Motor Provisions
in the Energy Policy Act of 1992

The U.S. Congress, as part of the Energy Policy Act of 1992 (U.S. Congress 1992), set minimum efficiency levels (see Table B-1) for motors falling under the following description: general purpose, T-frame, single-speed, foot-mounting, polyphase squirrel-cage induction motors of the National Electrical Manufacturers Association Designs A and B, rated for continuous duty and operating on 230/460 V and constant 60 Hz line power. Presently, EPAct lists energy efficiency levels for 1–200 hp electric motors. Since October 1997, all motors covered under EPAct (or "covered" motors) that are either manufactured alone or as a component of another piece of equipment must comply with EPAct efficiency levels and also be labeled with a certified efficiency value. This appendix will summarize the law and its implementation.

Covered Equipment

General purpose, one of several terms used to describe the type of motors covered, was not defined by EPAct. To avoid confusion, DOE's Office of Codes and Standards undertook the task of more clearly defining what motors are covered equipment as part of the final rule to implement EPAct directives concerning electric motors (Federal Register 1999). In general, DOE has interpreted *general purpose* to mean any motor that is designed in standard ratings, standard operating characteristics, and standard mechanical construction and can be used without restriction in a broad range of common applications. This

411

Table B-1

Allowable Minimum Full-Load Efficiencies for Motors Covered by EPAct

Number of Poles	Nominal Full-Load Efficiency					
	Open Motors			Enclosed Motors		
Motor Horsepower/ Standard Kilowatt Equivalent	6	4	2	6	4	2
1/.75	80.0	82.5	—	80.0	82.5	75.5
1.5/1.1	84.0	84.0	82.5	85.5	84.0	82.5
2/1.5	85.5	84.0	84.0	86.5	84.0	84.0
3/2.2	86.5	86.5	84.0	87.5	87.5	85.5
5/3.7	87.5	87.5	85.5	87.5	87.5	87.5
7.5/5.5	88.5	88.5	87.5	89.5	89.5	88.5
10/7.5	90.2	89.5	88.5	89.5	89.5	89.5
15/11	90.2	91.0	89.5	90.2	91.0	90.2
20/15	91.0	91.0	90.2	90.2	91.0	90.2
25/18.5	91.7	91.7	91.0	91.7	92.4	91.0
30/22	92.4	92.4	91.0	91.7	92.4	91.0
40/30	93.0	93.0	91.7	93.0	93.0	91.7
50/37	93.0	93.0	92.4	93.0	93.0	92.4
60/45	93.6	93.6	93.0	93.6	93.6	93.0
75/55	93.6	94.1	93.0	93.6	94.1	93.0
100/75	94.1	94.1	93.0	94.1	94.5	93.6
125/90	94.1	94.5	93.6	94.1	94.5	94.5
150/110	94.5	95.0	93.6	95.0	95.0	94.5
200/150	94.5	95.0	94.5	95.0	95.0	95.0

broad definition was clarified in the rule by a set of guidelines and examples of mechanical and electrical modifications, which can be used to determine whether a motor is considered covered equipment. Table B-2 displays the examples.

In addition, EPAct energy efficiency levels apply to electric motors that are rated in kilowatts or horsepowers other than those specified in NEMA MG 1, Table 10-4. Table B-1 shows the standard horsepower/kilowatt-equivalent ratings for metric motors. For other kilowatt-rated motors, the conversion to horsepower is done using the formula:

$$1 \text{ kilowatt} = (1/0.746) \text{ horsepower}$$

For motors with power ratings that fall between the values in the table, the appropriate nominal efficiency level is determined as follows: round up for a horsepower rating at or above the midpoint between two consecutive horsepowers and round down for a rating below the midpoint.

Determination of Efficiency

In order to comply with EPAct, the average full-load efficiency of each *basic model* of electric motor must be determined by either testing or the application of an alternative efficiency determination method (AEDM). *Basic model* refers to all units of a given type of motor that reflect the fundamental efficiency characteristics of a family of motors. Such characteristics are derived from the same general design and are anticipated to have similar efficiency values.

Testing

In general, efficiency is determined in accordance with NEMA Standard MG 1-1993, with Revisions 1–4 (NEMA 1993) and either IEEE Standard 112-1996, Test Method B, as amended (IEEE 1996) or CSA Standard C-390-93, Test Method 1 (CSA 1993).

Alternative Efficiency Determination Method

This method is the alternative to testing every basic model of motor for efficiency. It is based on testing a statistically valid sample of motors and applying the results to a mathematical model that represents the electrical, mechanical, and energy efficiency characteristics of a basic model. The accuracy and reliability of an AEDM must be substantiated before it can be used. In general, the tested losses must agree within 10 percent of the estimated losses.

Certification of Compliance

A manufacturer or private labeler must certify its electric motors are in compliance through either independent testing or a certification program nationally recognized in the United States.

If independent testing is used, the testing laboratory must be accredited by one of the following:

- The National Institute of Standards and Technology/National Voluntary Laboratory Accreditation Program (NIST/NVLAP)

(text continues on page 418)

413

Table B-2

Examples of EPAct Coverage of NEMA Products for Three-Phase Electric Motors, 1–200 hp, Based on Physical and Electrical Characteristics

MOTOR MODIFICATION		CATEGORY				EXPLANATION
		I	II	III	IV	V	
A: Electrical Modifications							
1	ALTITUDE	X					General purpose up to a frame series change
2	AMBIENT	X					General purpose up to a frame series change
3	MULTISPEED					X	EPAct applies to single speed only
4	SPECIAL LEADS	X					
5	SPECIAL INSULATION	X					
6	ENCAPSULATION				X		Due to special construction
7	HIGH SERVICE FACTOR	X					General purpose up to a frame series change
8	SPACE HEATERS	X					
9	WYE DELTA START	X					
10	PART WIND START	X					
11	TEMPERATURE RISE	X					General purpose up to a frame series change
12	THERMAL PROTECTION		X				Require resetting and third-party agency approval
13	THERMOSTAT/THERMAL SENSOR	X					
14	SPECIAL VOLTAGES					X	EPAct applies to motors operating on 230/460 V at 60 Hz
15	NONSTANDARD H/P		X				Round horsepower according to 10CFR 431.42 for efficiency

	I	II	III	IV	V	Notes
16 FREQUENCY					X	EPAct applies to motors operating on 230/460 V at 60 Hz
17 FUNGUS/TROP INSULATION	X					
B: Mechanical Modifications						
18 SPECIAL BALANCE	X					
19 BEARING TEMPERATURE DETECTOR	X					
20 SPECIAL BASE/FEET					X	Does not meet definition of T-frame
21 SPECIAL CONDUIT BOX	X					
22 AUXILIARY CONDUIT BOX	X					
23 SPECIAL PAINT/COATING	X					
24 DRAINS	X					
25 DRIP COVER	X					
26 GROUNDING LUG/HOLE	X					
27 SCREENS ODP	X					
28 MOUNTING F1, F2; W1–4; C1, C2	X					Foot-mounting, ridged base, and resilient base

Category I—General purpose electric motor as defined in EPAct
Category II—Definite purpose electric motor that can be used in most general purpose applications as defined in EPAct
Category III—Definite purpose electric motor as defined in EPAct
Category IV—Special purpose electric motor as defined in EPAct
Category V—Outside the scope of "electric motor" as defined in EPAct

continued on next page

Table B-2 (continued from previous page)

Examples of EPAct Coverage of NEMA Products for Three-Phase Electric Motors, 1–200 hp, Based on Physical and Electrical Characteristics

MOTOR MODIFICATION	I	II	III	IV	V	EXPLANATION
C: Bearings						
29 BEARING CAPS	X					
30 ROLLER BEARINGS		X				Test with a standard bearing
31 SHIELDED BEARINGS	X					
32 SEALED BEARINGS	X					Test with a standard bearing
33 THRUST BEARINGS				X		Special mechanical construction
34 CLAMPED BEARINGS	X					
35 SLEEVE BEARINGS				X		Due to special construction
D: Special End Shields						
36 C FACE	X					As defined in NEMA MG 1
37 D FLANGE	X					As defined in NEMA MG 1
38 CUSTOMER DEFINED				X		Special design for a particular application
E: Seals						
39 CONTACT SEAL	X					Includes lip seals and taconite seals—test with seals removed
40 NONCONTACT SEAL	X					Includes labyrinth and slinger seals—test with seals installed

No.	Type	I	II	III	IV	V	Comments
F: Shafts							
41	STANDARD SHAFTS AS DEFINED IN NEMA MG 1	X					Includes single and double, cylindrical, tapered, and short shafts
42	NONSTANDARD MATERIAL	X					
G: Fans							
43	SPECIAL MATERIAL	X					
44	QUIET DESIGN	X					
H: Other Special Types of Motors							
45	WASHDOWN	X					Test with seals removed
46	CLOSE-COUPLED PUMP		X				DIM and JP frame assignments
47	INTEGRAL GEAR MOTOR			X			Typically special mechanical design—not a T-frame; motor and gearbox inseparable and operate as one system
48	VERTICAL NORMAL THRUST (FOOTLESS)			X			EPAct covers foot-mounting
49	SAW ARBOR				X		Special electrical/mechanical design
50	TENV				X		Totally enclosed non-ventilated—special electrical/mechanical design
51	TEAO				X		Totally enclosed air-over requires airflow from external source—not integral to the motor
52	FIRE PUMP	X					When does not require safety certification
53	NONCONTIGUOUS				X		EPAct covers continuous ratings
54	INTEGRAL BRAKE MOTOR					X	Integral brake design factory built within the motor

Category I—General purpose electric motor as defined in EPAct
Category II—Definite purpose electric motor that can be used in most general purpose applications as defined in EPAct
Category III—Definite purpose electric motor as defined in EPAct
Category IV—Special purpose electric motor as defined in EPAct
Category V—Outside the scope of "electric motor" as defined in EPAct

Source: Based on DOE 1999

- An accreditation body having a mutual recognition arrangement with NIST/NVLAP
- An organization that has petitioned and is classified by DOE as an accreditation body; such a testing laboratory must be an *independent* facility in order to render test reports objectively and without bias.

If a certification program is used, the certification organization must meet certain criteria and submit a petition to DOE to be classified as nationally recognized.

Upon acceptance of a manufacturer's or private labeler's certification that its electric motors comply with the energy efficiency requirements contained in the EPAct final rule (Federal Register 1999), DOE will issue a Compliance Certification (CC) number to that manufacturer or private labeler.

Labeling

The nominal full-load efficiency, as determined by testing or use of an AEDM, must be marked on the motor's permanent nameplate. The CC number must be displayed on the permanent nameplate within 90 days of the number's issuance by DOE. In addition, a manufacturer or private labeler has the option to mark its complying electric motors with the encircled lowercase letters "ee," as in the following example, or with some comparable designation or logo.

Also, such energy efficiency information must be prominently displayed in motor catalogs and other materials used to market the motor.

Imported Motors

Any covered motor imported into the United States, whether it is manufactured alone or as a component of another piece of equipment, must meet the energy efficiency requirements prescribed by EPAct. These motors must also comply with certification and labeling requirements set forth in the EPAct final rule (Federal Register 1999). The importer would be responsible for certifying compliance if the covered motor had not already been certified for compliance by the manufacturer or private labeler.

Enforcement

Typically, DOE relies upon the marketplace to identify potential violations of the statutory requirements for electric motors. Upon receiving written information alleging that there has been a violation, DOE will investigate to determine whether a violation actually has occurred. DOE follows a prescribed procedure for enforcement as spelled out in the EPAct final rule (Federal Register 1999). As part of the enforcement process, the department can require the testing of motors under investigation. A violation can result in penalties, as provided under Section 431.128 in the final rule, and/or an order for "cessation of distribution of a basic model."

Glossary

Material for this glossary was taken in part from the following sources:

- *1981 Fundamentals*. Atlanta, Ga.: American Society of Heating, Refrigerating and Air-Conditioning Engineers (ASHRAE). 1981.

- *Cooling and Heating Load Calculation Manual*. Atlanta, Ga.: American Society of Heating, Refrigerating and Air-Conditioning Engineers (ASHRAE). 1979.

- *Dictionary of Mechanical Engineering*. Prepared by J.L. Nayler and G.H.F. Nayler. New York, N.Y.: Hart Publishing Company. 1967.

- "Energy Savings Potential in California's Existing Office and Retail Buildings." Staff Report. Sacramento, Calif.: California Energy Commission. 1984.

- *Glossary of Frequently Occurring Motor Terms*. Wallingford, Conn.: EMS, Inc. 1983.

- *Guide to HVAC Equipment*. Sacramento, Calif.: California Energy Commission. 1980.

- *Guidelines for Saving Energy in Existing Buildings: Building Owners and Operators Manual*. ECM 1. Washington, D.C.: Federal Energy Administration. 1975.

- *IEEE Standard Dictionary of Electrical and Electronics Terms*. New York, N.Y.: Institute of Electrical and Electronics Engineers. 1988.

- *Terminology of Heating, Ventilation, Air-Conditioning, and Refrigeration*. Atlanta, Ga.: American Society of Heating, Refrigerating and Air-Conditioning Engineers (ASHRAE). 1986.

Actuator: A device, either electrically, pneumatically, or hydraulically operated, that changes the position of a valve or damper.

Adjustable-speed drive (ASD): A motor accessory that enables the driven equipment (e.g., fan or pump) to be operated over a range of speeds. The two general categories of ASDs are mechanical units (installed between the motor and the driven load) and electronic units (installed in the electrical wiring to the motor).

Air transport system: A system that distributes air to the various spaces in a building, generally comprising fans, ducts, dampers, registers, etc. It is sometimes referred to as a ventilation system, but the air transport of warm or cool air for space conditioning may be separate from the mechanical ventilation system in some buildings.

Alternating current (AC): Electric current that is characterized by the electrons flowing back and forth along the conductors that constitute the circuit. Normal building wiring in the United States is alternating current with a frequency of back-and-forth flow of 60 cycles per second. See *direct current*.

Ambient: Surrounding (e.g., ambient temperature is the temperature in the surrounding space).

Amperes (amps): Equal to the flow of $6.25 \times 1,018$ electrons per second, or one coulomb per second.

> **Full-load amps (FLA)**: The amount of current the motor can be expected to draw under full-load (torque) conditions when operating at the rated voltage. Also known as nameplate amps.
>
> **Locked-rotor amps (LRA)**: The amount of current the motor can be expected to draw under starting conditions when full voltage is applied. Also known as starting inrush.
>
> **Service-factor amps**: The amount of current the motor will draw when it is subjected to a percentage of overload equal to the service factor on the nameplate of the motor. For example, many motors have a service factor of 1.15, meaning that the motor can handle a 15% overload.

See *current*.

Amps: See *amperes*.

Apparent efficiency: The product of a motor's efficiency and its power factor.

ASHRAE 90: Comprising voluntary building standards for new buildings, developed by the American Society of Heating, Refrigerating and Air-Conditioning Engineers. These standards include minimum equipment efficiencies, building envelope characteristics, and required control strategies for nonresidential buildings.

Average efficiency: See *nominal efficiency*.

Avoided cost: Cost to the utility of the marginal kilowatt-hour produced. When conservation or an alternative supply allows a utility to reduce its own power production, the savings to the utility is its avoided cost. This quantity (which includes avoided operations and maintenance, transmission and distribution, and capacity costs) varies depending on a wide range of factors, including fuel cost, generation type (which may vary over the course of the day and the year), etc.

Basic model: All units of a given type of covered equipment manufactured by a single manufacturer and, with respect to electric motors, that have the same rating and essentially identical electrical and efficiency characteristics (Federal Register 1999).

Bearings: The supports that hold a revolving shaft in its correct position. In the context of motors, the two rotor shaft bearings (mounted in the motor frame) allow rotary motion of the shaft relative to the enclosure while preventing axial or radial motion. Bearings come in a wide variety of types. Most integral-horsepower motors use ball bearings with rolling steel balls that contact the two main parts ("races") of the bearing to allow the relative motion. Many fractional-horsepower motors (especially the smallest sizes) use sleeve or journal bearings with a bearing lubricant to keep the spinning shaft from contacting the stationary bearing.

Belt: A band of flexible material (usually rubber or plastic reinforced with fabric or steel) for transmitting power from one shaft to another by running over flat, grooved, or toothed pulleys. See Figure 3-15 for illustrations. The common belt types include

Flat belts: Smooth belts with a flat cross-section, riding on corresponding smooth pulleys. Flat belts are thinner and wider than V-belts used in the same applications.

Synchronous belts: Belts with a flat cross-section and teeth formed in the inner belt surface. The belt teeth engage the teeth of the pulleys, preventing any slippage (hence the name).

V-belts: Belts with a V-shaped cross-section and a smooth or cogged inner belt surface. V-belts ride in pulleys (sheaves) with corresponding smooth, V-shaped grooves. The "cogged" V-belts are toothed with transverse grooves or notches. These notches do not interface with the notched sheaves but rather increase the contact force between the belt and the smooth sheave, reducing slippage while making the belt more flexible by reducing bending due to heating.

Bipolar transistor: Three-terminal electronic switch in which the current between two terminals (the collector and the emitter) is controlled by the third terminal (the base). The base current is typically 50–100 times smaller than the output current.

Brushes: Conductors, usually composed in part of carbon, serving to maintain an electrical connection between the stationary and rotating parts of a motor. Brushes contact either slip rings (in AC wound-rotor motors) or the contacts of the commutator (in DC motors).

Capacitor: A component containing a dielectric (nonconducting) material sandwiched between two metallic layers. Capacitors are widely used for power-factor compensation and filters. See *power factor*.

CEE premium-efficiency motor: A motor that meets or exceeds the minimum-efficiency level specified by the Consortium for Energy Efficiency. These levels represent an efficiency level above that of EPAct motors. This specification is used by many entities as the qualifying efficiency level for participation in motor programs. See *Consortium for Energy Efficiency and EPAct motors*.

Centrifugal chiller: A machine that produces cold water by using centrifugal action in its compressor to raise the pressure level of the refrigerant gas. Centrifugal chillers are commonly used in large commercial buildings to supply chilled water to cooling coils in the buildings' HVAC systems. Chiller unloading (operating at cooling loads below maximum) is generally regulated by varying the flow of the refrigerant gas with variable-inlet vanes on the input side of the compressor.

Centrifugal fan: A device for propelling air by centrifugal action. Forward-curved fans have blades that are sloped forward relative to the direction of rotation, while backward-curved fans have blades that are sloped backward and are generally more efficient at high pressures than forward-curved fans.

Chiller: A refrigeration machine that produces cooled water, generally at a temperature of 40–55°F. Types include reciprocating, screw, centrifugal (named for the type of compressor used in the motor-driven compression-expansion cycle), and absorption (for the heat-driven absorption cycle).

Chopper: A device that converts DC power into a square wave. When used with an output filter, a chopper can be used with a constant-voltage input to create a variable-voltage output by altering the ratio of on-time to off-time in the square wave.

Code letter: An indication of the amount of locked rotor (inrush) current required by the motor when it is started. See *amperes, locked-rotor*.

Coefficient of performance (COP): A measure of the efficiency of cooling or refrigeration equipment. COP is defined as the ratio of cooling output to energy input, with both quantities in the same units of measure (kilowatts or British thermal units per hour). Electric cooling equipment has COPs ranging between approximately 2 and 6. See *energy efficiency ratio*.

Compressor: A mechanical device that increases the pressure, and thereby the temperature, of a gas. Refrigerant compressors are the most common in building applications, followed by air compressors.

Condenser: A heat exchanger in which a refrigerant is condensed from a vapor to a liquid. Common types of condensers are air-cooled (either by natural air flow, as in the coil on the back of many residential refrigerators, or fan-forced, as in air conditioners); water-cooled (as in most large chillers for commercial buildings); and evaporative, where water is sprayed on the outside of the refrigerant tubes and a fan forces air to evaporate a portion of the water, providing a cooling effect.

Consortium for Energy Efficiency (CEE): A nonprofit organization located in Boston, Massachusetts, that develops and deploys market transformation programs for member utilities, government agencies, and public interest groups. CEE's motor committee has developed several motor system initiatives. See *market transformation*.

Cooling load: The heat and moisture that accumulate in a building and that must be removed in order to maintain comfortable temperature and humidity conditions.

Cooling tower: A device that cools water directly by evaporation and is typically used to reject heat from one or more condensers.

Covered motor: A motor, defined within EPAct regulations, to which EPAct specifications apply (see Appendix B for further explanation).

Current: The flow of electrons in an electrical circuit. Current is measured in amperes. See *amperes.*

Current signature: The unique distortions in the current profile caused by an operating electromechanical device.

Current-source inverter (CSI): A type of electronic ASD that works by converting the AC input to controlled-current DC and then synthesizing the variable-frequency AC output by using a DC-to-AC inverter. See *adjustable-speed drive, variable-frequency drive,* and *voltage-source inverter.*

Cycloconverter: An AC converter in which the AC supply from the grid is converted directly into another AC voltage waveform with a lower frequency, without an intermediate DC stage. The output frequency ranges between 0% and 50% of the input frequency.

Damper: A restrictive device used to vary the volume of air passing through an air outlet, inlet, or duct.

Demand charge: The amount charged by the utility per kilowatt of peak power used (demanded) by the customer. Demand charges are usually billed per month; the peak demand is measured by a special demand meter that records the highest average demand (typically over a 15- or 30-minute interval) during the month. The charge may be fixed or variable according to the time of day, season, and level of demand.

Demand-side management (DSM): These programs focus on reducing energy consumption by energy end-users and, in general, are operated by utilities, government, and public benefit entities. Projects may focus on education, incentives, or market transformation. See *public benefit fund* and *market transformation.*

Design: The design letter on a motor nameplate is an indication of the shape of the torque-speed curve. Figure 2-9 shows the typical shape of the most commonly used NEMA design letters (A, B, C, D, and E). Design B is the standard industrial-duty motor, which has reasonable starting torque with moderate starting current and good overall performance for most industrial applications. Design C is used for hard-to-start loads and is specifically designed to have high starting torque. Design D is the so-called high-slip motor, which tends to have very high starting torque with high slip at full-load torque. The motors are particularly suited for low-speed punch press, hoist, and elevator applications. Generally, the efficiency of Design D motors at full load is rather poor, and thus they are normally used on those applications where the torque characteristics are of primary importance. Design A motors are not commonly specified, but specialized motors used for injection molding applications have characteristics similar to Design A's. The most important characteristic of this type is that the pull-out torque is somewhat higher than Design B's; otherwise A and B are quite similar. Design E motors are comparable in specification to Design A's motors with high starting currents and limited pull-up torques but require special starters and are therefore predominately used in HVAC fan applications. See *slip, torque, pull-out* and *torque, pull-up*.

Direct current (DC): Electrical current characterized by electrons flowing in one direction only. See *alternating current*.

Discharge dampers: Dampers that regulate the flow of air on the outlet side of a fan in variable-air-volume systems. Dampers are the least efficient method of regulating air flow.

Drivepower: Energy consumed by motors and motor-driven equipment.

EASA-Q: A certification program developed by the Electrical Apparatus Service Association for quality motor repair practices. See *Electrical Apparatus Service Association* and *motor repair*.

ECM or ECPM: Electronically commutated permanent-magnet motor. See *permanent-magnet motors*.

Eddy (or eddy-current) losses: See *magnetic losses*.

Efficiency (motor): In general, this is the ratio of the mechanical power output to the electrical power input. See other efficiencies: *apparent, minimum*, and *nominal*.

Electrical Apparatus Service Association (EASA): A trade association representing many motor repair shops, principally in North America. EASA also develops standards for motor repair practices.

Electromagnetic interference (EMI): Impairment of a transmitted electromagnetic signal by an electromagnetic disturbance; it's particularly relevant to communications and data processing applications.

Energy charge: The amount charged by the utility for each kilowatt-hour of energy used by the customer. The energy charge may be fixed or variable, depending on the time of day, season, and level of usage.

Energy efficiency ratio (EER): A U.S. measure of cooling equipment efficiency, defined as

(cooling output in Btu/h)/(electric input in watts)

EER = COP × 3.412. See *seasonal energy efficiency ratio*.

Energy-efficient motor (EEM): A motor that meets or exceeds the minimum-efficiency levels specified in NEMA MG 1, Table 12-10. These levels correspond with the minimum-efficiency levels specified in the Energy Policy Act of 1992. See *EPAct motor* and *NEMA MG 1*.

Energy Policy Act of 1992 (EPAct): Federal legislation that amended the Energy Policy and Conservation Act of 1978. Among other actions, it established minimum-efficiency standards for integral-horsepower, general purpose, polyphase induction motors of 200 hp or less.

EPAct: See *Energy Policy Act of 1992*.

EPAct motor: A motor that complies with the minimum-efficiency levels specified in the Energy Policy Act of 1992. These motors also meet the NEMA definition of energy efficient. See *energy-efficient motor* and *Energy Policy Act of 1992*.

Explosion-proof (EXP): A type of motor package ("enclosure") designed to withstand the explosion of a specified gas or vapor within it and to prevent ignition of a specified external gas or vapor by sparks, flashes, or explosions that may occur within the motor casing.

First cost: The initial cost of a project, including design, procurement, equipment, and installation costs.

Forced commutation inverter: Inverter in which a special commutation circuit is required to turn off the thyristor, making the inverter design more complex. See *thyristor*.

Fractional-horsepower motor: A motor with a rated output power of less than 1 hp. See *horsepower* and *integral-horsepower motor*.

Frame size: Motors come in various physical sizes to match the requirements of the application. In general, the frame size gets larger with increasing horsepower or with decreasing speed. In order to promote standardization, NEMA prescribes standard frame sizes for certain horsepower, speed, and enclosure combinations. Frame size specifies the mounting and shaft dimensions of standard motors. For example, a motor with a frame size of 56 will always have a shaft height above the base of 3.5 inches. Frame sizes are usually listed as a combination of a number and a letter, with the number indicating the relative size and the letter the general frame type (such as T, U, etc.). See *frame type*.

Frame type: This is the general characteristics of a motor's size and mounting configuration, usually expressed by a letter. For example, NEMA T-frame motors (base-mount, single-ended shaft) are the most commonly made three-phase frame type; the similar but larger U-frame motors were most common until the 1960s. U-frame and T-frame motors have the same shaft size for the same power and speed. Another early design of the same type, A-frame motors, differs from T-frames in both motor size and shaft size. C- and J-frame motors are end-mounted and designed to be bolted directly to the driven equipment. L-frame motors are similar to C-frames except that they are designed to mount vertically above the load (usually a pump). Fractional-horsepower motors generally do not have a letter designation. See *frame size*.

Free rider: A participant in a promotional conservation program who would have performed the conservation action even without the program.

Frequency: The rate of oscillation of an alternating current, expressed in cycles per second (or hertz). In North America, the predominant frequency of AC power is 60 Hz.

Full-load speed: The approximate speed at which the motor will run when it is operating at full rated output torque or horsepower.

Gate turn-off thyristor (GTO): An electronic switch with the same properties as a thyristor, but possible to turn off by applying a small control signal in the gate. This is in contrast to standard thyristors, which must have the voltage across the main terminals brought close to zero in order to be turned off (requiring the use of such techniques as forced commutation). See *thyristor*.

Gears: A mechanical system for transmitting rotation through the use of toothed wheels in direct engagement. Gears are used to change the speed, direction, or orientation of rotation from one shaft to another. There are a great many types and combinations of gears; four of the most common types of gears are bevel, helical, worm, and spur gears. Helical and worm gears are shown and described in Figure 3-12. Spur gears are cylindrical gear wheels in which the teeth are parallel to the shaft and are used for transmitting power between parallel shafts. Bevel gears are beveled in order to transmit rotation between nonparallel shafts and are commonly used to transmit power at 90° to the output shaft, that is, between shafts with intersecting axes at right angles.

General purpose motors: NEMA defines a general purpose motor as an open or closed motor, 500 hp or less, rated for continuous duty, without special mechanical construction, that can be used in typical service conditions without restrictions to a particular application or type of application.

Harmonics: Electrical signals with frequencies that are integral multiples of the fundamental frequency. For example, in a 60 Hz application, a 180 Hz component is called the third harmonic.

Header: The manifold into which multiple pumps or compressors discharge.

Heating, ventilation, and air conditioning (HVAC) system: A system that provides one or more of the functions of heating, ventilation, and air conditioning (cooling) for a building.

Hertz (Hz): Frequency of AC power in cycles per second. The predominant frequency of power in North America is 60 Hz; in most other countries it's 50 Hz. See *frequency*.

High-inertia load: A load that has a relatively high flywheel effect (or moment of inertia). Large fans, blowers, punch presses, centrifuges, industrial washing machines, and similar loads can be classified as high-inertia loads. See *inertia*.

Horsepower (hp): A unit of power equal to 746 watts or 33,000 ft-lb/minute. In the United States, horsepower is used to indicate the rated output (shaft) power of a motor. One horsepower = torque (ft-lb) × speed (rpm)/5,252. In compressor sizing, it is the full-load output rating of the electric motor driving the compressor.

Hysteresis losses: See *magnetic losses*.

Inductance: The property of an electrical circuit by which an electromotive force is induced in it or in a nearby circuit by a change of current in either circuit.

Induction motor: The most common type of AC motor, in which a primary winding on one member (usually the stator) is connected to the power source and a secondary winding (in the case of wound-rotor induction motors) or a squirrel cage of metal bars (in the case of squirrel-cage induction motors). On the other member (usually the rotor), the induced current is carried. The changing magnetic field created by the stator induces a current in the rotor conductors, which in turn creates the rotor magnetic field. The interaction of the stator and rotor magnetic fields causes the motor to rotate.

Inductors: Generally, they are devices with a magnetic core around which windings of wire are wrapped, a construction that results in high inductance relative to the size of the device. An electromagnet is a type of inductor.

Industrial Best Practices: *Motors*, formerly *Motor Challenge*: See Chapter 9.

Inertia: That property of a body by which it tends to resist a change in its state of rest or uniform motion. Inertia is measured by mass (equivalent to weight) when linear accelerations are considered. In the context of motor systems where rotational acceleration is the primary concern, inertia is measured by the moment of inertia, about the axis of rotation. The moment of inertia is the Σmr^2, where m is the mass of a part of the rotating equipment and r is its perpendicular distance from the axis of rotation. That is, the moment of inertia depends on the weight of the rotating system and how far the weight is from the axis of rotation (the farther away it is, the more effect the same weight will have).

Inlet vanes: Variable vanes on the inlet side of a fan that regulate airflow in a variable-air-volume system. Inlet vanes are also used in centrifugal chillers.

Insulated gate transistor (IGT): A three-terminal electronic switch with an input stage that is an MOS transistor and an output stage that is a bipolar transistor. In this way, the IGT combines the best properties of both transistors (requires negligible input power to control the transistor and results in low losses in the conduction state when the IGT is fully on). See *MOS transistor*.

Insulation class: A measure of the resistance of the insulation components of a motor to their degradation from heat. The four major classifications of insulation used in motors are, in order of increasing thermal capabilities, Classes A, B, F, and H. Class A is no longer used in integral-horsepower motors; the designations C through E and G were never used.

Integral-horsepower motor: This motor has an output power rating of 1 hp or above. See *fractional-horsepower motor* and *horsepower*.

Inverter: A device or system that changes DC power to AC power.

Inverter drive: A type of adjustable-speed drive that varies the motor speed, changing the frequency of the motor input current.

See *adjustable-speed drive, variable-frequency drive,* and *voltage-source inverter.*

Inverter duty motor: A motor manufactured in conformance with NEMA MG 1, Part 31, with a higher class of insulation that allows the safe operation of inverter drives. See *inverter drive* and *NEMA MG 1.*

Isolation transformer: A transformer with primary and secondary windings physically separated, thus preventing primary circuit voltage from being forced onto the secondary circuits. Isolation transformers are often used with large ASDs to reduce the power quality degradation caused by the ASD.

Kilovolt-ampere (kVA): The product of the voltage (in volts) and current (in amperes) in an electrical circuit, divided by one thousand. In DC circuits, kilovolt-ampere equals kilowatt flowing. In AC circuits, the kilovolt-ampere equals the kilowatt if the power factor equals one; otherwise the kilovolt-ampere is higher than the kilowatt. See *kilowatt* and *power factor.*

Kilowatt (kW): A unit of (usually) electrical power equal to one thousand watts, or the flow of one thousand joules of energy per second. Equivalent to 3,412 British thermal units (Btus) per hour of thermal power or 1.34 hp. Other than in the United States, it is commonly used to indicate motor output (shaft) power. See *horsepower* and *kilovolt-ampere.*

Kilowatt-hour (kWh): A unit of electrical energy equal to one kilowatt of power flowing for one hour, i.e. 3,600,000 joules of energy. Equivalent to 3,412 Btus of thermal energy or 1.34 hp/hr. Kilowatt-hour is the most common unit used for metering electricity. See *kilowatt.*

Laminations: Thin steel sheets stacked together and used in electromagnetic devices. In motors, they form the core of the stator and rotor magnets. In inductors and transformers, laminations provide the magnetic core around which the windings of wire are placed.

Leakage reactance: The motor reactance associated with that fraction of the magnetic flux generated by the stator winding that does not cross the air gap and therefore does not reach the rotor (and vice versa, from the rotor to the stator). The leakage reactance is a trade-off value: for example, a high degree of leakage reactance results in lower starting current (a desirable result), but with undesirable reductions in steady-state motor performance. The leakage reactance increases with the air gap size and is also a function of other motor design parameters such as slot design, saturation of the magnetic circuit, and winding configuration.

Load profile: Distribution over time of the heating, cooling, ventilation, electrical, or any other loads of a building or process. Load profile is usually expressed on an hourly basis over a day but may also be expressed on a seasonal basis over a year.

Load types:

Constant-horsepower: Loads where the torque requirement decreases as the speed increases, and vice versa. Constant-horsepower loads are usually associated with applications such as traction (in electric vehicles, for example) and metal removal (e.g., drill presses, lathes, and milling machines).

Constant-torque: Loads where the amount of torque required to drive the machine is constant regardless of the speed at which it is driven. For example, most conveyors and many reciprocating compressors are constant-torque loads.

Variable-torque: Loads that require low torque at low speeds and increasing torque as the speed is increased. Centrifugal fans and pumps are typical examples of variable-torque loads.

Magnetic losses: When the iron core in the motor is subjected to a changing magnetic field, as it is during normal operation, there are two types of losses: eddy current and hysteresis. Eddy-current (or simply eddy) losses are due to the currents induced in the iron by the change in the magnetic flux, with losses growing with the square of the flux density and the square of the frequency. Eddy losses can be minimized by using thinner laminations and silicon steel with a higher electric resistivity. Hysteresis losses are due to the rotation of groups of iron atoms as they are excited by the changing magnetic field. Hysteresis losses are proportional to the square of the flux density and to the frequency. Hysteresis losses can be decreased by using high-performance silicon steel with high permeability and a narrow hysteresis cycle. Both types of magnetic losses can be decreased by using a lower magnetic flux density, which means using larger cross-sections in the magnetic circuit (i.e., more iron in the motor).

Market transformation: This concept involves programs and measures that seek to permanently change the market's structure or behavior to a desired goal (e.g., procurement of energy-efficient products). Strategies can involve education, targeted incentives, or formation of new market structures. These efforts are frequently carried out by government, public interest, public benefit, or utility entities. See *demand-side management.*

Mechanical cooling: Cooling by energy-using equipment such as chillers and air conditioners. Cooling accomplished through use of outside air or by evaporative coolers is generally not considered mechanical cooling.

MG 1: See *NEMA MG 1.*

Microelectronic: Electronic devices characterized by highly integrated circuits (many semiconductor devices on one chip of silicon) that are usually used for computation and control and generally operate at currents well below 1 ampere with voltages below 10 V. See *power electronic devices.*

Minimum efficiency: The minimum level of efficiency for a group of motors of the same specification. Up to 5% of motors can have an efficiency lower than the minimum efficiency. Minimum efficiency is sometimes guaranteed by the motor manufacturer. The NEMA minimum efficiency levels are set at two standard increments of efficiency below the NEMA nominal efficiency. See *nominal efficiency.*

Minimum-efficiency standard or specification: A standard or specification requiring a particular type of equipment to meet a minimum level of operating efficiency. In the case of motors, such standards generally set different minimum levels of nominal motor efficiency according to the motor size (in horsepower output rating). See *CEE premium-efficiency motor, efficiency, EPAct motor, NEMA MG 1, NEMA Premium Motor™,* and *nominal efficiency.*

MOS transistor: A three-terminal electronic switch in which the conduction between the two main terminals (the drain and the source) is controlled by the voltage applied between the third terminal (the gate) and the source. The input current in the gate is almost zero, and the input power required to control the transistor is negligible. This leads to simple control circuits and improved efficiency.

Motor (electric): A machine that converts electrical power into mechanical power in the form of a rotating shaft. See *induction motor* and *synchronous motor.*

Motor Challenge program, now *Industrial Best Practices: Motors:* See Chapter 9.

Motor repair: This area covers a range of services that involve the maintenance and repair of electric motors. These services can range from cleaning, preventive maintenance, and mechanical repair to the replacement of the electrical winding. See *Electric Apparatus Service Association* and *motor rewind*.

Motor rewind: This procedure involves the removal of the motor stator winding and replacing it with a new winding. A rewind is usually performed on a motor that has experienced an electrical failure. Rewinding usually also involves other mechanical and electric repairs such as cleaning and bearing replacement. See *motor repair*.

National electrical code: The standards document setting forth accepted sizing and installation practices for electrical equipment, used as a reference in setting local building codes.

Natural commutation: A circuit in which the voltage applied to the thyristors reverses in polarity, leading to the turnoff of the device when the voltage crosses zero.

NEMA: National Electrical Manufacturers Association.

NEMA MG 1: A standard issued by the Motor Generator Committee of NEMA that provides design, labeling, and application specifications for electric motors and generators. Table 12-10 provides the specification of energy-efficient motors that was incorporated in the EPAct Motor Standard. See *energy-efficient motor, EPAct, NEMA*, and *NEMA Premium Motor™*.

NEMA Premium Motor™: A minimum-efficiency specification for motors issued by NEMA. See *minimum-efficiency specification* and *NEMA*.

NEMA TP-1: A NEMA standard issued in 1996, entitled the *Guide for Determining Energy Efficiency for Distribution Transformers*. This standard specifies how cost of ownership for distribution transformers should be calculated and provides a default table, Table 4-2, of minimum-efficiency levels for different classes of transformers to be labeled "energy efficient." See *NEMA*.

Nominal efficiency: The average expected efficiency for a group of motors of the same specification. Half of the motors are expected to fall below the nominal value, and half above. NEMA's nominal efficiency (a rating indicating that the motor's nominal efficiency falls within a certain range) is now being stamped on the nameplate of most domestically produced integral-horsepower electric motors. See *minimum efficiency*.

Open drip-proof (ODP): A type of motor package ("enclosure") in which cooling is provided by an internal fan(s) forcing air through the motor. The ventilation openings are positioned to keep out liquid or solid particles falling at any angle from 0° to 15° from the vertical.

Participation rate: The fraction (or percentage) of the eligible customers taking part in a program.

Part-load ratio: The ratio of instantaneous output from a piece of equipment to the equipment's rated output. For example, if a piece of cooling equipment is exercising 60% of its full cooling capacity, the part-load ratio is 0.6.

Peak cooling load: The maximum rate of cooling that occurs in a building during the year.

Penetration rate: The degree to which a technology has become the standard in a marketplace. For example, if energy-efficient motors are sold for 10% of the general purpose motor applications, then they have achieved a 10% penetration rate in that market. The market context must be clarified for the penetration rate to be meaningful. For example, one needs to know if the target market is new applications or the existing stock.

Permanent-magnet (PM) motors: A family of motors in which a permanent magnet replaces the stator winding. In some small PM DC motors, the rotor is still fed by a conventional brush-and-commutator system. A more important type of PM motor has a stator with three windings producing a rotating field, as in induction and synchronous motors. The rotor consists of one or more permanent magnets that interact with the rotating field so as to align the poles in the rotor with the poles of the rotating field. The speed of the motor is the speed of the rotating field. Because there is no rotor current and the rotor magnetic field is constant, there are no losses in the rotor, helping to make PM motors more efficient (by five to ten percentage points in small sizes) than induction motors. The most common form of a PM motor is the brushless DC motor, also known as an electronically commutated motor (ECM).

Phase: The indication of the type of power supply for which the motor is designed. The two main categories are single-phase and three-phase (sometimes referred to as polyphase).

Poles: The ends of a magnet, which are always present in a pair consisting of a north and a south pole. Thus, the number of poles is always even. Poles may be located on permanent magnets or electromagnets. In AC motors, the synchronous speed is determined by the frequency of the power supply and the number of poles; four different motors operating at 60 Hz with two, four, six, and eight poles will have synchronous speeds of 3,600, 1,800, 1,200, and 900 rpm, respectively.

Positive displacement: A term used to describe mechanical equipment (such as compressors, pumps, and blowers) characterized by a reduction of the internal volume of a chamber, usually by a piston.

Power conditioning equipment: Electronic devices intended to correct power quality problems such as low power factor or harmonics.

Power electronic devices: Electronic devices used for the direct control of electrical power to various types of equipment, including motors. Power electronic devices are available with ratings up to about 5,000 V and 5,000 amperes, with a trend toward ever higher ratings.

Power factor: The ratio between the real power (measured in watts or kilowatts) and apparent power (the product of the voltage times the current measured in volt-amperes or kilovolt-amperes). Power factor is expressed either as a decimal fraction (zero to one) or a percentage (0% to 100%). In the case of pure sinusoidal waveforms (those not distorted by harmonics), the power factor is equal to the cosine of the phase angle between the voltage and current waves in an AC circuit. This value is known as the displacement power factor because it deals with the time displacement between the voltage and current. Since cosine values range from 0 to 1, the apparent power is always greater than or equal to the real power. If the power factor is less than 1, more current is required to deliver a unit of real power at a certain voltage than if the power factor were 1. In the case of waveforms that include harmonics, the harmonic current adds to the total current without contributing to the real power, so the power factor is reduced. Many power electronic devices (such as ASDs) have high displacement power factors (over 90%) but overall power factors that are significantly lower, depending on design and operating conditions (see *current* and *voltage*). This higher current is undesirable because the energy lost to heat in the wires supplying power is proportional to the square of the current. In motors and other inductive loads operating in AC circuits, the current wave lags behind the voltage wave. When a capacitive load is applied to an AC circuit, the voltage wave lags behind the current wave. Since these are opposite effects, they can be used to cancel each other. Thus, capacitors can be (and very commonly are) used to correct low power factor. In DC circuits, the power factor is always 1. See *kilovolt-ampere* and *kilowatt*.

Public benefit fund: A fund collected as a surcharge on energy sales that is used to sponsor activities that benefit the public such as conservation and efficiency programs, low-income energy programs, and energy research and development. See *demand-side management*.

Pulley: See *sheaves*.

Reciprocating compressor: A machine that uses positive displacement pistons for compression. The pistons move back and forth within their cylinders, much as in a standard automobile engine. Common applications of reciprocating compressors are refrigeration, air conditioning (including reciprocating chillers), and compressed-air systems.

Rectifier: A two-terminal (a positive anode and a negative cathode) electronic device that conducts a current in one direction with low resistance and blocks the current flow in the opposite direction. Rectifiers are mainly used to convert AC power into DC power. The most common rectifiers produced are solid-state silicon devices. In the past, mercury rectifiers (using liquid mercury in a vacuum tube) were commonly used in high-current applications. See *inverter*.

Regeneration capability (also called regenerative braking): This is the return of energy to the supply system when a motor is braking, in which case the motor is working as a generator. The input stage of the ASD must have the capability to work as an inverter to pump the energy back to the AC supply.

Resistance: A property of electrical conductors that, depending on their dimensions, material, and temperature, determines the current produced by a given voltage difference across the resistance. Resistance is the property of a material that impedes current and results in the dissipation of power in the form of heat. It is measured in ohms; one ohm is the resistance through which a voltage difference of one volt will produce a current of one ampere.

Resistor: A device connected to an electrical circuit to introduce a specified resistance.

Retrofit: To replace an operating piece of equipment with a more efficient product (in contrast to replace on failure).

Rewind damage: Damage to a motor resulting from improper repair practices such as overheating the motor core during winding removal.

Rewinding: See *motor rewind*.

Root mean square (RMS): The constant value of a periodic current or voltage that when applied to a resistance would produce the same amount of power. RMS is also known as equivalent DC. The RMS value of a periodic quantity is equal to the square root of the average of the squares of the instantaneous values of the quantity for the period. For example, the mathematical expression of the RMS value of a current is

$$I_{RMS} = \sqrt{\frac{1}{\tau} \int_{1}^{2} I^2(t)dt}$$

where
τ = the period of time for one cycle

t_1 = the time measurement starts

t_2 = the time measurement ends

$I^2(t)$ = the square of instantaneous value of the current at a time t between t_1 and t_2.

A similar expression applies to the RMS voltage and power values. If the quantity is a sine wave (the nominal form for voltage and current in AC circuits), the RMS value is 0.707 times the peak value of the wave.

Rotary compressor: A positive displacement compressor that changes the internal volume of its compression chamber(s) by the rotary motion of its positive displacement member(s). Two common types of rotary compressors are

Rolling-piston compressor: A small rotary compressor with its rotor aligned eccentrically within the stator; used in domestic refrigerators and some room air conditioners.

Screw compressor: A rotary compressor that produces compression with two intermeshing helical rotors. Applications include medium-to-large refrigeration and HVAC (including screw chillers) and compressed-air systems.

Rotor: The part of the motor that rotates.

SCR: See *thyristor*.

Screw compressor, screw chiller: See *rotary compressor*.

Seasonal energy efficiency ratio (SEER): A U.S. rating measure for unitary air conditioning equipment. Measured in a standard test that averages across different part-load ratios of equipment throughout a simulated cooling season. See *energy efficiency ratio*.

Self-commutation: Circuits that use electronic devices, such as transistors and gate-turnoff thyristors, that turn off with the application of a small control signal at their input.

Service factor: The service factor is a multiplier that indicates the amount of overload a motor can be expected to handle. For example, a motor with a 1.0 service factor cannot be expected to handle more than its nameplate horsepower on a continuous basis. Similarly, a motor with a 1.15 service factor can be expected to safely handle continuous loads of 15% beyond its nameplate horsepower.

Servodriver: See *servomotor*.

Servomotor: A low-power electric motor that performs a positioning function. Examples include actuators for dampers, valves, and adjustable pulleys.

Shaded-pole motor: The shaded-pole motor, a type of single-phase induction design, is most commonly used in packaged equipment applications below 0.17 (1/6) hp. Although shaded-pole motors are cheaper than single-phase squirrel-cage motors, their efficiency is poor (below 20%) and their use should be restricted to low-power applications with a limited number of operating hours.

Sheaves: Grooved wheels attached to the motor shaft and to the shaft of the driven equipment, such as a fan. Sheaves transmit mechanical power by means of one or more belts that ride in the grooves of the pair of sheaves. Another name for sheave is pulley.

Silicon-controlled rectifier (SCR): See *thyristor*.

Slip: The difference between motor operating speed and synchronous motor speed, expressed either directly in revolutions per minute or as a percentage of synchronous speed (see *synchronous speed*). For example, an 1,800 rpm motor operating at a full-load speed of 1,725 rpm is running at a slip of 75 rpm, or 4.2%. Most standard induction motors run at a full-load slip of 2% to 5%.

Slip rings: In an AC motor, they are a set of metal rings that are mounted on the rotor shaft and conduct current into or out of the rotor through stationary brushes.

Space conditioning loads: A building's heat losses and gains that need to be counteracted by heating or cooling in order to maintain a comfortable temperature and humidity.

Squirrel-cage induction motor: A type of induction motor with a squirrel-cage winding consisting of a number of conducting bars connected at each end by metal rings that are located in slots in the rotor core. The bars are parallel to the motor shaft; the rings are concentric with the axis of the shaft. This motor is the most common type in use. In order to deliver torque to a load, its shaft must run with slip, or below synchronous speed. See *induction motor, slip,* and *synchronous speed.*

Stator: The nonrotating magnetic section of a motor. In most induction motors, the stator contains the windings.

Synchronous motor: An AC motor in which the speed of operation is exactly proportional to the frequency of power to which it is connected (the motor operates with no slip). Synchronous motors generally have the rotor electromagnets supplied with DC power through slip rings. Since these motors produce little torque except at speeds near to the synchronous speed, they need special methods for starting.

Synchronous speed: The speed at which the motor's magnetic field rotates. It approximates the speed of no-load operation. A four-pole motor running on 60-cycle-per-second power will have a synchronous speed of 1,800 rpm; a two-pole motor at the same frequency will have a synchronous speed of 3,600 rpm. See *slip.*

TEFC (totally enclosed fan-cooled): A type of motor package ("enclosure") in which there is no air exchange between the inside and outside of the motor. The fan is located in a cover opposite the driving (power output) shaft and is driven by an extension of the motor shaft through the housing.

Temperature, ambient: The maximum safe room temperature surrounding the motor if it is going to be operated continuously at full load. In most cases, the standardized ambient temperature rating is 40°C (104°F). Certain types of applications, such as ships and boiler rooms, may require motors with a higher ambient temperature capability such as 50°C or 60°C. Note that this definition is specific to motors, in contrast to the general definition of *ambient.*

Temperature rise: The amount of temperature increase that can be expected within the windings of the motor when going from nonoperating (cool condition) to full load and continuous operation. Temperature rise is normally expressed in degrees Celsius.

Throttle: A device that regulates the flow of a gas or liquid by directly restricting the flow. Discharge dampers, inlet vanes, and valves can all be throttles.

Thyristor (also called silicon-controlled rectifier [SCR] or phase-controlled rectifier): Electronic devices that have both the same capabilities as rectifiers and a third terminal (the gate). The gate allows conduction control from 0% to 100% when the polarity applied to the main terminals is positive. If the polarity is negative, the thyristor blocks the current like a rectifier.

Time rating: Most motors are rated for continuous duty, meaning that they can operate at full-load torque continuously without overheating. Motors used in certain applications (such as waste disposers, valve actuators, hoists, and other intermittent loads) will frequently be rated for short-term duty such as 5 minutes, 15 minutes, 30 minutes, or 1 hour.

Torque: The twisting force exerted by the motor shaft on the load. Torque is measured in units of length times force in foot-pounds or inch-pounds (or, for small motors, inch-ounces). For an illustration of the following types of torque, see Figure C-1.

> **Breakdown torque**: See *pull-out torque*.
>
> **Full-load torque**: The rated continuous torque that the motor can support without overheating within its time rating.
>
> **Peak torque**: Many types of loads, such as reciprocating compressors, have cycling torques, where the amount of torque

Figure C-1

Typical Torque-Speed Curve

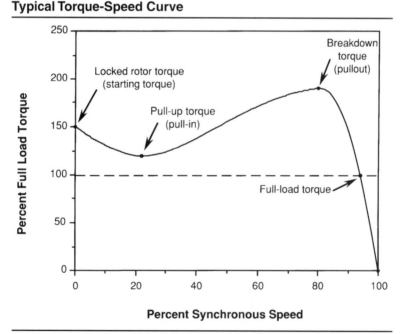

Percent Synchronous Speed

required varies depending on the position of the machine. The actual maximum torque requirement at any point is called the peak torque requirement. Peak torques are involved in types of loads (such as punch presses) that have an oscillating torque requirement. A motor's pull-up torque must be greater than the load's peak torque requirement to prevent stalling the motor.

Pull-out torque: The maximum amount of torque that is available from the motor shaft when the motor is operating at rated voltage and running at full speed. Also known as breakdown torque.

Pull-up torque: The lowest point on the torque-speed curve for a motor accelerating a load up to speed. Pull-up torque limits a motor's ability to accelerate its load and to meet a load's peak torque requirement. Some motor designs (typically NEMA Designs A and B) do not have a separate value for pull-up torque because the lowest point may occur at the locked rotor (starting) point. In this case, pull-up torque is the same as starting torque.

Starting torque: The amount of torque the motor produces when energized at full-rated voltage with the shaft locked in place. It is the amount of torque available when the motor is energized to break the load away (start it moving) and begin accelerating it up to speed. Also known as locked-rotor torque.

TP-1: See *NEMA TP-1*.

Transformer: The most common form of transformers is a device to increase or decrease the voltage in an AC system. The primary side of the transformer is connected to the source of power, the secondary side to the load. A step-down transformer (the most common type in transmission and distribution systems) reduces the primary voltage to the secondary voltage. A step-up transformer (used, for example, at power plants to increase the generation voltage to the transmission voltage) increases the primary voltage to the secondary voltage. Transformers work by using the current in the primary winding to create a changing magnetic field, which is used to induce a voltage (and thus current when connected to a load) in the secondary winding. Another common transformer type is the isolation transformer. Efficient transformers are specified in the NEMA TP-1 standard.

Transistor: See *bipolar transistor, insulated gate transistor*, and *MOS transistor*.

Variable-air-volume (VAV): An HVAC system in which the amount of cooling is controlled by changing the air flow rate; VAV heating systems are also used, as well as VAV controls of room pressurization.

Variable-frequency drive (VFD): Another name for the most common type of electronic adjustable speed drive. This type of drive uses an electronic package between the fixed-frequency AC input and the motor. The speed is varied by supplying the motor with synthesized AC power of changing frequency. See *adjustable-speed drive*.

Variable-speed drive: See *adjustable-speed drive*.

Ventilation: The introduction of fresh air into a building specifically for the purpose of maintaining good air quality. Air is usually drawn from outdoors but can also be purified, recirculated air. Often, the term *ventilation* is used loosely to include transport of any air, not just of fresh air. See *air transport system*.

Venturi: A constricted throat in an air passage creating a vacuum.

Voltage: The rated voltage under which a motor or related electrical equipment is designed to operate. In general, voltage is the electrical potential at any point relative to some reference point in a circuit. The voltage represents the energy level of a quantity of electrical charge (electrons) at that point in the circuit. In the Système Internationale system of measurement, the unit of voltage is the volt, which equals one joule of energy per one coulomb of charge (see *current*). When there is a flow of charge at a given voltage, this stream of energy is electrical power. This power is measured in watts (joules per second) and is equal at any instant to the product of the voltage and the current in the circuit.

Voltage-source inverter (VSI): A type of electronic ASD that converts the AC input to controlled-voltage DC and then synthesizes the variable-frequency AC output by using a DC-to-AC inverter. See *adjustable-speed drive, current-source inverter,* and *variable-frequency drive.*

Watt: A unit of (usually) electrical power equal to one joule of energy flowing per second. See *kilowatt*.

Windage: Motor loss resulting from the aerodynamic drag of the spinning motor rotor.

Winding: Windings are the turns of insulated wire (usually copper) wrapped around the core of steel laminations in motor stators, transformers, inductors, and electromagnets. The stator windings are generally connected to the power supply. In squirrel-cage motor rotors, the windings are several bars of uninsulated aluminum or copper, arranged in a cylinder and connected together at both ends by rings of the same material. The windings of wound-rotor motors are similar to those of the motor stator. When a motor is rewound, the insulated wire is removed and replaced with new wire.

Wk²: The symbol used for moment of inertia and measured in lb-ft². See *inertia*.

Wound-rotor induction motors: This class of motors features insulated copper windings in the rotor similar to those in the stator. The rotor windings are fed with power using slip rings and brushes.

Equipment Manufacturers and Associations

This appendix is divided into three sections:

1. Motor and drive manufacturers
2. Manufacturers of motor system test and repair equipment
3. Trade and professional associations related to motors and drives

This appendix is also available on the ACEEE.org Web site, where it will be regularly updated.

1. Motor and Drive Manufacturers

Most manufacturers are members of the National Electric Motors Association. Their contact info is listed on the NEMA Web site, where it is updated regularly. Please contact NEMA for the most recent information on manufacturers.

National Electrical Manufacturers Association
1300 North 17th Street, Suite 1847
Rosslyn, VA 22209
(703) 841-3200
(703) 841-5900 (fax)
www.nema.org

One notable absence from NEMA is Baldor Electric:

Baldor Electric Company
P.O. Box 2400
Fort Smith, AR 72902
(501) 646-4711
(501) 648-5792 (fax)
www.baldor.com

2. Manufacturers of Motor System Test and Repair Equipment

Abbreviations for product categories are as follows:

  (c)   core loss testers
  (d)   dynamometers
  (k)   kW meters
  (m)   megohmmeters
  (p)   power quality analyzers
  (r)   rewind equipment
  (t)   tachometers

Note that while digital multimeters and surface thermometers are also commonly used for motor testing, they are widely available and thus not listed here.

AEMC Corporation (m)
99 Chauncey Street
Boston, MA 02111
(617) 451-0227
www.aemc.com

Ametek (t)
37 North Valley Road, Building 4
P.O. Box 1764
Paoli, PA 19301
(610) 647-2121
(610) 296-3412 (fax)
www.ametek.com

Amprobe (m)
630 Merrick Road
Lynnbrook, NY 11563
(516) 593-5600
(516) 593-5682 (fax)
www.amprobe.com

AVO Instruments (m)
4271 Bronze Way
Dallas, TX 75237
(800) 723-2861
(214) 467-7341 (fax)
www.avointl.com

AW Dynamometer (d)
P.O. Box 428
Colfax, IL 61728
(800) 447-2511
(309) 723-4951 (fax)
www.awdynamometer.com

Dranetz/BMI Technologies (k, p)
1000 New Durham Road
P.O. Box 4019
Edison, NJ 08818-4019
(800) 372-6832
(201) 287-8627 (fax)
www.dranetz.com

Dreisilker Electric Motors (r)
352 Roosevelt Road
Glen Ellyn, IL 60137
(312) 469-7510
(312) 469-3474 (fax)
www.dreisilker.com

Eaton Corp. (d)
1111 Superior Avenue
Cleveland, OH 44114-2584
(216) 523-5000
www.eaton.com

Esterline Angus Instrument Corp. (k)
P.O. Box 24000
Indianapolis, IN 46224
(800) 543-0829
(317) 247-4749 (fax)

Inductor, Inc. (d)
5821 5th Avenue
Kenosha, WI 53141
(414) 657-0984
(414) 657-1200 (fax)
www.inductor.com

Lexington Sales and Engineering (Lexseco) (c)
4740 Allmond Avenue
Louisville, KY 40209
(502) 367-4393
(502) 386-3377 (fax)
www.lexseco.com

Monarch Instrument (t)
15 Columbia Road
Amherst, NH 03031
(603) 883-3390
(603) 886-3300 (fax)
www.monarchinstrument.com

NLB (r)
29830 Beck Road
Wixom, MI 48393-2824
(248) 624-5555
(248) 624-0908 (fax)
www.nlbcorp.com

WOMA (r)
Raritan Center
95 Newfield Avenue
Edison, NJ 08837
(732) 417-0010
(732) 417-0015 (fax)
www.woma.de

Yokogawa Corp. (t, m)
2 Dart Road
Newnan, GA 30265-1018
(404) 253-7000
(404) 251-2088 (fax)
www.yca.com

3. Trade and Professional Associations Related to Motors and Drives

Air Conditioning and Refrigeration Institute (ARI)
1501 Wilson Boulevard, Suite 600
Arlington, VA 22209
(703) 524-8800
(703) 528-3816 (fax)
www.ari.org

Air Movement and Control Association
30 West University Drive
Arlington Heights, IL 60004
(847) 394-0150
(847) 253-0088 (fax)
www.amca.org

American Chain Association
6724 Lone Oak Boulevard
Naples, FL 34109
(941) 514-3441
(941) 514-3470 (fax)
www.americanchainassn.org

American Gear Manufacturers Association
1500 King Street, Suite 201
Alexandria, VA 22314
(703) 684-0211
(703) 684-0242 (fax)
www.agma.org

American Society of Mechanical Engineers (ASME)
345 East 47th Street
New York, NY 10017
(212) 705-7800
www.asme.org

Association of Energy Engineers (AEE)
4025 Pleasantdale Road, Suite 420
Atlanta, GA 30340
(770) 447-5083
(770) 446-3969 (fax)
www.aeecenter.org

Bearing Specialists Association
Building C, Suite 20
800 Roosevelt Road
Glen Ellyn, IL 60137
(630) 858-3838
(630) 790-3095
www.bsahome.org

Electric Power Research Institute
3412 Hillview Avenue
Palo Alto, CA 94304
(415) 855-2000
www.epri.com

Electrical Apparatus Service Association
International Headquarters
1331 Baur Boulevard
St. Louis, MO 63132
(314) 993-2220
(314) 993-1269 (fax)
www.easa.com

Fluid Power Distributors Association
P.O. Box 1420
Cherry Hill, NJ 08034-0054
(856) 424-8998
(856) 424-9248
www.fpda.org

Institute of Electrical and
Electronics Engineers
445 Hoes Lane, P.O. Box 1331
Piscataway, NJ 08855-1331
(732) 981-0060
(732) 981-1721 (fax)
www.ieee.org

Mechanical Power Transmission
Association
6724 Lone Oak Boulevard
Naples, FL 34109
(941) 514-3441
(941) 514-3470 (fax)
www.mpta.org

National Electrical
Manufacturers Association
1300 North 17th Street,
Suite 1847
Rosslyn, VA 22209
(703) 841-3200
(703) 841-5900 (fax)
www.nema.org

National Fluid Power
Association
3333 North Mayfair Road,
Suite 311
Milwaukee, WI 53222
(414) 778-3344
www.nfpa.com

National Industrial Belting
Association
N19 W24400 Riverwood Drive
Waukesha, WI 53188
(262) 523-9090
(262) 523-9091 (fax)
www.niba.org

National Lubricating Grease
Institute
4635 Wyandotte Street
Kansas City, MO 64112
(816) 931-9480
(816) 753-5026 (fax)
www.nlgi.com

Power Transmissions
Distributors Association
250 South Wacker Drive,
Suite 300
Chicago, IL 60606-5840
(312) 876-9461
(312) 876-9490 (fax)
www.ptda.org

Rubber Manufacturers
Association
1400 K Street NW
Washington, DC 20005
(202) 682-4800
www.rma.org

Society of Tribologists and
Lubrication Engineers
840 Busse Highway
Park Ridge, IL 60068-2376
(847) 825-5536
(847) 825-1456
www.stle.org

Annotated Bibliography

This Annotated Bibliography describes some of the books, reports, journals, software tools, and Web sites that are most useful for obtaining additional information on motor systems.

Books and Reports

Andreas, John. 1982. *Energy-Efficient Electric Motors: Selection and Application.* New York, N.Y.: Marcel Dekker.

This reference, written in simple language, provides guidelines for selecting and applying electric motors on the basis of life-cycle costs. Particular emphasis is given to single- and three-phase motors from 1 to 125 hp. The book covers the economics of energy-efficient motors in detail and discusses some of the interactions between the power supply and the motor. There is a brief section on adjustable-speed drives.

Arthur D. Little, Inc. (ADL). 1980. *Classification and Evaluation of Electric Motors and Pumps.* Report DOE/TIC-11339. Prepared for the U.S. Department of Energy, Office of Industrial Programs. Springfield, Va.: National Technical Information Service.

This study, based on data from the late 1970s, represents the first attempt to describe the motor and pump markets and to analyze whether efficiency standards and labeling requirements for motors and pumps were desirable. The data in this report served as the basis for the development of the EPAct motor rule. The report contains many detailed breakdowns on the motor and pump populations, but the accuracy of some of the numbers is questionable (due to limitations in the underlying data). The report contains a politically biased conclusion that neither efficiency standards nor labeling requirements are desirable. An earlier version of the report (DOE/CS-1047, same title and publisher) concluded that efficiency standards and labeling might be advantageous. This earlier version also contained some data that did not make it into the final report.

Arthur D. Little, Inc. (ADL). 1999. *Opportunities for Energy Savings in the Residential and Commercial Sectors with High-Efficiency Electric Motors, Final Report*. Prepared for the U.S. Department of Energy. Washington, D.C.: Arthur D. Little, Inc.

This study, commissioned by DOE's Office of Energy Efficiency and Renewable Energy, is the most complete study of motor use in the commercial and residential sectors. The report profiles motor technologies and applications found in the commercial and residential sectors, current motor populations, energy use, and savings potentials, and also identifies barriers to increased use of efficient motors. In contrast to the XENERGY (1998) study of motors in the industrial sector, this report is based on secondary data sources. However, when combined with the XENERGY (1998) study, these studies offer the most comprehensive picture of motor use in the United States currently available.

Bensch, Ingo. 1999. *POS Evaluation: Looking Back on the Performance Optimization Service Program, Report Summary*. Madison, Wis.: Energy Center of Wisconsin.

This ten-page report discusses the results and lessons from the Wisconsin Performance Optimization Service program, which is one of the most extensive programs to promote system optimization in North America. This study is useful reading for program planners and implementers interested in encouraging systems optimization since the Wisconsin program implementers had some notable successes but also learned some important lessons that will need to be addressed by future programs of this type. An in-depth report is also available for those wanting further details.

Consortium for Energy Efficiency (CEE). 1996. *Premium Efficiency Motor Initiative*. Boston, Mass.: Consortium for Energy Efficiency.

CEE's premium-efficiency motor initiative was inspired by a desire to define a new efficiency point for manufacturers to use as a target when they designed their new product lines in response to EPAct. The description discusses the motivation behind the initiative, how the levels in the specification were arrived at, and possibilities for the specification to be used in market transformation programs.

Dreisilker, Henry. Undated. "Safe Stator and Rotor Stripping Method." 10-page typescript. Glen Ellyn, Ill.: Dreisilker Electric Motors.

Henry Dreisilker, president of a large motor distribution and repair business, has waged a one-man campaign for 30 years against the use of burnout-oven stripping. He maintains that conventional motor repair practice damages motors and that the low-temperature, mechanical technique he uses and markets does a better job without damaging the motors. Dreisilker has an extensive collection of testimonials in support of his method and case studies of the damage caused by conventional practice.

E Source, Inc. 1999. *Drivepower Technology Atlas Series, Volume IV.* Prepared by B. Howe, A. Lovins, D. Houghton, M. Shepard, and B. Stickney. Boulder, Colo.: E Source, Inc.

This volume is the third edition of one of the standard motor efficiency references that built upon the 1989 report *The State of the Art: Drivepower* (Lovins et al. 1989). E Source's series was among the first motor energy efficiency technical references and helped establish the credibility of this topic. This encyclopedic work provides in-depth technical information on motor and related technologies; motor systems; and motor selection, operation, and maintenance. The book focuses in particular on the state-of-the-art technologies and practice and provides information not readily available from other sources.

Easton Consultants. 1996. *National Market Transformation Strategies for Industrial Electric Motor Systems: Volume II, Market Assessment.* DOE/PO-0044. Washington, D.C.: U.S. Department of Energy.

This report is the result of a multi-funder research effort to characterize the opportunities for energy efficiency in key segments of the original equipment manufacturer motor marketplace. The structure of the motor drive, pump, fan, and compressed-air industry are described, and key market players are identified.

Easton Consultants. 2000. *Market Research Report: Variable Frequency Drives.* Report #00-054. Portland, Oreg.: Northwest Energy Efficiency Alliance.

This study builds upon the above report to provide a characterization of the ASD market and represents the most current market analysis available. This report projects the market for ASDs, estimates the installed base, characterizes how the market functions, provides current cost data, and identifies potential implementation problems associated with drives.

Easton Consultants and XENERGY. 1999. *Opportunities for Industrial Motor Systems in the Pacific Northwest*. Portland, Oreg.: Northwest Energy Efficiency Alliance.

Identifies opportunities for reducing motor system energy use by measure (e.g., motor efficiency upgrade, pump system efficiency improvement, etc.) and sector (e.g., pulp and paper, irrigation, etc.). This work is largely based on a previous study by XENERGY for DOE (XENERGY 1998). It also evaluates the sectors and measures on specific criteria and identifies five major program "opportunity clusters"—a motor package (efficiency, proper rewinding, and downsizing), an equipment package (primarily fan and pump systems), a compressed-air package, irrigation pumping, and refrigeration in the food-processing industries. The approach and packages may be appropriate for other regions.

Electrical Apparatus Service Association (EASA). 1985. *Core Iron Study*. St. Louis, Mo.: Electrical Apparatus Service Association.

This widely cited study from the 1980s, prepared by the trade association of motor repair shops, sought to resolve the question of whether conventional burnout-oven stripping degrades motor cores. Although the study concluded that no damage should occur when burnout ovens are set no higher than 650°F, the data from EASA's tests do show some damage and suggest that lower temperature limits may be warranted. Much of the more recent work, especially by the Washington State University (Schueler, Leistner, and Douglass 1994) build upon this report to address this question.

Friedman, R., C. Burrell, J. DeKorte, N. Elliott, and B. Meberg. 1996. *Electric Motor System Market Transformation*. Washington, D.C.: American Council for an Energy-Efficient Economy.

This report, the foundation for many of the motor system programs developed in recent years (such as the *Compressed Air Challenge*), remains an important reference on motor system markets. The study identified and characterized the major motor systems market segments and provided market structures for each segment, identifying key players. Opportunities for transforming the markets were identified, and intervention strategies were proposed for each. Based on this analysis, the various strategies were ranked based on their energy savings and likelihood of success. The report was prepared by ACEEE under contract to DOE's *Motor Challenge* program and was also published by DOE (DOE/PO-0044, Volume I, 1996) along with a companion study, *National Market Transformation Strategies for Industrial Electric Motor Systems: Volume II, Market Assessment* (Easton Consultants 1996).

Lawrence Berkeley National Laboratory and Resource Dynamics Corporation. 2001. *Improving Motor and Drive System Performance: A Sourcebook for Industry*. Washington, D.C.: U.S. Department of Energy, Office of Industrial Technologies.

This new compilation assembles much of the key reference information into a single volume. The book provides an overview of motor technology basics and includes eight fact sheets on key motor topics and a list of resources available to motor users. The report also includes the motor repair documents prepared by the Washington State University.

Nailen, Robert. 1987. *Motors, Volume 6, Power Plant Reference Series*. Palo Alto, Calif.: Electric Power Research Institute.

This motor manual is directed mainly at power plant engineers, although most of the information is useful in other fields. Techniques for matching a motor to an application are described in relation to the load characteristics, environment, and power systems. Motor industry standards and maintenance practice are also covered.

National Electrical Manufacturers Association (NEMA). 1999. *Motors and Generators*. NEMA Standards Publication No. MG 1-1998. Rosslyn, Va.: National Electrical Manufacturers Association.

One of the most important technical references for understanding the energy efficiency of electric motors. This document is the primary technical standard by which motors are designed and specified in North America and is widely referenced throughout the world. The standard specifies allowable ranges for key operating parameters for different "designs" of motors and generators. MG 1 allows for the interchangeability of motors of a given design among different manufacturers. Also provided are guidelines for labeling motors, including regarding energy efficiency. This standard is updated on approximately a 2-year cycle.

Schueler, V., P. Leistner, and J. Douglass. 1994. *Industrial Motor Repair in the United States*. Portland, Oreg.: Bonneville Power Administration.

The first major independent study of the motor repair industry and motor repair techniques. The study persuasively makes the case regarding why repair is an important energy issue, provides a profile of the motor repair industry, discusses the repair process and its possible impacts on efficiency, identifies the major barriers to quality repair, and proposes strategies to encourage quality repairs. Appendices provide tools to assist end-users in managing their motor repairs.

Seton, Johnson, & Odell, Inc. 1987. *Lost Conservation Opportunities in the Industrial Sector*. Portland, Oreg.: Bonneville Power Administration.

This report examines opportunities for obtaining efficiency improvements at low cost when new equipment is purchased or existing equipment is being replaced. It discusses several motor-related industrial energy efficiency opportunities, including motors, pumps, and piping. Extensive data on motor sales, costs, and efficiencies are included.

U.S. Department of Energy (DOE). 1998. *Improving Compressed Air System Performance: A Source Book for Industry*. Washington, D.C.: U.S. Department of Energy.

This practical reference provides guidance for engineers and compressed-air system operators on opportunity identification and system performance improvements. The guide was developed by DOE's *Motor Challenge* program, in cooperation with the *Compressed Air Challenge*. The leading experts in the compressed-air industry contributed to and reviewed this guide. The volume is organized into three parts:
1. An overview of compressed-air systems that describes types of compressors and other system components and discusses uses for compressed air
2. A set of 11 fact sheets covering the main performance opportunities
3. A reference guide that directs the reader to additional resources for assistance in compressed-air system optimization and operation

XENERGY. 1998. *United States Industrial Electric Motor Systems Market Opportunities Assessment*. Prepared for Oak Ridge National Laboratory and DOE's Office of Industrial Technologies. Washington, D.C.: U.S. Department of Energy, Office of Energy Efficiency and Renewable Energy.

This study, perhaps the most important new data source on electric motors to become available in the past decade, was commissioned by DOE and undertook a systematic review of all data sources in order to characterize the industrial motor marketplace. This review was supplemented by extensive field assessment of manufacturing facilities. This research provided an accurate characterization of the number, size, and application of electric motors in industry and yielded important insights into how these assets are managed by plants and the sizes and locations of major efficiency opportunities. The main report, which is available online, is an important reference; however, the data available in the appendices of the complete report represent an even more valuable research resource. These include profiles of motor use and savings opportunities in selected industries, methodological information, and a stock adjustment model to project changes in the motor marketplace. Unfortunately, the complete report has become difficult to obtain.

XENERGY. 2000. *Assessment of the Market for Compressed Air Efficiency Services*. Prepared for DOE and the *Compressed Air Challenge*. Burlington, Mass.: XENERGY.
This study characterizes the compressed-air marketplace from both demand and supply sides. The report characterizes the knowledge of compressed-air system users and suppliers, provides a picture of the market structure, estimates energy use, and identifies the magnitude of efficiency opportunities. Market barriers on both sides are identified, and strategies for addressing these barriers are suggested.

Journals and Periodicals

ASHRAE Journal, ASHRAE Transactions. American Society of Heating, Refrigerating and Air-Conditioning Engineers, 1791 Tullie Circle, N.E., Atlanta, Ga. 30329.
The *Journal*, the monthly magazine of ASHRAE, covers topics related to their mission in articles, advertisements, and product listings and is of primary interest to mechanical engineers designing or retrofitting HVAC and refrigeration systems. *Transactions* is published twice each year and contains the research papers presented at the two annual ASHRAE meetings.

Consulting-Specifying Engineer, Design News, Plant Engineering. The Cahners Publishing Company, 275 Washington Street, Newton, Mass. 02158.
Consulting-Specifying Engineer is published monthly and is aimed at mechanical and electrical engineers working in the building construction industry. The articles, advertising, and product listings cover a wide range of technologies, including those related to motors. *Design News* is published twice monthly and is written for mechanical and electrical engineers designing components and systems for buildings, industry, and transportation. It includes articles, advertising, and product listings. *Plant Engineering* is also published twice monthly and includes articles, advertising, and product listings. It is written for engineers working in industry.

Electrical Construction and Maintenance. Intertec Publishing, 888 7th Avenue, 38th Floor, New York, N.Y. 10106.
This monthly magazine covers the installation, maintenance, and repair of a range of electrical technologies. Each issue includes "Motor

Facts," which covers a variety of issues in the selection, installation, and care of motors.

Energy Engineering. Association of Energy Engineers, 700 Indian Trail, Lilburn, Ga. 30247.

This bimonthly publication is the journal of the Association of Energy Engineers. Each issue concentrates on a single topic, such as motor systems, energy management control systems software, etc., and includes several articles plus a product directory.

Energy User News. The Chilton Company, 7 East 12th Street, New York, N.Y. 10003.

This monthly magazine is targeted at facility managers in commercial and institutional buildings. It presents case studies, interviews, and surveys on energy use practices and reports on trends in energy costs. The magazine includes advertising and product directories on a variety of energy-efficient technologies, including motors and drives.

Engineered Systems. Business News Publishing Company, P.O. Box 7016, Troy, Mich. 48007.

This magazine, published bimonthly, "provides information to assist people who specify, install, buy, and maintain commercial, industrial, and institutional HVAC/R systems." Articles, advertisements, and product directories cover a wide range of topics in the areas of both mechanical and electrical technologies, including motors and motor systems.

Heating, Piping, and Air Conditioning. Penton Publishing, Inc., 1100 Superior Avenue, Cleveland, Ohio 44114.

This monthly magazine is addressed to mechanical engineers working in the building trade. Articles, advertising, and product listings cover a variety of topics, including pumps, fans, piping, and ductwork.

IEEE Transactions on Industry Applications, IEEE Transactions on Power Systems. Institute of Electrical and Electronics Engineers, 345 East 47th Street, New York, N.Y. 10017-2394.

Industry Applications, published six times a year, covers a variety of motor-related technologies of interest to industry (including recent developments in adjustable-speed drives and their applications) and includes papers presented at conferences of the IEEE Industry Applications Society. *Power Systems* focuses on topics of interest to electric utilities, including new types of motors and the interaction of motor systems with utilities. This quarterly publication contains papers presented at conferences of the IEEE Power Engineering Society.

Software Tools

Jacobs Engineering Group, Inc. 1996. *ASDMaster*™. Palo Alto, Calif.: Electric Power Research Institute.

ASDMaster™ is a software tool developed by EPRI to assist with the analysis, application, and specification of adjustable-speed drives. *ASDMaster*™ provides the end-user with a screening tool that aids in identifying ASD applications, evaluating the economics, selecting the right ASD to suit the application and environment, developing a purchase specification, and locating manufacturers of suitable ASDs. The companion user's guide provides an excellent overview of ASD technologies, application considerations, economic evaluation, and several case studies.

Office of Industrial Technologies. 2000. *Pumping System Assessment Tool*. Washington, D.C.: U.S. Department of Energy, Office of Industrial Technologies.

The *Pumping System Assessment Tool* (PSAT) is a software program developed by DOE's Office of Industrial Technologies to assist engineers and facility operators in performing assessments of pumping system energy usage. PSAT is also well suited for consultants or plant engineers performing plant energy usage surveys. End-users in the field will find PSAT easy to use since it was carefully designed to require only the minimum essential operation data (or requirements) to perform its analysis. Although PSAT does not specify recommendations for measures to improve systems, it does prioritize likely opportunities for efficiency improvement and allows the user to broaden or narrow searches for improving efficiency.

Washington State University Energy Program. 1999. *MotorMaster+*®, Version 3.01. Olympia, Wash.: Washington State University.

This program is the latest incarnation of what started as the Washington State Energy Office's Motor Database in the late 1980s. The core feature of this tool is a database of catalog data for integral-horsepower motors from all major manufacturers. This database allows the user to compare different motors and evaluate repair/replace decisions. The database has also proven to be a valuable research tool, allowing the range of available products to be investigated. In the *plus* version,

461

a robust motor inventory and management function and a life-cycle cost calculator have been added to the compare function. While there are limitations with the database, as discussed in Chapter 2, the large database and the ability to download for free make this tool a must-have.

Washington State University Energy Program. 2001. *AirMaster+®*. Olympia, Wash.: Washington State University.

AirMaster+® is a stand-alone Windows-based software tool used to analyze industrial compressed-air systems. It is intended to enable auditors to model both existing system operation and future improvements and evaluate savings from energy-efficiency measures with relatively short payback periods. *AirMaster+®* provides a systematic approach to assessing compressed-air systems, analyzing collected data, and reporting results. Available upon request by e-mailing the OIT Clearinghouse at Clearinghouse@ee.doe.gov or calling 800-862-2086.

Web-Based Resources

ACEEE Motors Web Page
http://aceee.org/motor

ACEEE maintains updated material and links to Web resources on motors, motor-driven equipment, and motor systems.

Air Movement and Control Association International, Inc.
http://www.amca.org

AMCA is the association of the fan and blower industry. Information on AMCA-certified products is available on the site, along with downloadable Adobe PDF versions of their newsletter *TechSpecs*.

Compressed Air and Gas Institute
http://www.cagi.org

CAGI is the manufacturers' association of the air-compressor industry. Their Web site has a "toolbox" including data sheets on compressed-air equipment performance and systems, a glossary, formulas, and a discussion of the European Union's *Pressure Equipment Directive*.

Compressed Air Challenge
http://www.compressedairchallenge.org

This Web site provides an online version of *Improving Compressed Air System Performance: A Source Book for Industry*, as well as information on training offered by the CAC.

Consortium for Energy Efficiency's Industrial Programs
http://www.cee1.org/ind/ind-main.php3

CEE's Industrial Programs maintains a site that provides updated information on its motor system initiatives, including their *premium-efficiency* motor initiative. The program has also assembled a "toolkit" of Web-based resources to help end-users and market transformation programs improve energy efficiency of motor systems.

Hydraulic Institute
http://www.pumps.org

The Hydraulic Institute, the association of the pump industry, has been working with OIT to develop materials on energy efficiency in pump systems, much of which is available in the pump resources section of their Web site. Information on pump manufacturers is also available at the site.

Industrial Best Practices: Motors
http://www.oit.doe.gov/bestpractices/motors

DOE's Office of Industrial Technologies maintains a Web site with many technical publications and some excellent software tools available for downloading. An online version of the *MotorMaster+*® software is available. The site also maintains a calendar identifying training opportunities related to motors and motor systems.

Office of Industrial Technologies Clearinghouse
http://www.oit.doe.gov/clearinghouse/

DOE's Office of Industrial Technologies also offers a clearinghouse where trained staff answer questions on OIT's products and services, including motor programs. The clearinghouse can be contacted at clearinghouse@ee.doe.gov or 800-862-2086.

Productive Energy Solutions, LLC
http://www.ProductiveEnergy.com/

Productive Energy Solutions is a consulting engineering firm focusing on motor systems optimization. The site provides material that can assist plant staff in assessing energy savings opportunities from motor systems optimization. The site also includes an online motor system cost calculator.

Washington State University's Energy Program
http://www.energy.wsu.edu/index/industrial.cfm

WSU's Energy Program has developed many of the technical content and software tools for OIT's Industrial Best Practices program, many of which are available for downloading from this Web site. The site also has numerous research reports and technical assistance guides available both online and as Adobe PDF files.

Index